MRINAL K. DAS

EVOLUTION OF
FISH SPECIES FLOCKS

EVOLUTION OF
FISH SPECIES FLOCKS

EDITED BY
ANTHONY A. ECHELLE
OKLAHOMA STATE UNIVERSITY

IRV KORNFIELD
UNIVERSITY OF MAINE AT ORONO

University of Maine at Orono Press, Orono, Maine 1984

Oklahoma State University Museum of Cultural and Natural History, Stillwater, Oklahoma

Migratory Fish Research Institute, Orono, Maine

The cover illustration is an untitled charcoal by Nina Sutcliff permanently housed in the Art Collection, the University of Maine at Orono.

Library of Congress Catalog Card Number
84-051502

ISBNO-89101-058-0 (cloth)
ISBNO-89101-057-2 (paper)

CONTRIBUTORS

CLYDE D BARBOUR, Department of Biological Science, Wright State University. Dayton Ohio 45435 USA

PERETI BASASIBWAKI, Uganda Freshwater Fisheries Research Organization. P.O. Box 343 Jinja Uganda

KENT E CARPENTER, East-West Center and Department of Zoology, University of Hawaii. Honolulu Hawaii 96822 USA

BARRY CHERNOFF, Department of Ichthyology, Academy of Natural Sciences of Philadelphia. Philadelphia Pennsylvania 19103 USA

WALLACE J DOMINEY, Museum of Zoology and Division of Biological Sciences, University of Michigan. Ann Arbor Michigan 48109 USA

ALICE F ECHELLE, Department of Zoology, Oklahoma State University. Stillwater Oklahoma 74078 USA

ANTHONY A ECHELLE, Department of Zoology, Oklahoma State University. Stillwater Oklahoma 74078 USA

W NOEL GRAY, Department of Surveys. P.O. Box 349 Blantyre Malawi

P HUMPHRY GREENWOOD, Department of Zoology, British Museum (Natural History). Cromwell Road London SW7 England

JULIAN M HUMPHRIES, Museum of Zoology, University of Michigan. Ann Arbor Michigan 48109 USA

LESLIE S KAUFMAN, New England Aquarium. Central Wharf, Boston Massachusetts 02110 USA

IRV KORNFIELD, Migratory Fish Research Institute and Department of Zoology, University of Maine. Orono Maine 04469 USA

KAREL F LIEM, Museum of Comparative Zoology, Harvard University. Cambridge Massachusetts 02138 USA

PAUL V LOISELLE, 6707 North 77th Street No 6. Milwaukee Wisconsin 53223 USA

ERNST MAYR, Museum of Comparative Zoology, Harvard University. Cambridge Massachusetts 02138 USA

AMY R McCUNE, Section of Ecology and Systematics, Cornell University. Ithaca New York 14853 USA

KENNETH R McKAYE, Cape McClear Fisheries Research Station and Duke University Marine Laboratory. Pivers Island, Beaufort North Carolina 28516 USA

PAUL E OLSEN, Lamont-Doherty Geological Observatory, Columbia University. Palisades New York 10964 USA

LYNNE R PARENTI, Department of Ichthyology, American Museum of Natural History. Central Park West at 79th Street, New York New York 10024 USA

ANTHONY J RIBBINK, J.L.B. Smith Insititute of Ichthyology. Grahamstown 7140 South Africa

RICHARD D SAGE, Museum of Vertebrate Zoology, University of California. Berkeley California 94720 USA

GERALD R SMITH, Museum of Zoology, University of Michigan, Ann Arbor Michigan 48109 USA

RICHARD E STRAUSS, Museum of Zoology, University of Michigan, Ann Arbor Michigan 48109 USA

KEITH S THOMSON, Department of Biology, Yale University. New Haven Connecticut 06511 USA

THOMAS N TODD, Great Lakes Fishery Laboratory, U.S. Fish and Wildlife Service. 1451 Green Road Ann Arbor Michigan 48105 USA

ALLAN C WILSON, Department of Biochemistry, University of California. Berkeley California 94720 USA

FRANS WITTE, Zoologisch Laboratorium, University of Leiden. Kaiserstraat 63 Postbus 9516, 2300RA Leiden, The Netherlands

CONTENTS

PREFACE

The last general treatment of the evolution of species flocks of fishes was John Brooks' paper "Speciation in ancient lakes" published almost 25 years ago. Since then, new flocks have been discovered and other flocks have proved more speciose and phylogenetically more diverse than previously suspected. The relatively recent development of molecular techniques to probe genealogical relationships and objective methodologies to reconstruct evolutionary histories along with the maturation of evolutionary and ecological theory, have provided an impetus to study these systems in more detail. In preparing this volume, our purpose was to consolidate contemporary perspectives about species flocks, including reassessments of old questions in the context of new data and current thought. This book grew out of a symposium we organized which convened at the annual meetings of the American Society of Ichthyologists and Herpetologists, Tallahassee, Florida, 21-22 June 1983.

The book begins with a commentary by Ernst Mayr (Chapter 1). Species flocks, including fishes, figured significantly in his book, *Systematics and the Origin of Species.* Mayr was the first to provide a definition for the term *species flock* (= species swarm), but 42 years later there is no consensus on exactly what a species flock is. The symposium participants roughly fall into two camps summarized by the contrasting views of Greenwood (Chapter 2) and Ribbink (Chapter 3). From our perspective, the term *species flock* in its most general usage encapsulates the range of evolutionary questions presented in this volume, and acts as a semaphore to alert biologists to the uniqueness of endemic lacustrine faunas.

The first half of the book deals with non-cichlid flocks. McCune et al. (Chapter 4) examine fossil flocks of holostean fishes and suggest that evolutionary dynamics in lake-adapted faunas may have a partial explana-tion in celestial mechanics. Particularly interesting is their treatment of diversification and extinction episodes in the fossil record, thus providing a valuable temporal perspective for current research on extant systems. Smith and Todd (Chapter 5) present a hypothetico-deductive assessment of modes of speciation in north temperate flocks. Intralacustrine diversification, possibly by competitive speciation, figures prominently in some of the systems examined. Kornfield and Carpenter (Chapter 6) and Parenti (Chapter 7) reassess, with new tools, the evolutionary relationships and biogeography of two groups classically referred to as species flocks: the cyprinids of Lake Lanao (Philippines) and the cyprinodontoid genus *Orestias* (Lake Titicaca). The conventional scenario for the origin of the Lanao flock is supported by new field studies and biochemical comparisons. Parenti's cladistic reassessment of South American *Orestias* shows convincingly that the endemic fishes of Titicaca do not constitute a single species flock. Echelle and Echelle (Chapter 8) explore the evolution of silversides from the Mexican Plateau. Exceedingly high genetic similarities found among some species suggest a very recent radiation in ephemeral lakes. Barbour and Chernoff (Chapter 9) examine monophyly, hybridization and morphological differentiation in a subset of the silversides on the Mexican plateau, the pescados blancos. The evolutionary role of differential allometric growth figures significantly in their discussion of diversification. Humphries (Chapter 10) presents the genetic structure of a surprising, recently discovered species flock of Mexican pupfishes. He suggests that these endemic forms arose by intralacustrine speciation, perhaps via competitive speciation.

The spectacular assemblages of haplochromine cichlids of the African Great Lakes, synonymous with *fish species flock* in the minds of many biologists, have been the subject of very active investigations. Correspondingly, about half of the book deals with these and other cichlid fishes. Greenwood (Chapter 11) places the 200 + cichlids of Lake Victoria in the context of current debates on punctuated equilibrium. He emphasizes the point that lines within this flock provide examples of cladistic, rather than phyletic gradualism. Witte (Chapter 12) discusses ecological segregation in cichlids of southern Lake Victoria. Comparing his findings with the assumptions of earlier workers, he concludes that ecological characteristics and evolutionary processes are very similar among the three African Great Lakes. McKaye and Gray (Chapter 13) describe field experiments in Lake Malawi on intralacustrine barriers and the origins of species flocks. They find evidence for extrinsic barriers, but in contrast to earlier work, also suggest that intrinsic isolating mechanisms may be important in the evolution of some groups. Sage et al. (Chapter 14) explore morphological versus genetic differentiation in Lake Victoria cichlids. Because of the apparent tempo of speciation, they propose that behavioral drive may be an important evolutionary force in cichlid systems. Liem and Kaufman (Chapter 15) describe functional morphology and feeding experiments with two morphs of the unique cichlid of Cuatro Cienegas, Mexico. They consider the general role of intraspecific polymorphism in the production of species flocks and macroevolution. Although the Cuatro Cienegas cichlid is a single biological species, the reader may be surprised to learn that there may be some justification for considering it a species flock (see Chapter 9)! Strauss (Chapter 16) examines mechanisms associated with divergence in different, trophically associated structures of African cichlids. The simple relationship he finds between trophic differentiation and body size in a portion of the Victoria flock provides a significant alternative mechanism for the control of diversification. Against the background of recent evolutionary thought, Dominey (Chapter 17) compares mating systems and life history traits among African cichlids and Hawaiian *Drosophila* to explore common features that might explain "explosive speciation." He presents a particularly strong case for the role of sexual selection in the speciation

of endemic faunas.

Finally, we discuss natural and anthropogenic aspects of extinction in fish species flocks, emphasizing the need for students of these unique assemblages to become actively involved in their preservation (Chapter 18),

Explicitly indicated or implied in most contributions are suggestions for future research. There is an obvious need to obtain more information about temporal and spatial patterns of resource utilization and population structuring, life history information and the finer details of genealogical relationships within species flocks. However, one theme which pervades all the contributions presented here should perhaps be emphasized above all else: quantitative genetics. Given the apparent rapid radiation in many species flocks, it is of primary importance to undestand the genetic control of morphololgy and coloration, those characters permitting recognition of endemic species. Knowledge of the magnitude of phenotypic plasticity and the number of effective factors (or loci) segregating within and between species are essential to distinguish among alternative hypotheses for speciation and radiation.

The success of any symposium depends on the caliber and cooperation of its participants; in this respect we were extremely fortunate. The authors provided timely submissions and indulged us in the delays associated with the editorial process. We express our appreciation to those who served as extramural reviewers of the contributions. We wish to thank Bryan Glass of the Oklaholma State University Museum of Cultural and Natural History, Jerry Wilhm of the OSU Zoology Department and Gary Menchen of the University of Maine at Orono Press for providing funds and making concessions that enabled us to produce this book rapidly and to offer it at a price that hopefully will make it generally available. We also extend our thanks to Ralph Yerger, Florida State University, for assisting in the organization of the symposium. Finally we gratefully acknowledge the type-setting and paste-up efforts of "retired" Professor Don Allen who worked arduously for little recompense except our thanks.

Anthony A Echelle   **June 1984**
Irv Kornfield

EVOLUTION OF
FISH SPECIES FLOCKS

EVOLUTION OF FISH SPECIES FLOCKS: A COMMENTARY

ERNST MAYR

That freshwater lakes, particularly old ones, sometimes have very species-rich endemic faunas of fishes, snails, crustaceans, and other invertebrates has long been known. Seemingly monophyletic groups of closely related species coexisting in the same area are often referred to as species flocks (Greenwood 1984b). There is no need to define this term of convenience too rigidly. To hold that the term "is independent of taxonomists' assessments of species boundaries ... and that the smallest species flock is a single polymorphic species" (Barbour, Chernoff 1984) would require applying the term *species flock* to all polymorphic species, in conflict with the traditional meaning of the term. Common descent and coexistence in the same lake are the earmarks of lacustrine species flocks.

The existence of species flocks has raised a number of interesting questions, two of which have generated controversies that have not yet been muted: 1) Can the presence of scores, if not hundreds, of closely related species in a single lake be explained by allopatric speciation? 2) Do these species flocks refute the principle of competitive exclusion? Or, stated differently, how can so many sympatric species partition the available resources in such a way as to avoid too intense competition? This symposium volume provides partial answers to these questions and shows what problems remain to be studied. I will give a short summary of findings and discuss critically some open problems.

MODE OF SPECIATION

Being unable to discover any major geographic barriers within lakes supporting species flocks, most early authors postulated some process of sympatric speciation. This is not the place to describe the manifold difficulties of the sympatric model which ignores the genetic difficulties (Mayr 1976) and is not

substantiated by any incipient cases (Greenwood 1984a). Such a model is difficult to visualize, since speciation is a populational process and therefore "niche space," always having a spatial dimension, is at least micro-allopatric.

Several papers in this volume demonstrate the fact that some sympatric sibling species of fishes have only slight differences in their allozymes. However, this can not be considered evidence for sympatric speciation. It is now quite evident that the allozymes, isozymes or electromorphs have little or nothing to do with the speciation event (consensus in Barigozzi 1983). The amount of difference in allozymes between two species is a rough indication of the time that has passed since their phyletic lines split apart from each other. In rapidly speciating groups perfectly good species may have adequate isolating mchanisms and yet show no electrophoretic differences at all. Allozyme differences among sibling species may have evolved subsequent to the speciation event, and even subsequent to the reestablishment of sympatry. What is now needed is a careful comparison of closely related sympatric and allopatric species to determine the average amount of allozyme difference between these two groups of species. The genus *Labidochromis* in which many geographically isolated species occur would seem particularly well suited for such a comparison.

The discovery of two markedly different trophic morphs in *Cichlasoma minckleyi* has revived the theory of speciation by disruptive selection. If assortative mating should develop among the members of two morphs, the papilliform and the molariform, it would represent a case of sympatric speciation. How probable is such a development? The hypothesis of disruptive speciation in polymorphic species must be regarded from the viewpoint of natural selection. We must ask which of two individuals would in the long run leave more offspring: individual *A* which can produce several morphs, each partialy removed from competing with its brothers and sisters; or individual *B* which has only one kind of offspring restricted to a single niche? My guess is that individual *A* would leave more offspring, because it would be far better able to cope with resource fluctuations and changes in the environment. A narrowing down of the food niche through increased specialization would be favored by selection only when the range of the polymorphic population were invaded by a related and competing species. If the latter were equally polymorphic, each species would presumably be the superior competitor in a different subniche. If the new invader were already specializing in one of the subniches, it would almost surely displace the resident polymorphic species from that subniche. The occupation of several subniches by an ecologically polymorphic species thus is like hedging a bet: it permits an instantaneous retreat to a subniche not occupied by an invading competitor. Certain papers in this volume propose speciation by disruptive selection owing to the sympatric disruption of a polymorphic population (e.g., Humphries 1984), but I did not find the arguments particularly convincing.

Rensch (1933) was apparently the first to attempt to explain the rich species flocks in freshwater lakes by various modes of allopatric speciation. This interpretation was adopted by Mayr (1942) and during the 25 years after 1949 most authors endeavored to show that the species flocks in freshwater lakes could be explained by allopatric speciation (Brooks 1950; Fryer, Iles 1972; Greenwood 1974). However in the 1970's theories of sympatric speciation again found favor owing to the discovery of the distinct trophic morphs in Cuatro Cienegas (Sage, Selander 1975) and the unexpectedly high number of sibling species of cichlids in East African lakes. The question was raised again: Can such rich species flocks be explained by the process of allopatric speciation? While investigating this question, it is particularly important to remember two aspects of these species flocks: other higher taxa in the same lakes have not evolved species flocks, and

the sister groups of lacustrine species flocks in nearby river systems likewise have not evolved species flocks. It is thus evident that there must be some aspect in the life history and perhaps the genome of certain taxa which permit them to speciate rapidly inside of lakes.

Most contributors to this volume fully realize that the solution of the problem of explosive speciation in freshwater lakes is to be looked for in the behavior and ecology of the speciating taxa. Alas, we will probably never learn the answer for the 18 endemic species of cyprinids of Laka Lanao on Mindanao because 15 or 16 of these species are now extinct (Kornfield, Carpenter 1984) and nothing can be said about their niche occupation and behavior. Nor can one speculate on the contribution of behavior or ecology to the speciation of the 25 or so endemic species of the cyprinodont *Orestias* of Lake Titicaca in the Andes because there are little or no life history data. Many of the 43 species of this genus occur in Andean stream systems of Peru, Chile, and Bolivia, and several colonizations of the Lake Titicaca Basin are evident (Parenti 1984). Nevertheless there are monophyletic species assemblages inside the lake which evidently originated through intralacustrine speciation.

Brooks (1950) reports on the early history of the problem of speciation in freshwater lakes. Rensch (1933:38), in an attempt to refute Woltereck's (1931) hypothesis of sympatric speciation, proposed three mechanisms to account for the richness of the lake faunas: 1) Repeated colonizations from rivers leading into and out of the lakes; 2) allopatric speciation in different portions of the lake; and 3) an accumulation of species through a fusion of previously existing bodies of fresh water.

Truly intralacustrine speciation was minimized in this scheme and the same was true for Mayr's (1942) presentation that was largely based on Rensch's. Our knowledge of species flocks in freshwater lakes has increased enormously in the past 30 years, and the need for new analysis of the problem of species flocks in freshwater lakes is apparent. Does the new evidence substantiate the occurrence of sympatric speciation and, if not, what is the relative importance of various allopatric speciation processes?

The reason why no definitive answers to

these questions have yet been reached in spite of the excellent research carried out in recent years are the following:

a) There is still great uncertainty concerning the age of the species flocks and of the lakes in which they occur. The age of Lake Victoria, for instance, is given as 750,000 years by some, but only as 100,000 years by others, and still others suggest that the lake basin had almost entirely dried up during one phase of the Pleistocene. Or to take another example, the nearest relatives of *Orestias* of Lake Titicaca are still uncertain, and the origin of the genus would have to be placed in the early Mesozoic if the unlikely suggestion were true that the Anatolian Cyprinodonts are the sister group of *Orestias* (Parenti 1984). Similar uncertainties exist for nearly all the lakes and the taxa they contain. For instance, the age of the Laka Lanao species flock evidently is far greater than the 10,000 years postulated in the earlier literature (Kornfield, Carpenter 1984). Even if the maximum age is accepted for Lake Victoria, this is much shorter than the time intervals normally available to the student of fossils, and it is a time span seemingly totally inadequate to explain the origin of a flock of 250 haplochromine species.

b) Taxonomic analysis is still incomplete. Sibling species abound in most species flocks and many earlier type series consist of two or three sibling species. Yet, as Greenwood (1984a) points out, these are unquestionably good biological species. There is no intergradation among them and where their breeding habits were studied, they have been found to be completely isolated reproductively. Genetically they are good species, no matter how similar they may be morphologically. The problems escalate in the case of fossil species. McCune et al. (1984), on the basis of rather few characters, recognize 21 sympatric species of similar body form and without any noticeable trophic specializations in semionotid fishes of the Mesozoic of eastern North America. Are these 21 morphotypes truly species? How could they partition the available resources, considering their extreme similarity? One is faced with numerous unanswered questions.

c) Limitations to the cladistic technique. Groups of closely related species have exceedingly similar genotypes. There is a great potential of such species to develop independently the same synapomorph characters. Not surprisingly Lewis (1982) during a cladistic analysis of the species of *Labidochromis* from Lake Malawi, came to the conclusion that these species do not form sister groups. That such pluripotentiality is a major obstacle to the application of the cladistic method to close relatives has long been known. Other problems with applying cladistic methods to species flocks are discussed in this volume (Barbour, Chernoff 1984; Echelle, Echelle 1984; McCune et al. 1984).

These various classes of difficulties must be carefully kept in mind during the analysis of potential patterns of speciation.

It is now rather generally agreed that there are three possible major causes for species flocks in lakes: 1) multiple colonization; 2) amalgamation of smaller lakes owing to a rise in water level; 3) intralacustrine speciation by one or another of several possible mechanisms (Smith, Todd 1984). Although Rensch and Mayr thought that the first-mentioned two causes make the major contribution to the size of the species flocks, it is now evident, particularly for the immense species flocks of African cichlids, that intralacustrine speciation was largely responsible. Even though multiple colonization is not entirely excluded from having contributed to the species flocks in the African lakes, this process was apparently more important for temperate zone lakes, for Lake Titicaca, and for atherinids in lakes on the Mesa Central of Mexico (Echelle, Echelle 1984). A fusion of smaller lakes might have contributed to the size of the species flock in Lake Victoria (Greenwood 1951). On the other hand the Great Basin lakes in North America, in spite of a series of Pleistocene lake level fluctuations, have not produced species flocks (Smith, Todd 1984).

As far as intralacustrine allopatric speciation is concerned, three possibilities have been suggested:

1) Lake basins. A giant lake like Lake Baikal (636 km length) has several partly separated basins and would permit some extent of allopatric speciation within the lake, particularly if there were major fluctuations of lake level. Good basins are also accepted for Lake

Tanganyika (Fryer, Iles 1972). However, major basins are not present in most other lakes although the five Great Lakes in North America are actually five basins of one great lake. There is indeed some indication of incipient speciation in the ciscoes (*Coregonus*) of the Great Lakes of North America (Smith, Todd 1984).

Allopatric speciation within a single lake may be possible even in lakes not divided into separate lake basins. Most species in the larger lakes occur only in a portion of the lake, and such spotty distribution favors intralacustrine speciation. Only about a quarter of the haplochromines of Lake Victoria seem to have a lakewide distribution. The cichlids of Lake Malawi seem to be even more localized.

2) Substrate localization. The frequently made assumption that a lake is a rather homogeneous habitat is quite erroneous. In nearly all lakes, and this is particularly true for the large East African lakes, there is an alternation of rocky, sandy, and muddy stretches of sublittoral substrates. Contrary to earlier claims, this is also true for Lake Victoria (Witte 1984). Many species of cichlids are completely restricted to one kind of habitat. Such substrate dependence is greatest in algal grazers, less so in piscivores and omnivores. Many species of the rock-dwelling genus *Labidochromis* in Lake Malawi are known only from a single rock outcrop (Lewis 1982). Nearly all haplochromines are mouth-brooders, and have always been assumed to be, on the whole, poor dispersers. As a result, many closely related species are still allopatric (Lewis 1982). It appears from available descriptions that habitats in a lake that are suitable for a given species are more or less scattered and isolated like islands in an archipelago. Intralacustrine speciation has therefore been sometimes compared to archipelago speciation in terrestrial organisms (Brooks 1950; Mayr 1963:507). Rock-inhabiting species found on offshore islands illustrate the process of peripatric speciation particularly well. Of 150 mbuna species on the Malawi coast of Lake Malawi only three are recorded for the entire area.

Of particular interest are the experiments of McKaye and Gray (1984) of building artificial ''reefs'' in sandy areas. These reefs were colonized by more than 20 species, most of them plankton feeders and not found elsewhere in surveys of sandy habitats. There is a seeming and not yet resolved contradiction between this capacity for dispersal and the fact that these plankton feeders belong to the most speciose of all the haplochromine cichlids.

c) Spawning site isolation. Discontinuities between spawning sites, even in the absence of visible barriers, might serve as the means by which the isolation is achieved which is the prerequisite for the acquisition of isolating mechanisms. It must be stressed, since this was misrepresented in certain contributions to this volume, that there are two classical theories concerning this process. According to one theory, that of Darwin, H. J. Muller, and Mayr, the isolating mechanisms are acquired during spatial isolation as a byproduct of the genetic changes of the isolated population. According to the other theory, that of A. R. Wallace and Dobzhansky, the isolating mechanisms are formed during isolation only very incompletely, and are perfected through natural selection (owing to the inferiority of the hybrids) only after the secondary encounter of the neospecies. The almost complete absence of hybrids in the lake cichlids is convincing evidence for the validity of the Darwin theory as far as the species flocks in lakes are concerned.

Two sets of factors, then, must be demonstrated in order to explain any instance of allopatric speciation. These are 1) the isolating factors which safeguard the integrity of the incipient species during the period of its genetic reconstruction, and 2) the factors responsible for the building up of isolating mechanisms before there is any contact with the species against which the isolating mechanisms eventually will be employed.

As far as spatial isolation is concerned, there are many indications that such coarse substrate criteria as rock, sand, or mud are not sufficient to explain the exuberant speciation of cichlids in the East African lakes. There is, however, one aspect of intralacustrine speciation that has not been sufficiently emphasized. While speciation in terrestrial organisms is constrained by their restriction to a two-dimensional space, lake inhabitants have a third available dimension. There are good indications that this additional dimension facilitiates speciation. But there are still other factors. Foremost among these is the fact that the water mass of a lake is not homogeneous, and this is particularly true for feeding conditions. As is well known, mortality in fishes is highest among the freshly hatched fry. If they hatch in an area of abundant food supply there will be high survival. This fact sets up a selection force for the appropriate site finding (homing) and timing behavior of the spawning parents (Smith, Todd 1984). Individuals not spawning at the right time and place face the risk of lower reproductive success. This will lead to a local clustering of spawning populations and to a continuing increase of philopatry.

But how will such spawning clusters acquire isolating mechanisms against each other? This has always been a weakness of the Darwin theory as interpreted by recent evolutionists. Mayr (1963:551) and others thought that the isolating mechanisms arose as a byproduct of the ecological divergence of the incipient species, but this interpretation was not very convincing, owing to the slightness of the niche differences in many incipient species, illustrated by the spawning clusters of lake fishes.

The probable causal factors in the origin of isolating mechanisms has recently been discovered in sexual selection. I find Dominey's (1984) explanation, largely based on Thornhill and Alcock (1983) and West-Eberhard (1983), completely convincing. Local fashions in mate preference can apparently change rather rapidly in small isolated populations, and these can develop into behavioral isolating mechanisms. Furthermore, they serve as recognition features of the individuals that belong to such a spawning cluster, and thus reinforce the spatial isolation. Dominey rightly stresses the important role played by the mating system of cichlids in this process of speciation. Although the distances among the spawning clusters may be small, the isolation nevertheless is spatial, and the process can be justly designated as micro-allopatric.

One of the consequences of this model of intralacustrine speciation (Smith, Todd 1984) is that the first characters to diverge should relate to courtship, while characters relating to competition for shared resources will evolve later, particularly after sympatry is established. The frequency of male color differences in morphological sibling species is consistent with this prediction. By contrast, the morphological consequences of sympatric speciation would be that characters related to resource partitioning (food utilization) should be the first to diverge. Eventually there will be character displacement wherever such newly evolved species come into competition with each other: morphological differences which reinforce resource partitioning will evolve.

Much of intralacustrine speciation, according to this model, is thus due to an interaction of two sets of factors, one having to do with the spawning site (appropriate territoriality and philopatry), while the other concerns the development of genetic mechanisms for localized recognition characters which isolate the spawners from straying individuals of other populations. Such a recognition pattern (behavioral isolation) could presumably be acquired by very few genetic changes. Once acquired, its strength seems to be considerable, as documented by the virtual absence of hybrids in lake cichlids. The complex courtship presumably prevents cases of mismating.

The Smith-Todd-Dominey model is at this time not much more than a working hypothesis, even though it fits the known facts extremely well. It will require a considerable amount of field work to test the validity of its various assumptions. How permanent are the spawning areas, and how reliable is the food supply for the hatched fry? How much dispersal is there among the offspring of a given female? How selective are females when faced with a choice of two kinds of male? Life history studies will presumably shed far more light on speciation in lakes than attempts at a genetic analysis. It is unfortunately true that we still know virtually nothing about the actual genetic factors responsible for speciation even in *Drosophila* and other genetically well-known terrestrial species (Barigozzi 1983).

The above postulated process of intralacustrine speciation is applicable primarily to the cichlid species flocks in the lakes of tropical East Africa. By contrast, intralacustrine speciation in temperate zone lakes is dominated by other factors. Here, seasonal phenomena are apparently utilized to produce isolation. Time of spawning as well as spawning migrations seem to play a far greater role than in tropical lakes, and some of the spawning migrations may have been a byproduct of multiple colonization (Smith, Todd 1984).

FISH FLOCKS AND EVOLUTIONARY THEORY

The study of fish species flocks in freshwater lakes is making considerable contributions to evolutionary theory, as is evident from the papers in this volume. For instance, the two morphotypes in the Cuatro Cienegas cichlid remind us that the term *gradual evolution* has two independent meanings that are almost invariably confounded. Populational gradualism was most important for Darwin in his fight against essentialism. However, this is not the same as phenotypic gradualism, since a single population can be polymorphic for sharply distinct phenotypes. Phenotypic discontinuities are not necessarily incompatible with populational gradualism.

There is an interesting correlation in the East African lakes between the age of the lake and the average morphological distance among the species. In the youngest of these lakes, Lake Victoria, which is at most 750,000 years old, there are no extreme morphotypes, and related species are connected by intermediate species forming a morphocline (Greenwood 1984a). By contrast, some rather diverse types have evolved in Lake Tanganyika, which is at least two million years old, with a much higher average morphological distance between species. Lake Malawi, apparently younger, is intermediate in the morphological distance between species. The evidence would seem to indicate that natural selection continues to push species toward a more extreme divergence and that intermediate types are more vulnerable to extinction, presumably by competition.

The specializations in the East African lakes illustrate the opportunism of natural selection. Parallel utilization of the same resources (trophic types) has evolved in the various lakes, and correlated with it parallel morphological specializations. Apparently, all the major potential resources are utilized in all the lakes. Different kinds of organisms may evolve

along rather different pathways, even under similar conditions. Such evolutionary pluralism, as rightly stressed by several recent authors, is abundantly illustrated by the fish flocks. The cichlids and coregonids, for instance, exhibit rather different evolutionary processes from other fish with which they share their habitat as well as from freshwater inhabiting invertebrates. To mention another instance of pluralism, founder populations may sometimes acquire rather different phenotypes or else evolve as sibling species with hardly any morphological change. Even though most of the species must have originated through peripatric speciation, they nevertheless have retained an average level of heterozygosity and the same polymorphism may turn up in rather distantly related species (Sage et al. 1984). It must be remembered that heterozygosity does not necessarily have to be largely eliminated during peripatric speciation provided the period of the bottleneck is very short (Nei et al. 1975).

Cichlid species provide many illustrations of mosaic evolution. Trophic specialization, for instance, and reproductive isolation would seem to evolve independently. In *Cichlasoma minckleyi* trophic polymorphism evolved without reproductive isolation. In the numerous sibling species in the East African lakes reproductive isolation evolved without trophic specialization.

Cichlid fishes have long provided by far the best illustration for rapid speciation (Mayr 1967). There is much evidence for speciation within three to four thousand years, as documented for instance by the cichlids of Lake Nabugabo (Greenwood 1965). In view of this rapidity the slightness of the morphological change is not surprising. What is surprising, perhaps, is the rapidity by which the presumed isolating mechanisms are acquired (Dominey 1984).

It has been suggested that Lake Victoria might have completely dried up during one of the arid periods of the Pleistocene, about 14,000 years ago. The production of over 200 endemic species of haplochromines since that recent date would seem utterly improbable. However, if we were to assume that each bout of speciation would require about 2800 years, and that the founder species of the new Lake Victoria gave rise by peripatric speciation to three daughter species in different parts of the new lake, and that this process was repeated every 2800 years, there could be 243 species after 14,000 years.

I do not know of any other organism for which such a rapid rate of speciation could even be considered. Actually, Lake Victoria presumably never dried up completely, but was merely reduced to a series of ponds in which at least some of the previous fish fauna survived. Under such an assumption, the length of individual speciation bouts could be increased considerably beyond 2800 years.

What all this teaches us, and this is in line with the admonitions of several leading evolutionists, is that there is a great diversity of evolutionary processes, and that the parameters that are valid for one group of organisms may not necessarily be at all applicable to another group. There are few, if any, universal laws in evolutionary biology (Mayr 1967).

ADAPTIVE RADIATION AND COMPETITIVE EXCLUSION

The coexistence of hundreds of closely related species in the same lake poses some fundamental questions concerning competition and resource utilization. To what extent, if any, is the existence of fish flocks in freshwater lakes in conflict with the concept of competitive exclusion? Greenwood's contribution to this volume gives us a superb analysis of the problem. The fact that competition must exist is self-evident. That it sets up selection forces is indicated by the fact that there is more morphological divergence among species in Lake Tanganyika, the oldest of the African lakes, than in the relatively recent Lake Victoria. The severity of the competition is also indicated by the fact that the haplochromines utilize every potentially available resource, and that every truly lacustrine habitat is occupied by one or several haplochromine species. Nevertheless, there are far more species than trophic types; for instance, of the 11 trophic types among 250 + species of haplochromines in Lake Victoria, 30 to 40 percent are piscivorous.

What factors permit the coexistence of so many competitors and prevent them from exterminating each other? The adaptive radiation was evidently facilitated by the seeming flexibility of the exploratory tendencies in haplochromines. This agrees with the observation that most organisms have a rather open genetic program concerning habitat and food utilization (Mayr 1974). Not surprisingly, in different lakes where similar diversities of resources are available, parallel adaptations have evolved. In such cases, I would prefer not to speak of "the deterministic action of selection" (Greenwood 1984a) since I can not see anything deterministic in the opportunism of natural selection. If those individuals of a certain population that can better crush snails leave more offspring than other members of their population, one would expect that natural selection would favor those aspects of behavior or structural modifications that facilitate use of snails as a food resource.

What, then, are the factors that mitigate competition? First, many species are localized, and even though there may be 250 haplochromines in Lake Victoria, they will not all coexist at the same locality. Further, syntopy is often reduced by exclusion in time or space. A preference for different water depth of potential competitors often reduces competition (Marsh et al. 1981). Witte (1984) has made a particularly careful analysis of niche differences among similar species. For instance, in the zooplanktivores of Lake Victoria each case of asserted "total interspecific overlap" that was studied in detail revealed niche segregation. A thorough study of the ecology of these species is beginning only now, and little is so far known about components of the life cycle, seasonal differences in food utilization, and other aspects that play a role in reducing competition.

One of the surprising findings is that some structurally rather specialized trophic types are nevertheless facultative feeders, utilizing a variety of different resources. Greenwood calls attention to the seeming paradox that the specialists among the cichlids usually can also exploit the more generalized food niches while the generalists cannot do the reverse. Hence, widespread assumptions notwithstanding, it is the specialists who have the potential for the exploitation of a wider range of food niches. Indeed, as Liem and Kaufman (1984) point out, the prey capture apparatus in cichlids is built in such a way that it can meet multiple problems. Profound functional shifts can be made without any major structural alteration of jaws and associated muscles. And a seemingly rather drastic difference, such as between the molariform and the papilliform morphs in *Cichlasoma minckleyi*, can be effected by a very small shift in regulatory genes.

The recent works, particularly on the haplochromines in East Africa, are shedding considerable light on the relation between behavior, structure, speciation, and food niche. The fact that most closely related species show no morphological differences at all or only very minor ones in trophic adaptations indicates that morphological differentiation is neither a prerequisite nor even a necessary corollary of speciation. The motto that behavior is the pacemaker of evolution is massively supported by the evolutionary changes in the cichlid species flocks. Also interesting is the degree to which evolutionary change is restricted to trophic structures. "There is remarkably little diversity in body form" (Greenwood 1984a). The cichlids thus have experienced a major ecological evolution without a corresponding morphological one. This parallels the situation in the songbirds (Oscines) where one also finds remarkably little divergence except for coloration and features of bill and feet.

There are, of course, many unanswered questions, and additional puzzling aspects of evolution will surely be discovered during the further study of the ecology and behavior of the cichlid species flocks. Greenwood (1984a) echoes a question often raised in the history of Darwinism when he states, "It is difficult to imagine how selection can be invoked to explain the evolution of highly derived features (dental specialization) when functionally and morphologically intermediate stages are still extant and present in species occurring syntopically with the more derived forms." In a way this is merely a new version of the question raised by anti-Darwinians in the last century: "Why would higher organisms evolve when simple ones like bacteria and protists are so eminently successful?" A single component of the phenotype, like dental specialisation, is of course not the whole of the adaptive value of an organism. Each newly evolving species presumably has some character, by no means necessarily a

structural one, which gives it the capacity to compete successfully, and thus to coexist, with other species that may have acquired a structural specialization.

The present symposium shows beautifully how many important biological problems are being illuminated by the study of even so specialized a subject as that of fish species flocks in freshwater lakes. And yet, the analysis of this intriguing evolutionary phenomenon has only begun. There is every reason to believe that further studies will result in raising tantalizing new questions but also contribute to the solution of still open problems. What is most important now is to push these studies as hard as possible as long as these diversified faunas still exist. Lake Lanao and other lakes where the endemic faunas have been largely exterminated are an ominous warning of what the future might have in store.

REFERENCES

Barbour C D, B Chernoff 1984 Comparative morphology and morphometrics of the pescados blancos (Genus *Chirostoma*) from Lake Chapala, Mexico. 111-128 A A Echelle, I Kornfield eds. *Evolution of Fish Species Flocks*. Univ Maine at Orono Press

Barigozzi C 1983 *Mechanisms in Speciation*. Liss. New York

Brooks J L 1950 Speciation in ancient lakes. *Quart Rev Biol* 25:30-176

Dominey W J 1984 Effects of sexual selection and life history on speciation: species flocks in African cichlids and Hawaiian *Drosophila*. 231-250 A A Echelle, I Kornfield eds. *Evolution of Fish Species Flocks*. Univ Maine at Orono Press

Echelle A A, A F Echelle 1984 Evolutionary genetics of a "species flock": Atherinid fishes on the Mesa Central of Mexico. 93-110 A A Echelle, I Kornfield eds. *Evolution of Fish Species Flocks*. Univ Maine at Orono Press

Fryer G, T D Iles 1972 *The Cichlid Fishes of the Great Lakes of Africa. Their Biology and Evolution*. Oliver, Boyd. Edinburgh

Greenwood P H 1951 Evolution of the African cichlid fishes. The haplochromine species flock in Lake Victoria. *Nature* 167:19-20

———— 1965 The cichlid fishes of Lake Nabugabo, Uganda. *Bull Brit Mus Nat Hist (Zool)* 12:315-357

———— 1974 The cichlid fishes of Lake Victoria, East Africa: The biology and evolution of a species flock. *Bull Brit Mus Nat Hist (Zool)* Suppl 6:1-134

———— 1984a African cichlids and evolutionary theories. 141-154 A A Echelle, I Kornfield eds. *Evolution of Fish Species Flocks*. Univ Maine at Orono Press

_____ 1984b What *is* a species flock? 13-20 A A Echelle, I Kornfield eds. *Evolution of Fish Species Flocks.* Univ Maine at Orono Press

Humphries J M 1984 Genetics of speciation in pupfish from Laguna Chichancanab, Mexico. 129-140 A A Echelle, I Kornfield eds. *Evolution of Fish Species Flocks.* Univ Maine at Orono Press

Kornfield I, K Carpenter 1984 Cyprinids of Lake Lanao, Phillipines: Taxonomic validity, evolutionary rates and speciation scenarios. 69-84 A A Echelle, I Kornfield eds. *Evolution of Fish Species Flocks.* Univ Maine at Orono Press

Lewis D S C 1982 A revision of the genus *Labidochromis* from Lake Malawi. *Zool J Linn Soc* 75:189-265

Liem J F, L Kaufman 1984 Intraspecific macroevolution: Functional biology of the polymorphic cichlid species *Cichlasoma minckleyi.* 203-216 A A Echelle, I Kornfield eds. *Evolution of Fish Species Flocks.* Univ Maine at Orono Press

Marsh A C, A J Ribbink, B A Marsh 1981 Sibling species complexes in sympatric populations of *Petrotilapia* Trewavas (Cichlidae, Lake Malawi). *Zool J Linn Soc* 71:253-264

Mayr E 1942 *Systematics and the Origin of Species.* Columbia Univ Press. New York

_____ 1963 *Animal Species and Evolution.* Harvard Univ Press. Cambridge

_____ 1967 Evolutionary challenges to the mathematical interpretation of evolution. 47-58 P S Moorhead, M M Kaplan eds. *Mathematical Challenges to the neo-Darwinian Interpretation of Evolution.* Wistar Institute Symposium. Monograph No 5. Wistar Institute Press. (Reprinted in Mayr 1976:53-63)

_____ 1974 Behavior programs and evolutionary strategies. *Amer Scientist* 62:650-659 (Reprinted in Mayr 1976:694-711)

_____ 1976 *Evolution and the Diversity of Life.* Harvard Univ Press. Cambridge

McKaye K R, W N Gray 1984 Extrinsic barriers to gène flow in rock-dwelling cichlids of Lake Malawi: macrohabitat heterogeneity and reef colonization. 169-184 A A Echelle, I Kornfield eds. *Evolution of Fish Species Flocks.* Univ Maine at Orono Press

McKaye K R, T Kocher, P Reimtal, I Kornfield 1982 A sympatric sibling species complex of *Petrotilapia* Trewavas from Lake Malawi analyzed by enzyme electrophoresis (Pisces, Cichlidae). *Zool J Linn Soc* 76:91-96

McCune A R, KS Thomson, P E Olsen 1984 Semionotid fishes of the Mesozoic Great Lakes of North America. 27-46 A A Echelle, I Kornfield eds. *Evolution of Fish Species Flocks.* Univ Maine at Orono Press

Nei M, T Mayurama, R Chakraborty 1975 The bottleneck effect and the genetic variability in populations. *Evolution* 29:1-10

Parenti L R 1984 The species flock concept as it relates to the phylogeny and biogeography of the Andean killifish *Orestias.* 85-92 A A Echelle, I Kornfield eds. *Evolution of Fish Species Flocks.* Univ Maine at Orono Press

Rensch B 1933 Zoologische Systematik und Artbildungsproblem. *Verh Deutsch Zool Ges* 1933:19-83

Sage R D, P Loiselle, P Basasibwaki, A C Wilson 1984 Molecular versus morphological change among cichlid fishes of Lake Victoria. 185-202 A A Echelle, I Kornfield eds. *Evolution of Fish Species Flocks.* Univ Maine at Orono Press

Sage R D, R K Selander 1975 Trophic radiation through polymorphism in cichlid fishes. *Proc Nat Acad Sci* 72:4669-4673

Smith G R, T N Todd 1984 Evolution of species flocks of fishes in north-temperate lakes. 47-68 A A Echelle, I Kornfield eds. *Evolution of Fish Species Flocks.* Univ Maine at Orono Press

Thornhill R, J Alcock 1983 *The Evolution of Insect Mating Systems.* Harvard Univ Press. Cambridge

West-Eberhard M J 1983 Sexual selection, social competition, and speciation. *Quart Rev Biol* 58:155-183

Witte F 1984 Ecological differentiation in Lake Victoria haplochromines: comparison of cichlid species flocks in African lakes. 155-168 A A Echelle, I Kornfield eds. *Evolution of Fish Species Flocks.* Univ Maine at Orono Press

Woltereck R 1931 Beobachtungen und Versuche zum Fragenkomplex der Artbildung I. *Biol Centralblatt* 51:231-253

WHAT *IS* A SPECIES FLOCK?

P H GREENWOOD

Reviewing the relevant but not very extensive literature on this topic soon discloses a varied interpretation of the phrase *species flock,* and its synonym, *species swarm.* To some authors it is simply a descriptive term used to embrace a number of taxonomically allied and endemic species occurring in a geographically circumscribed area. For others, there are evolutionary implications, especially with regard to the processes involved in a flock's formation. To yet others there are phylogenetic or genealogical implications to be considered with or without reference to the evolutionary processes involved. Almost the only common factors in those different approaches are that members of a flock should have some taxonomic cohesion, should occupy a very restricted area, usually a lake or island, and that the species should be endemic to that area. It is difficult to discover a unified species flock concept. It would seem to be more a descriptive term than a concept, an idea which, like Harriet Beecher Stowe's Topsy, "one 'spects just growed." Also reminiscent of Topsy, it is difficult to get much information on the term's actual birth, although that child's comment that she ".. was raised by a speculator" seems readily applicable to some of the ideas on how a species flock might have evolved.

Despite its origin, discussed below as something little more than a descriptive phrase, the idea of a species flock soon became entangled with, and almost absorbed by, various ideas concerned with modes of speciation. That situation persisted until the early 1940's when, for the first time, there appeared to be some attempt to formulate a species flock concept as a particular aspect of evolutionary theory.

Ironically, during the pre-1940 period the term was not employed by Regan, Herre and Trewavas, three ichthyological pioneers in the study of what were later to become some of the classical examples of species flocks, namely the African lake cichlids and the cyprinids of Lake Lanao, Philippines. Neither was the term used by other investigators working on different lakes and on islands, the faunas of which could be or were later identified as species flocks.

The last half century or so has seen little progress towards identifying what precisely is implied when the term *species flock* is used. In particular, there has been no trenchant statement on what levels of phyletic interrelationship limit the use of the term to describe an assemblage of related taxa as a species flock. These difficulties must be faced and solved if the term is to lose its currently rather vague meaning, and its at times confusing application to situations which, at first sight, are quite "obviously" examples of species flocks, but which on closer examination fail to reveal just what was so "obvious."

When attempting to reconstruct the history of the species flock idea and the usage of the phrase *species flock* one encounters several problems. Not least of these is the fact that the term was rarely used during the first half of this century, the period when the concept itself, vague though it may be, was beginning to develop (see especially: Moore 1903; Plate 1913; Regan 1922a,b; Woltereck 1931; Rensch 1933; Herre 1933; Myers 1936; Worthington 1937). Indeed, the same problem is encountered in the writings of later workers as well. Thus, Trewavas (1949), Stankovic (1935), and Kozhov (1963) described and commented on situations which others by then or later had designated as species flocks.

To overcome this difficulty I have brought into the discussion relevant ideas, comments and observations used by those authors who did not employ the term *species flock* themselves, but were studying situations later recognized by others as species flocks. These situations include, in particular, the geospizine finches of the Galapagos Islands, drepanidid honeycreepers, drosophilid flies, gastropods and weevils of Hawaii, the gammarid Crustacea and cottid fishes of Lake Baikal (see extensive reviews by Berg 1928, Brooks 1950, Kozhov 1963), and the coregonid fishes of northern Europe.

As far as I can discover, the word *flock* was first used in the context we are considering by two botanists, Sargent and Peck (1906). They wrote: "The particular tendency of *Crataegus* is to flock together.. It is rare to find any large area occupied by a single species." A year later a trio of botanists, MacDougal, Vail and Schull (1907), working on *Oenothera,* described a situation in terms that clearly foreshadow those with which we are now familiar in the literature on species flocks: ".. under the name *biennis* is included a swarm of elementary species very closely related, but easily distinguishable when grown side by side, having no intermediate forms." Here, at least implicitly, and associated with the words *flock* and *swarm* are the factors of close interspecific relationship and relative geographical restriction.

A few years earlier, in 1903, J.E.S. Moore, writing about the fish fauna of Lake Tanganyika, east Africa, commented on the high level of endemicity to be found there (76 of the 87 then known species were endemic). He said that ".. the extraordinary large number of endemic forms present can only be viewed either as the result of the formation and multiplication of species through natural selection and other similar causes, in the lake itself, or as a survival in this lake of some old fauna which was rich in such types of fishes" (Moore 1903:134).

The precise implication of Moore's second alternative explanation is rather vague, since it might suggest that the lake was a refuge for species that had evolved outside Lake Tanganyika. It is clear, however, that he considered the Lake Tanganyika cichlids and some other taxa as well, to have been derived from marine ancestors. Moore's first suggestion, on the other hand, clearly encapsules the essential features of the species flock phenomenon in its broadest sense. Moore was apparently content to note the phenomenon, albeit as an outstanding one worthy of detailed comment, but coined no names for it.

Regan (1922a,b), some nineteen years later, was also impressed with the high endemicity and large numbers of African lake cichlids. By then, these fishes were known to be a characteristic feature of several African lakes, especially Lakes Victoria and Malawi, and to involve even greater numbers of species and even higher levels of endemicity. Like Moore, Regan let the phenomenon pass unnamed, but he did speculate on certain questions of phylogeny, and on the possible evolutionary processes involved. As an evolutionary explanation, he favored *habitudinal segregation* (Gulick 1905) rather than natural selection (Regan 1922b:159).

As a result of Regan's taxonomic revisions of the African cichlid genera, the nature of this lake flock phenomenon was more sharply etched than before. The distinct cichlid flocks in Lakes Victoria and Malawi were shown to be confined mainly to the genus *Haplochromis*, whereas in Lake Tanganyika a greater number of genera, belonging in Regan's opinion to two distinct lineages, were involved, each genus being represented by far fewer species than was generally the case in Lakes Victoria and Malawi. For all three lakes, Regan stressed the point that, with one or two exceptions, the constitutent species of a group, irrespective of its size, evolved within the lake.

Both explicitly and implicitly, Regan emphasized the notion of close phyletic intrarelationship between members of what later were to be called the *Haplochromis* species flocks of Lakes Victoria and Malawi. It is not clear from his writings whether Regan was proposing a monophyletic or an oligophyletic origin for the *Haplochromis* species of each lake. On some occasions he refers to a single ancestral species, and on others he refers to as many as half a dozen ancestors (Regan 1922a,b). Be that as it may, it is certainly obvious that Regan considered the half-dozen (or fewer) ancestors as intimately related, at least in the sense that they closely resembled one another morphologically and anatomically. He also emphasized that each flock was what he termed a *natural group* (Regan 1922b:158).

Thus for the first time in an ichthyological context three criteria were recognized for what was later to be called a species flock. That is: a high level of endemicity amongst its constituent species, their close phyletic relationship, and their geographical circumscription. The first and last of these features, it will be recalled, were noted by Moore (1903) in his discussion of the Lake Tanganyika fishes, cichlid and otherwise, but the context in which they were mentioned by Moore was less precisely delimited than that discussed by Regan, and the question of phyletic relationships was treated in a much more general way.

In his papers on the Victoria and Malawi cichlid species, Regan (1922a,b) did not make any reference to the temporal factors involved in their evolution. Moore (1903), however, obviously considered Lake Tanganyika to be of great age (Jurassic at least), and the cichlids probably marine relics which had evolved within and with the lake.

At a more theoretical level, at least with regard to the evolutionary processes involved, Woltereck (1931) used the cichlids of Lakes Tanganyika and Victoria, together with the gammarid Crustacea of Lake Baikal and the *Achatinella* species (Gastropoda) of Hawaii, as models for his theory of *schitzotypische Artsplitterung*, now generally anglicized as *explosive speciation*. Essential elements in his thesis are that speciation be sympatric and that it result from ecological specialization in localities where numerous niches become available for occupation and exploitation. This in essence is a theory virtually identical with that of Gulick's *habitudinal segregation* used by Regan to explain the evolution of the Lake Victoria *Haplochromis* flock (Gulick 1905; Regan 1922b). Like other workers of the time, Woltereck did not use the term *species flock* to describe the outcome of *explosive speciation*, but in later years, the two concepts were closely associated (Mayr 1942).

Woltereck's very explicit use of the sympatric or intralacustrine speciation model to account for the origin of species flocks led to several decades of debate on the importance and even the feasibility of sympatric speciation (see discussion in Fryer and Iles 1972). In many ways, this preoccupation with process rather than pattern inhibited the development of anything approaching precise formulation of a species flock concept. To a large degree the debate about sympatric speciation was more semantic than biological, devolving on the spatial limits implied by the prefixes *sym-* and *allo-*.

Of more direct application to the concept of a species flock, since it involves problems of genealogy, are questions relating to the level and kinds of spatial separation envisaged when the term *intralacustrine speciation* is used. For example, if in its history, a lake as we know it today originated from the union of several smaller lakes, can one reasonably say that its present population of species evolved by a process of intralacustrine speciation?

Questions of intralacustrine and sympatric speciation, coupled with or exclusive of multiple invasions, have long beffled discussions about the origin of Lake Baikal's very numerous and taxonomically diverse species flocks. See Kozhov (1963) for an excellent summary and a review of the literature. As far as I can determine, none of the authors contributing to the debate used the term *species flock* to describe any of the species aggregates. However, it is interesting to note that Dorogostaisky (1923) indicated, at least implicitly, a monophyletic origin for several aggregates that later were to be labeled as species flocks. Regrettably, considering the great amount of information available about Lake Baikal and its biota, students of evolution in that lake have been little concerned with the species flock concept *per se*. Instead, the term *species flock* was applied only in retrospect, and in a purely descriptive way to the lake's various endemic species groups.

Mayr (1942:213) discussed at some length the role of sympatric speciation in the origin of lacustrine species flocks, and used that term to describe the species aggregates he was considering. Although Mayr specifically mentioned the close relationship of species in a flock, he was no supporter of their origin by sympatric or intralacustrine speciation. Rather, he favored an origin through what might be described as multiple colonization, especially for the African lake flocks. Mayr's views were summarized in his statement: "The accumulation of large numbers of related species in these lakes is apparently partly due to the fact that a number of different river systems have 'washed' their own endemic species into the lake, and partly to the fact that some of these lakes originated by the amalgamation of a series of smaller lakes. Ecological specialization helps now to preserve the discontinuities between the species, but is not responsible for their creation. Species flocks in fresh-water lakes cannot be considered as proof of explosive speciation." He was using *explosive speciation* in Woltereck's sense.

I find the views expressed in that statement rather confusing. If different river systems contributed their "own endemic species" to a developing lake, it is probable that the endemic species of any one river would be more closely related to each other than to the endemic species of other rivers. Is it reasonable to conclude that all or even the majority of present species in the lake flock are closely related? As with Regan's statements, there is an element of uncertainty about what level of relationship is implied. The same question is raised when one considers the species derived from a series of smaller lakes, which in Mayr's argument, were amalgamated during the formation of the presentday lake.

Mayr's concept of a species flock seems to differ from Regan's in one very important respect. Where Regan postulated a truly intralacustrine origin for the consitituent species, Mayr was arguing for virtually an extra-lacustrine one. In terms of the flock's phylogenetic intrarelationships, the difference is important, and indeed, is central to the question of whether or not one can speak of a species flock concept and not merely a descriptive use of the term. It could be argued that, on the basis of Mayr's 1942 model, a species flock is little more than a zoogeographical accident.

With the passage of time and the increased information available about African lake flocks, Mayr's views shifted, considering his paper of 1947 and the comments he added later when it was reprinted in 1976. His discussion of the problem in 1963 included the description of a species swarm as ".. a considerable number of closely related species .. confined to a narrowly circumscribed area, no close relatives occuring elsewhere" (Mayr 1963 464). This viewpoint was rather closer

to Regan's than were his ideas of 1942, especially since he now apparently had dropped the idea of the flock's origin through multiple colonization. But there are still problems associated with precisely what is meant by such expressions as *closely related species*, and *no close relatives.*

A clearer statement on those points was made by Kosswig (1963) who, when considering modes of speciation in African cichlids, wrote: "In the first line, attention must be drawn to the fact that the cichlids in each lake can be divided into a *series of species flocks,* all species of one flock being more intimately interrelated with one another than with any cichlid species outside that lake" (italics added). At least to a phylogeneticist, Kosswig appears to be stipulating that the members of a species flock should be of monophyletic origin, and that the monophyly should be at a low level of universality, no higher than that of sister species.

Kosswig's statement is of particular interest since he is no longer using the term *species flock* to embrace almost the entire complex of cichlid species in a lake, as had Mayr, and as had Regan with respect to the *Haplochromis* of Victoria and Malawi. Instead he would regard each lake as harboring a number of species flocks. That Kosswig chose his example of a cichlid species flock, the *Lamprologus* species, from Lake Tanganyika may explain partly his thinking on the subject. The cichlids of Lake Tanganyika, unlike those of Victoria and Malawi, are not dominated by a single and speciose genus. I am using the taxonomy of 1963, disregarding my recent breakup of the genus *Haplochromis* (Greenwood 1979; 1980), because we are dealing with the history of the species flock concept.

Kosswig's views are also of interest in relation to Mayr's 1942 perspectives on the origin of speciose lake faunas, but his interpretation of a species flock is quite different from Mayr's. Whereas Mayr apparently interpreted closeness of relationship on a fairly broad basis, Kosswig is employing a much finer level of discrimination. At this finer level, and accepting monophyly as an essential element in the definition of a species flock, Mayr's postulated history of a lake's faunal origins through multiple invasions and accretions from different sources would still be an acceptable scenario. It would most probably result in an aggregate of several flocks, however, with the ancestors of the various flocks stemming from the different sources involved.

An earlier recognition of monophyly as an essential element in the origin of a species flock is apparent in Herre's (1933) paper on the evolution of the endemic cyprinid species in Lake Lanao, Philippines. Although Herre did not use the term *species flock,* he described a situation which later authors were to treat as a classical example of that phenomenon (Mayr 1942; Brooks 1950; Myers 1960). According to Herre the 18 or so endemic Lanao cyprinids, placed in four endemic and one non-endemic group, evolved in the lake from a single ancestral species, extant in the lake and widely distributed elsewhere from Thailand to Singapore and throughout Sumatra, Java, and Borneo. Speciation and morphological differentiation, Herre argued, were intralacustrine, and had occurred since the lake first formed, at a time later fixed at about 10,000 years ago, according to Myers (1960), but thought to be during the later Tertiary by others (Frey 1969).

Herre (1933) briefly outlines the interrelationships of the several endemic species, but unfortunately, gives few indications of the characters on which these are based. He is more precise, however, in suggesting that the flock differentiated in response to changing ecological conditions and niche availability in the developing lake, which is an example of *schitzotypische Artsplitterung* (*sensu* Woltereck 1931, a paper not cited by Herre). Myers (1960) fully supported Herre's views on the origin, relationship and evolution of the Lanao cyprinids, which he identified as a species flock, and provided a detailed scenario to account for the presence of the presumed ancestral species on Mindanao Island.

More recently, Kosswig and Villwock (1965) and Reid (1980) have argued against a monophyletic origin for the Lanao cyprinids, and thus the identity as a species flock. Those authors also criticized the validity of certain morphological features on which Herre based-some of his genera, and argued against the presumed intralacustrine origin of the so-called flock. Regrettably, neither Kosswig, Villwock, nor Reid has carried out a detailed

phyletic analysis of the species involved, so the questions of relationship and monophyly remain unresolved, as therefore does the question of the Lanao cyprinids being or not being a species flock, if Kosswig's criterion of strict monophyly is observed.

Two comments on Reid's paper are called for to clarify an already confused situation. Reid's apparent intention is to show that the concept of *explosive speciation* is a fallacy. But since he does not mention Woltereck's (1931) original concept of that phenomenon, nor the grounds on which Woltereck's model was based, I suspect that Reid is really aiming his criticisms at the descriptive usage of that term; that is, rapid speciation in a geographically restricted area. Secondly, Reid appears to have overlooked Herre's 1933 paper on the Lanao fishes, since he obviously believes Herre made no claims to the evolutionary importance of the Lanao phenomenon. He could not have entertained this idea had he read Herre's paper which, indeed, is titled: "The fishes of Lake Lanao: a problem in evolution."

Arguments supporting the identification of the Titicaca cyprinodonts as a true species flock, another classical example of the phenomenon, were also marshalled by Kosswig and Villwock (1965).

The importance of Kosswig and Villwock's (1965) arguments, and those of Reid (1960) lies in their underlining the question posed earlier: How does one identify a species flock? The other papers discussed here also serve to stress the confusion surrounding the problem, and in particular, the tangled issues of pattern and process. From all that has been said, is it possible to formulate a species flock concept, and if so, what is its significance in the broader realms of evolutionary theory?

This bifaceted problem is illustrated by the views of Myers (1960) and Herre (1933), on the one hand, and of Kosswig and Villwock (1965) on the other, regarding the status of the Lanao cyprinids. It is also clearly seen in the case of the haplochromine species of Lakes Victoria and Malawi (cf Greenwood 1974; 1979; 1980).

Both Brooks and Greenwood referred to each endemic species aggregate as a *Haplochromis* species flock because all the taxa involved were then placed in the genus

Haplochromis (Brooks 1950; Greenwood 1951-1974). The four monotypic genera endemic to Lake Victoria were also included because of their assumed close relationship to certain *Haplochromis* species in that lake.

On reflection it became apparent that one cannot establish the monophyletic origin of the Lake Victoria *Haplochromis* species, and that the genus itself is not a natural one (Greenwood 1979; 1980). Preliminary examinations also suggest that a situation like that in Victoria exists among the Malawi species previously referred to as *Haplochromis*, and that the features which Regan (1922a,b) used to establish the supposed common ancestry of all the Malawi *'Haplochromis'* species are of questionable value for that purpose (Greenwood 1983). In short, it might be argued that one should not speak about a single *Haplochromis* species flock in either Lake Malawi or Lake Victoria.

The Lake Victoria haplochromines, that is, species formerly referred to *Haplochromis* and the allied monotypic genera, together with haplochromine species from Lakes Edward, George, Kivu and Nabugabo, can, I believe, be broken down into a number of polyspecific but monophyletic lineages, members of which occur in more than one lake. Each lineage is, as yet, of indeterminate interrelationship with any other lineage in or outside of Lake Victoria (Greenwood 1979; 1980).

In many respects the situation in Lake Victoria, and probably Malawi now resembles that described above for Lake Tanganyika, and the situation postulated by Kosswig and Villwock (1965) for the cyprinid "flock" in Lake Lanao. For Lake Tanganyika, both Myers (1960) and Kosswig (1963), apparently independently, identify only one cichlid species flock: that of *Lamprologus*, a unit of 36 species. In Lake Lanao, Kosswig and Villwock (1965) argue against the endemic cyprinid species of that lake being a species flock. Myers (1960) identified the Lanao cyprinids as a true species flock, and like Kosswig and Villwock (1965), he accorded that status to the *Orestias* species of Lake Titicaca as well. Since in both Lanao and Titicaca, the species involved, apart from the so-called ancestral species in Lanao, are endemic and restricted to a narrowly circumscribed area, these differences in opinion depend on whether or not

the authors consider them to be of monophyletic origin.

Clearly then, the central issue in any attempt to formulate a species flock concept and identify species flocks must hinge on the monophyly of its members and the level of universality at which that monophyly is determined. In other words, an aggregate of several species should be identified as a flock only if its members are endemic to the geographically circumscribed area under consideration and are each others' closest living relatives. It is this element of immediate shared common ancestry which principally distinguishes a species flock from other species aggregates occupying geographically circumscribed areas. It is also an important indicator of the particular circumstances, both biotic and abiotic, leading to the flock's genesis, circumstances which are likely to differ substantially from those underlying the origin of other kinds of localized species aggregates.

Seen in this light, the importance of species flocks lies not in their being some near mystical evolutionary phenomenon, but in the empirical evidence which can be derived from them and applied in a broader evolutionary context. If periods of intensive speciation are common in phylogeny, as postulated in the punctuated equilibrium theory of evolution, there may be nothing extraordinary about a species flock, even if the flock's size is more in keeping with that of a Scottish crofter's than an Australian sheep farmer's.

REFERENCES

Berg L S 1928 Die Fauna des Baikalsees und ihre Herkunft. *Arch Hydrobiol* Suppl 4:479-526
Brooks J L 1950 Speciation in ancient lakes. *Quart Rev Biol* 25:30-60, 479-526
Dorogostaisky V C 1923 The vertical and horizontal distribution of fauna in Lake Baikal. *Rec de trav des proff de L'Univ d'Etat a Irkoutsk. Sci Nat* 4:103-131
Fryer G, T D Iles 1972 *The Cichlid Fishes of the Great Lakes of Africa.* Oliver & Boyd. Edinburgh
Frey D G 1969 Limnological reconnaissance of Lake Lanao. *Verh Int Verein Limnol* 17:1090-1102
Gulick J T 1905 Evolution: racial, habitudinal. *Publ Carnegie Inst* Washington 25:1-269
Greenwood P H 1951 Evolution of the African cichlid fishes: the *Haplochromis* species-flock in Lake Victoria. *Nature* 167:19-20
_____ 1975 Cichlid fishes of Lake Victoria, East Africa: the biology and evolution of a species flock. *Bull Brit Mus Natur Hist (Zool)* Suppl 6:1-134

Greenwood P H 1979, 1980 Towards a phyletic classification of the 'genus' *Haplochromis*, (Pisces, Cichlidae) and related taxa. Parts I, II. *Bull Brit Mus Natur Hist (Zool)*: Pt I, 35:265-322; Pt II, 39:1-101

_____ 1983 On *Macropleurodus, Chilotilapia* (Teleostei, Cichlidae) and the interrelationships of African cichlid species flocks. Ibid 45:209-231

Herre A W C T 1933 The fishes of Lake Lanao: a problem in evolution. *Amer Natur* 67:154-162

Kosswig C 1963 Ways of speciation in fishes. *Copeia* 1963:238-244

Kosswig C, W Villwock 1965 Das Problem der intra-lakustrischen Speciation im Titicaca- und im Lanosee. *Verh Deut Zool Ges Keil 1964 Zool Anz Suppl* 28:95-102

Kozhov M 1963 *Lake Baikal and its Life*. Junk. The Hague

MacDougal D T, A M Vail, G H Schull 1907 Mutations, variations, and relationships of the Oenotheras. *Carnegie Inst Publ* 81:1-92

Mayr E 1942 *Systematics and the Origin of Species*. Columbia Univ Press. New York

_____ 1947 Ecological factors in speciation. *Evolution* 1:268-288

_____ 1963 *Animal Species and Evolution*. Belknap. Cambridge Mass

_____ 1976 Sympatric speciation. *Evolution and Diversity of Life: Selected Essays*. Belknap. Cambridge Mass

Moore J E S 1903 *The Tanganyika Problem*. Hurst Blackett. London

Myers G S 1936 Report on the fishes collected by H C Raven in Lake Tanganyika in 1920. *Proc U S Nat Mus* 84:10-15

Myers G S 1960 The endemic fish fauna of Lake Lanao and the evolution of higher taxonomic categories. *Evolution* 14:323-333

Plate L 1913 *Selektionsprinzip und Probleme der Artbildung*. Leipzig. Berlin

Reid G McG 1980 'Explosive speciation' of carps in Lake Lanao (Philippines) - fact or fancy? *Syst Zool* 29:314-316

Regan C T 1920 Classification of the fishes of the family Cichlidae - I. The Tanganyika genera. *Ann Mag Natur Hist* 9:5, 33-53

_____ 1922a The cichlid fishes of Lake Nyassa. *Proc Zool Soc London* 1921:675-727

_____ 1922b Cichlid fishes of Lake Victoria. *Proc Zool Soc London* 151-191

Rensch B 1933 Zool Systematik und Artbildungsproblem. *Verh Deut Zool Ges* 1933:19-83

Sargent C S, C H Peck 1906 Species of *Crataegus* found within twenty miles of Albany. *Bull N Y State Mus* 105:44-74

Stankovic S 1955 Sur la speciation de lac d'Ohrid. *Verh Int Ver Limnol* 12:478-506

Trewavas E 1949 Origin and evolution of the cichlid fishes of the great African lakes, with special reference to Lake Nyassa. *13th Int Cong Zool* 1948:365-368

Woltereck R 1931 Wie entsteht eine endemische Rasse oder Art? *Biol Zentralbl* 51:231-235

Worthington E B 1937 On the evolution of fish in the great lakes of Africa. *Int Rev Gesamten Hydrobiol Hydrogr* 35:304-317

IS THE SPECIES FLOCK CONCEPT TENABLE?

A J RIBBINK

INTRODUCTION

The term *species flock* is frequently used with reference to island-type faunas such as the Galapagos finches, Hawaiian *Drosophila* and Honeycreepers, and the faunas of lakes Baikal, Titicaca, Lanao, Victoria, Tanganyika and Malawi. The species flock concept has vague origins (Greenwood 1984) and has not had the benefit of a scientific definition. It is thus subject to various interpretations, but notions of what a species flock might be can be guided by empirical usage. Four important criteria emerge from the literature. A species flock should: 1) be monophyletic; 2) be speciose; 3) be endemic to a circumscribed area; and 4) show, relative to the size of the occupied area, a disproportionately high number of species compared with other areas. These criteria are applied here for the haplochromine fishes, particularly the *Mbuna* group, of Lake Malawi. Strict application of these criteria should demonstrate whether they are useful and whether the concept can be used in a narrow or a general sense.

Cichlidae of Lake Malawi

Lake Malawi probably supports 400-500 species of cichlid fish (Ribbink et al. 1983), of which only 4 are not endemic (Trewavas 1935). Seven species are of the tilapiine lineage and the remainder comprise the haplochromine lineage. The haplochromines include the *Cyrtocara* and the Mbuna groups, each with about 200 member species, several smaller groups including the genera *Lethrinops* with about 70 species and *Aulonocara*, *Trematocranus* and *Rhamphochromis*, each with about 15 species, and finally the di- and monotypic genera. The Mbuna group differs from the other groups in that it embodies 10 genera; all others represent a single genus, though revision of the genus *Cyrtocara* probably would result in several separate genera. The Mbuna genera

are "more closely related to each other than to any other genus" (Trewavas 1935). Fryer (1959) suggested that the Mbuna group is so distinct that it merits recognition as "at least a tribe."

EXAMINATION OF THE CRITERIA

Monophyly. Earlier it was assumed that the endemic haplochromines of Lake Malawi are monophyletic (Regan 1921; Trewavas 1935) and that the group constitutes a species flock (Fryer, Iles 1972; Greenwood 1974). Recently, however, Greenwood (1979, 1984) suggested that Lake Malawi haplochromines are oligophyletic, and therefore that they do not qualify as a single flock. Greenwood (1984) states that "the central issue in any attempt to formulate a species flock concept and identify species flocks must hinge on the monophyly of its members and the level of universality at which that monophyly is determined."

Research on the fishes of the Great Lakes of Africa is in its infancy. The poorly known taxonomy of these fishes coupled with the lack of data on their behavior, ecology and interrelationships makes it impossible to trace phylogenies in a manner which delineates all of the constituent species groups. As knowledge increases, the speciose cichlid fauna of each lake is no longer viewed as an amorphous mass of closely related morphologically similar species. Monophyletic groups are now being recognized. The trend is clearly demonstrated by Greenwood (1979, 1980) who created 25 genera from the single genus *Haplochromis*. Regarding the haplochromine species flock of Lake Victoria, he concluded that "if the component species cannot be shown to stem from a single and fairly recent ancestor" then, in the strictest sense, the term *species flock* should not be used (Greenwood 1980).

Since the species flock concept draws

attention to unexpectedly high numbers of related species which are assembled within a narrowly circumscribed area and not to the phylogenies of the various member species, the criterion of recent monophyly is inappropriate. This does not imply that monophyly has no part in the species flock concept. It certainly has. For the member species of a flock to be closely related they must share a common ancestry, even if the ancestor lived in the distant past.

For recognition of a species flock the most appropriate level of universality extends to the early ancestry or stem species from which the extant species now constituting a flock arose, rather than to the more recent patterns of shared ancestry considered important by Greenwood (1980, 1984). It is the high number of cichlid species in the African Great Lakes, the high number of cyprinid species in Lake Lanao, and the high number of crustacean species in Lake Baikal, *relative* to the numbers of species in surrounding areas that led to recognition of species flocks, and not concern about recency of monophyletic origin. Indeed, the criterion of monophyly implies that a species flock cannot be so designated until monophyly has been established. Yet flocks were recognized and the species flock concept was in use well before the taxonomic philosophies of Hennig became fashionable. Since a species flock is not a taxonomic category, it is not necessary that a flock be monophyletic in the cladistic sense (Nelson 1971). Therefore, I conclude that recent monophyly is not a general criterion of species flocks.

An examination of similarities among contemporary species enables one to recognize a hierarchy of subflocks, or as Kosswig (1963) noted, a series of species flocks within the main flock. This enables one to refer to a subsection of a flock without the need to enumerate all the component species within the subgroup.

Hierarchy of Cichlid Flocks in Lake Malawi:
Level 1. The cichlid flock.
Level 2. The tilapiine and haplochromine flocks.
Level 3. The Mbuna flock; i.e., the Mbuna genera. Similarly, the *Cyrtocara* group including its distinct subgoups.

It is conceivable that the haplochromine flock originated from several different ancestral forms (Greenwood 1979), but it does not really matter whether the flock is the product of one or several ancestors entering Lake Malawi. The collection of closely related haplochromine species within the confines of Lake Malawi has the qualities of a species flock.

Any further subdivision of the haplochromine species flock is of questionable value. For example, the *Lethrinops* species flock may also be called the *Lethrinops* species, *Lethrinops* species group, or *Lethrinops* species complex. The same is true of other speciose genera among the haplochromines of Lake Malawi. Perhaps the value of using the term *flock* in preference to other terms is that it implies speciosity.

Speciosity. A flock must be speciose, but how many closely related species are required to constitute a flock? The smallest recognized cichlid flock comprises the four *Oreochromis* species of the *saka-squamipinnis* group in southern Lake Malawi (Lowe-McConnell 1975). Similarly, a flock of five species of the genus *Cyprinodon* inhabit Laguna Chichancanab in Mexico (Humphries 1984). Therefore there is no doubt that groups with larger numbers of member species readily qualify as species flocks. However, I shall argue that it is the *relative* number of species, and not necessarily the absolute number within a circumscribed area which dictates whether a species assemblage constitutes a species flock.

Endemicity. Species flocks are most readily recognized among invasive faunas which apparently have moved into vacant island-type habitats where they underwent rapid, often spectacular, radiation and speciation. Hence a feature of extant species flocks is that they are confined to narrowly circumscribed areas, and since most if not all contemporary members evolved within the confines of such areas, a very high degree of endemicity is characteristic. There probably is general acceptance of the criterion of endemicity and sympatry, though problems may arise if the interpretation is too rigid. All Mbuna are endemic to Lake Malawi, but only two of the 196 species studied by Ribbink et al. (1983) have a lake-wide distribution. All other species have narrow distributions, some being confined to a small island, or part of the rocky shore on an island, or a small section of mainland coast. Indeed, several species are limited to areas less than 5000 m^2. As a consequence of the very narrow distribution of the vast majority of species, each island, rocky outcrop, and every portion of the mainland coastline is inhabited by a different assemblage of species. Each of these assemblages could be considered a species flock, such as the Mbuna flock of 40 species at Likoma Island or of 18 species in Monkey Bay. A further effect is that this fragments the species flock in a manner which is contrary to the sense in which the concept is generally used. Hence, such a rigorous use of the criterion of endemicity would erode the concept. A species flock is endemic to the whole of the circumscribed area invaded by its ancestors and should not be broken into constituent parts.

Disproportionate Speciosity.
The most important quality of a species flock is its disproportionate abundance of closely related species within a geographically circumscribed area (David H Eccles pers comm). Usually, a species flock is represented by more species than would be expected from empirical knowledge of the group in the regions which surround the area in which the flock developed. Further, the rate at which adaptive radiation and speciation of the invasive fauna took place to give rise to species flocks is greater than expected from studies of faunas derived from the same or similar stocks but which remained in areas from which colonization originated. By inference, the quality of disproportionate speciosity suggests unusually high rates of evolution and speciation of a particular group within a circumscribed area. Therefore, the term *species flock* is comparative. The species diversity in an island-type situation is high relative to that of an adjacent continental area. Similarly, the rate of speciation and degree of adaptive radiation of the invasive fauna is higher than that of the related, continental fauna in adjacent regions. Other forms of comparison are also made. Among the Cichlidae, the haplochromine group is highly speciose, but the tilapiine group is not. Thus, the four tilapiines of the *saka-squamipinnis* group of the genus *Oreochromis* can be considered

a species flock (Lowe-McConnell 1975) because, relative to *Oreochromis* occurrences elsewhere, the presence of four sympatric species of recent common ancestry is unexpectedly high.

DISCUSSION
The quality of disproportionate speciocity is the strongest attribute of the species flock concept. It is apparent that a strict application of the criteria of monophyly and endemicity tend to subdivide a flock, and thus erode the concept. There is no doubt that as knowledge inceases, and differences between fish groups become clearer, strict application of these criteria would result in an increasing tendency to subdivide these flocks further and thus accelerate the demise of the concept. Therefore, monophyly and endemicity should not be strictly applied criteria.

Although it is not necessary for the members of a flock to be products of recent monophyly, it is necessary that the members be fairly closely related. It seems incorrect to conclude, as did Greenwood (1980), that a flock which comprises a number of different recent monophyletic lineages is not strictly a flock. In my opinion, the haplochromines of lakes Victoria and Malawi do consititute flocks regardless of whether each flock arose from one or more closely related invasive species.

There has been a historic tendency to recognize subflocks within flocks, such as Brooks' (1950) mention of five cichlid flocks in Lake Malawi: *Tilapia, Haplochromis, Lethrinops, Mbuna* and *Rhamphochromis*. Kosswig (1963) recognized a series of species flocks among the cichlids of Lake Tanganyika, and Greenwood (1979) divided the *Haplochromis* flock of Lake Victoria into subportions. However, the taxonomic categories embraced by the species flock concept are highly variable with little conformity among authors so that it is difficult to accommodate the views of all in a useful definition. One might refer to the crustacean flock of Lake Baikal or to its flocks of Isopoda, Ostracoda or Amphipoda. The Amphipoda constitute the most remarkable feature of the Baikalian fauna, comprising about 300 species assigned to 30 genera (Brooks 1950). These have been called a gammarid flock and speciose genera have also been called flocks.

Similarly, the molluscan flock of Lake Baikal comprises the prosobranch, pulmonate and pelecypod subflocks, as well as subflocks of the speciose genera. In Lake Lanao the cyprinid flock comprises five genera, but Myers (1950) apparently does not recognize the group as a flock. In his view, 13 species of one genus constitute a flock, while the other genera with very few species are not flocks. The flock of finches in the Galapagos Islands comprises 14 species and 6 genera (Lack 1947), but few of the genera would constitute species flocks. Similarly, the *Drosophila* species flock of Hawaii is represented by numerous genera (Kaneshiro 1974). Should every speciose genus be recognized as a species flock? With such a wide spectrum of taxonomic ranks encompassed by its usage, the term *species flock* obviously is loosely applied.

If one is to be guided by empirical usage then it is most difficult to narrow the term *species flock*. The definition should be sufficiently explicit to indicate with some precision the nature of the flock, but not so detailed as to duplicate without advantage all the information contained in a currently used taxonomic categorization of the group. Thus, referring to "teleost flocks" of the Great Lakes of Africa, or to the "crustacean flocks" of Lake Baikal is of little value. Reference to "cichlid flocks" and "amphipod flocks" provides more specific information, as does "Mbuna flock." In contrast, there is little value in being so specific as to portray the generic subdivisions of the Mbuna as flocks. Yet there are instances where species of a single genus have been correctly classed as a species flock (Brooks 1950; Lowe 1952; Myers 1960; Greenwood 1974). Clearly, the term cannot be narrowly defined in a manner which limits its use to particular taxonomic categories.

As one proceeds up the hierarchy from the general, all-encompassing flock to specific constituent flocks, recent monophyly is likely to become a more apparent characteristic of the flocks. However, by following this route the quality of disproportion is attenuated.

By referring to subgroups as *species flocks* an impression of speciosity is conveyed, but since a flock may have from four to several hundred member species, it is necessary to include a qualifying statement regarding the number of species in the flock. Accordingly, the advantage of additional information imparted by referring to these groups as flocks is questionable. Since "Mbuna" is a collective term, there is merit in referring to this speciose group as a species flock, but little advantage accrues from referring to genera or species complexes within the group as flocks. Nevertheless, if a definition of species flocks is to be developed, it may be necessary for it to embrace even these hierarchical levels.

Since a species flock cannot be narrowly defined, and since it has few criteria which can be strictly applied, is there any value in retaining this loose term? The answer is a definite *Yes*! The species flock concept portrays the following biological information: 1) Compared with related species elsewhere, the group includes a disproportionately high number of closely related species. 2) The members of the group apparently are the products of an unusually high rate of speciation, resulting from the *explosive* evolutionary responses of an invasive fauna. 3) The distribution of the member species is limited to a circumscribed island-type area to which most or all species are endemic. Thus, a species flock is an assemblage of a disproportionately high number, relative to surrounding areas, of closely related species which apparently evolved rapidly within a narrowly circumscribed area to which all the member species are endemic.

SUMMARY

An examination of the species flock concept suggests the following conclusions: 1) A species flock is an assemblage of a disproportionately high number of closely related species which evolved rapidly within a narrowly circumscribed area to which all or almost all the member species are endemic. 2) Although recent monophyly is a characteristic of some flocks, it is not a general criterion of a species flock, since flocks are not taxonomic categories. 3) Hierarchies of subflocks may be recognized in those flocks which exhibit considerable species diversity. It is concluded that the species flock concept is useful and should be retained.

REFERENCES

Brooks J L 1950 Speciation in ancient lakes. *Q Rev Biol* 25:30-60, 131-176

Fryer G 1959 The trophic interrelationships and ecology of some littoral communities of Lake Nyasa with especial reference to the fishes and a discussion of the evolution of a group of rock-frequenting Cichlidae. *Proc Zool Soc Lond* 132:153-281

Fryer G, T D Iles 1972 *The Cichlid Fishes of the Great Lakes of Africa. Their Biology and Evolution.* Oliver Boyd. Edinburgh

Greenwood P H 1974 Cichlid fishes of Lake Victoria East Africa: the biology and evolution of a species flock. *Bull Br Mus Nat Hist (Zool)* Suppl 6:1-134

_____ 1979 Towards a phyletic classification of the 'genus' *Haplochromis* (Pisces Cichlidae) and related taxa. Part 1. *Bull Br Mus Nat Hist (Zool)* 35:265-322

_____ 1980 Towards a phyletic classification of the 'genus' *Haplochromis* (Pisces Cichlidae) and related taxa. Part 2. The species from Lakes Victoria, Nabugabo, Edward, George, and Kivu. *Bull Br Mus Nat Hist (Zool)* 39:1-101

_____ 1984 What *is* a species flock? 13-20 A A Echelle, I Kornfield eds. *Evolution of Fish Species Flocks*. Univ Maine at Orono Press

Humphries J M 1984 Genetics of speciation in pupfish from Laguna Chichancanab, Mexico. 129-140 A. A. Echelle, I Kornfield eds. *Evolution of Fish Species Flocks*. Univ Maine at Orono Press

Kaneshiro K Y 1974 Phylogenetic relationships of Hawaiian Drosophilidae based on morphology. 102-110 M J D White ed. *Genetic Mechanisms and Speciation in Insects*. Australia New Zealand Book Co. Sydney

Kosswig G 1963 Ways of speciation in fishes. *Copeia* 1963:238-244

Lack D 1947 *Darwin's Finches*. Cambridge Univ Press. Cambridge England

Lowe-McConnell R H 1975 *Fish Communities in Tropical Freshwaters*. Longman. London

Mayr E 1963 *Animal Species and Evolution*. Harvard Univ Press. Cambridge

Myers G S 1960 The endemic fish fauna of Lake Lanao and the evolution of higher taxonomic categories. *Evolution* 14:323-333

Nelson G J 1971 Paraphyly and polyphyly: redefinitions. *Syst Zool* 20:471-472

Regan C T 1921 The cichlid fishes of Lake Nyasa. *Proc Zool Soc Lond* 1971:675-727

Ribbink A J, A C Marsh, C C Ribbink, B J Sharp 1983 A preliminary survey of the cichlid fishes of rocky habitats in Lake Malawi. *S Afr J Zool* 18:149-310

Tewavas E 1935 A synopsis of the cichlid fishes of Lake Nyasa. *Ann Mag Nat Hist* 16:65-118

SEMIONOTID FISHES FROM THE MESOZOIC GREAT LAKES OF NORTH AMERICA

AMY REED MCCUNE, KEITH STEWART THOMSON, and PAUL ERIC OLSEN

INTRODUCTION

Fishes of the "holostean" family Semionotidae have been found throughout the Mesozoic in both freshwater and marine sediments, although generally the nature and extent of their diversity has not been recognized (McCune 1982; Olsen et al. 1982). We will show that, like African cichlids, American semionotids were extremely diverse, dominating a system of large, deep, stratified, rift valley lakes, the geological record of which forms part of the Newark Supergroup of eastern North America (Olsen 1978). While the historical record of African cichlids is fragmentary at best (Van Couvering 1982), the fossil record of the semionotid species complex in Newark lake deposits is exceptional. In Newark sediments, semionotids are extremely abundant, often whole, and fully articulated. Also, because of the cyclic nature of the sedimentary record, we are able to sample the fishes from a temporal succession of discrete lakes within a given lake basin. Within a portion of each lake history the stratigraphic resolution is in yearly intervals, not thousands or tens of thousands of years —within the range of a genetical-historical time scale (Stebbins 1982), and far superior to the usual resolution in the "best" stratigraphic sequences (Williamson 1981; Schindel 1982; Dingus, Sadler 1982).

This unusual combination of paleobiological and sedimentological circumstances offers a unique opportunity to examine the historical development of an "adaptive radiation" of fishes. Owing to the high diversity of both cichlid and semionotid fishes and the similarities of their respective physical and environmental settings, study of these two unrelated groups may be reciprocally illuminating. Most biological information of the kind available for living cichlids will always be lacking for semionotid fishes. For that kind of information, we are limited to inference and analogy based on living fishes. On the other hand, direct historical studies of semionotids may be more informative about possible long term factors in the evolution of lake species than inferences based on living cichlids.

Various questions have been asked about the conditions that allow or initiate "explosive" radiation (e.g., Greenwood 1981; Fryer, Iles 1972). In general these conditions fall in two categories: those intrinsic to the organism, and those that are environmental. What are the initial conditions for "explosive" radiation, and what sort of interplay is necessary between evolutionary potentials of the organism and environmental opportunities? In semionotids, the overall patterns of variation that we have charted point to an intrinsic but not obviously adaptive component to diversification. Beyond that, long term environmental cyclicity—the geological evanescence of Newark lakes—may have repeatedly provided these fishes new ecological opportunities.

GEOLOGICAL BACKGROUND

The Late Triassic to Early Jurassic Newark Supergroup includes the lacustrine sediments deposited in ten major and several minor basins in eastern North America (Figure 1). These basins are part of a rift valley system, much like the rift valleys in Africa today, formed during the tectonic prelude to the opening of the Atlantic Ocean (Manspeizer et al. 1978). Intermittently, for about 45 million years during the Late Triassic and Early Jurassic, these basins were filled by very large deep lakes, some of which may have been as large as modern lakes Tanganyika, Malawi, and Baikal (Olsen 1980b). For simplicity, we give a

Figure 1. Exposures of the Newark Supergroup in eastern North America. 1. Wadesboro Basin (Chatham Group). 2. Sanford Basin (Chatham Group). 3. Durham Basin (Chatham Group). 4. Davie County Basin. 5. Dan River Basin. 6. Scottsburg Basin. 7. Basins south of Farmville Basin. 8. Farmville Basin. 9. Richmond Basin. 10. Taylorsville Basin. 11. Scottsville Basin. 12. Culpeper Basin. 13. Gettysburg Basin. 14. Newark Basin. 15. Pomperaug Basin. 16. Hartford Basin and Cherry Brook Outlier. 17. Deerfield Basin. 18. Fundy Basin (Fundy Group). 19. Chedabucto Basin (Olsen et al. 1982)

generalized picture of Newark paleolimnology, taken mainly from Olsen et al. (1978) and Olsen (1980b).

Newark lake sediments are cyclic with a regularly repeating sequence of physically different sediments. These different sedimentary features, produced by sedimentation in different depths of water, reflect cyclic changes in lake level. Thus, understanding the deposition of these sedimentary cycles provides significant insight into the environmental context of semionotid evolution. Each sedimentary cycle records the history of one manifestation of a lake in the basin concerned. As lakes formed and reformed in the same basin through time, new cycles of sediment were laid down. A cycle or lake history can be divided into three units which represent 1) the expansion of a shallow lake over swamps and dry land, 2) a period of maximum lake transgression when the lake was large, deep, and stratified, and 3) the contraction and eventual evaporation of the lake (Figure 2).

The lower unit (Division 1) deposited during lake transgression is poorly bedded and may contain mudcracks, current bedding, and fossils such as plants (including roots), pollen, spores, dinosaur footprints and stromatolites. This basal unit grades upward into rock that is very finely bedded or microlaminated (Division 2). These microlaminated sediments, deposited in very large deep stratified lakes, contain fully articulated fish, reptiles and invertebrates —sometimes even insects. At this stage of maximum lake transgression, the lakes in some basins were in excess of 8000 square kilometers in area and more than 100 meters deep. The uppermost unit (Division 3) is defined by the breakdown of the microlaminated structure of the sediments. As in Division 1, the sediments of Division 3 were deposited in shallow water, but they reflect a contracting, not expanding lake. Here we find current bedding, salt casts from evaporation, mudcracks, in situ roots, and dinosaur footprints (Figure 2). Although there are fossils throughout the cycles, they are preserved best in the middle microlaminated sediments laid down in deep stratified lakes. The hypolimnion of these stratified lakes was anerobic, and thus hostile to bioturbating organisms and scavengers, allowing preservation of both fine sedimentary structures and articulated fossils like those pictured in Figure 3.

The microlaminated sediments in Division 2 comprise alternating layers of organic-rich silt and carbonate-rich silt analogous to the

GENERALIZED NEWARK SEDIMENTARY CYCLE

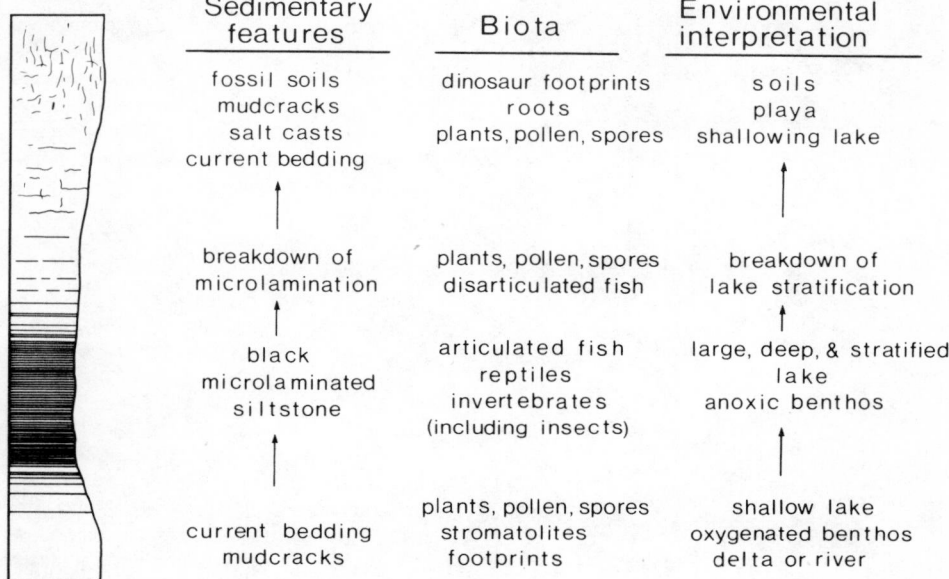

Sedimentary features	Biota	Environmental interpretation
fossil soils mudcracks salt casts current bedding	dinosaur footprints roots plants, pollen, spores	soils playa shallowing lake
↑		↑
breakdown of microlamination	plants, pollen, spores disarticulated fish	breakdown of lake stratification
↑		↑
black microlaminated siltstone	articulated fish reptiles invertebrates (including insects)	large, deep, & stratified lake anoxic benthos
↑		↑
current bedding mudcracks	plants, pollen, spores stromatolites footprints	shallow lake oxygenated benthos delta or river

Figure 2. Generalized Newark sedimentary cycle. A schematic stratigraphic section of a single Newark sedimentary cycle is pictured at left. The corresponding progression of certain sedimentarly features characteristic of a cycle is listed in the first column. Characteristic fossils and their state of preservation at corresponding positions are in the "biota" column. Together, the sedimentary features and character of the biota form the basis for the environmental interpretation of the stages of lake development (3rd column). One sedimentary cycle corresponds to the formation and evaporation of a single lake.

annual varves (each couplet = one year of deposition) deposited in modern perennially stratified lakes (Olsen 1980b; Davies, Ludlam 1973; Livingstone 1965). The analogy is important because it means that within the microlaminated portion of a given cycle, we can place individual fish in a year-by-year chronology.

As remarkable as it may seem, these great deep lakes periodically evaporated completely, as shown by the presence of salt casts, mudcracks, reptile footprints, in situ roots, and fossil soils in the upper divisions of cycles. Periodicity of lake formation and evaporation was first estimated at about 21,000 years by varve counts (Van Houten 1969; Olsen 1980b) and corroborated by an independent estimate based on radiometric dates and a stratigraphic correlation derived from vertebrate, palynological and paleomagnetic data (Olsen 1983). This periodicity has been linked to the climatic changes induced by the precession

cycle of the equinox and the cycle of eccentricity of the earth's orbit (Olsen 1983; Van Houten 1969).

SEMIONOTID FISHES

Semionotid fishes were first reported in North America early in the 19th century (Hitchcock 1819; Agassiz 1833-44). Since then, many thousand individuals have been collected from over 200 fish localities in Newark Supergroup sediments (McDonald 1983). It may seem remarkable that "species flocks" could go unrecognized for more than 160 years in such a well-collected fauna. Certainly early workers, recognizing that *Semionotus* was both abundant and diverse, described numerous species, although their descriptions were rarely diagnostic. Also, much variability remained hidden by poor preservation or lack of preparation. Confusion in early descriptions led most 20th century workers to believe that these Newark semionotids were much more

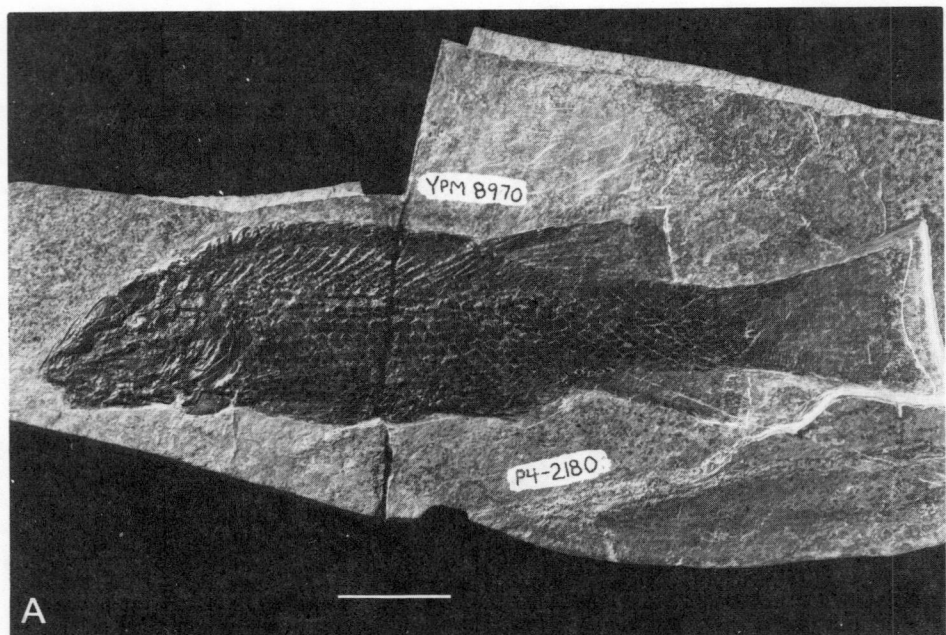

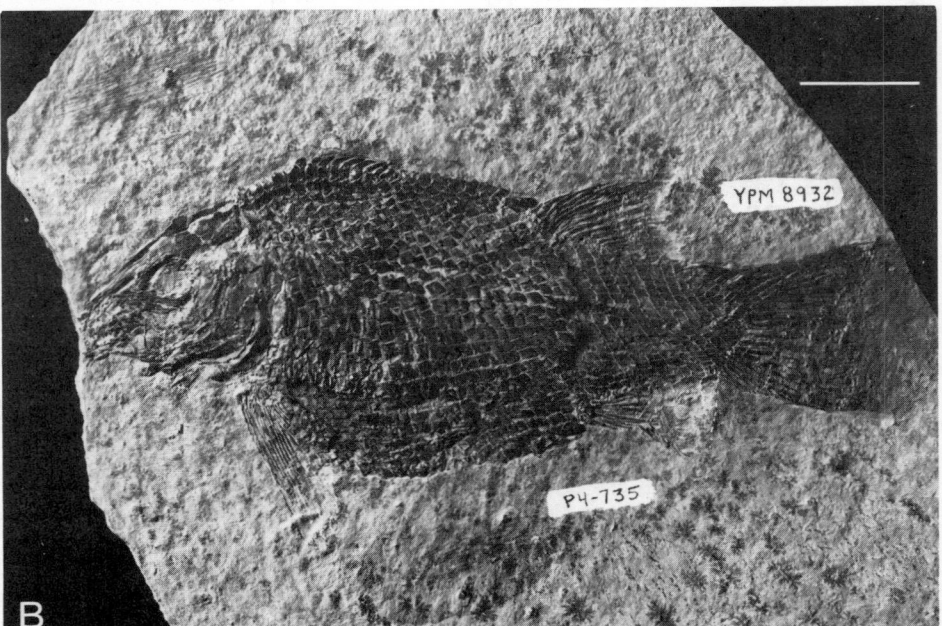

Figure 3. Exemplars of two semionotid species showing the preservational quality of fishes in microlaminated sediments. (Descriptions in McCune 1982). Both specimens are from cycle P4 (division 2) of the early Jurassic Towaco Formation of the Newark Basin. A: Member of species "p" in Figures 6, 7. B: Member of species "r" in Figures 6, 7. Scale bars = 2 cm.

diverse in name than in reality (eg., Schaeffer 1967).

Our revised assessment of semionotid diversity in Newark lakes is based on the many thousands of specimens that we collected from several large excavations in New Jersey and Connecticut, as well as by extensive prospecting throughout a broad range of other Newark Supergroup deposits (Olsen et al. 1982). To collect complete fish, and preserve the most detailed stratigraphic data possible, we removed sheets of rock 1-2 cm thick, collected the fish in each layer, and recorded the microgeographic positions of individual fish (McCune 1982). In addition to the advantages afforded by the resulting large samples of complete specimens, application of relatively recent advances in fossil preparation (e.g., negative acid preparation and high resolution casting compounds; see Rixon 1976) has enabled us to see and characterize variation in a way earlier workers could not.

Variation, Diversity, Distribution

Semionotus, Lepidotes, and two undescribed American genera form a monophyletic group, the Semionotidae, defined by 1) the presence of dorsal ridge scales, 2) a large posteriorly directed process on the epiotic, 3) premaxillae with prominent rostral processes sutured to the frontals, and 4) serially subdivided lacrymals (Olsen, McCune ms; McCune 1982). Semionotids, exemplified by the Newark semionotid illustrated in Figure 4, exhibit a suite of primitive characters such as hemiheterocercal tail, rhomboid shaped scales covered with ganoine, fins comprised of lepidotrichia and fringed by paired fulcra, a vertical suspensorium, a small mouth, and ordinary conical teeth. Most of the observed morphological diversity among Newark semionotids is confined to two complexes of characters —the modified scales along the dorsal midline (dorsal ridge scales) and the overall form of the body. There seems to be little diversity in the morphology of feeding structures. A few meristic characters are useful to define particular species.

Dorsal ridge scales are convex with posteriorly directed spines in the primitive state for semionotids (Figure 5a; Olsen at al. 1982). Deviations from this design are almost unknown in semionotids outside the Newark system (for exceptions see McCune 1982,

Olsen et al. 1982), but Newark semionotids exhibit an impressive array of morphological diversity (Figure 5). The dorsal ridge scales may be concave, often decorated by small tubercles anteriorly (Figure 5g) as described for the "*S. elegans* group" (Olsen et al. 1982). They may be laterally undercut and only partially covered by ganoine (Figure 5d). The spines may be abbreviated and the size of the scales reduced relative to the body (Figure 5c) as described for the "small scale group" (Olsen et al. 1982), or they may be dorsoventrally enlarged in the anterior region (Figure 5e,f), the more angular variety corresponding to scales described for the "*S. tenuiceps* group" (Olsen et al. 1982). There is a gradient of modification of scale structure according to position along the anterior-posterior axis of the series. In all cases, the anterior scales in the series are the most distinctive. Differences between types of dorsal ridge scales cannot be explained as ontogenetic stages. Even the most extreme forms of dorsal ridge scale types are found in both very small and very large individuals.

The adaptive significance of the dorsal ridge scale series is difficult to assess. Mate recognition is an unlikely adaptive explanation, as groups of species share the same dorsal ridge scale morphology. While there might also have been different locomotor consequences for different morphologies, there is no association of a particular body form with a particular dorsal ridge scale morphology.

Diversity of body form is the most prominent aspect of variability among Newark semionotids. Different shapes have been recognized by an iteration of pairwise geometrical comparisons of individuals, standardized by size. These groups were then described by bivariate plots and by a discriminant analysis of 13 measurements (McCune 1982). A sample of the variation in body shape for semionotids from a single lake cycle sharing the primitive dorsal ridge scale type is illustrated (Figure 6C a-g).

Because variation in dorsal ridge morphology and body shape is not concordant, the diversity of distinguishable forms is very high. For example, not all deep-bodied fish have robust scales, and both slender and deep-bodied fish may have robust scales. Though there are obvious difficulties in using

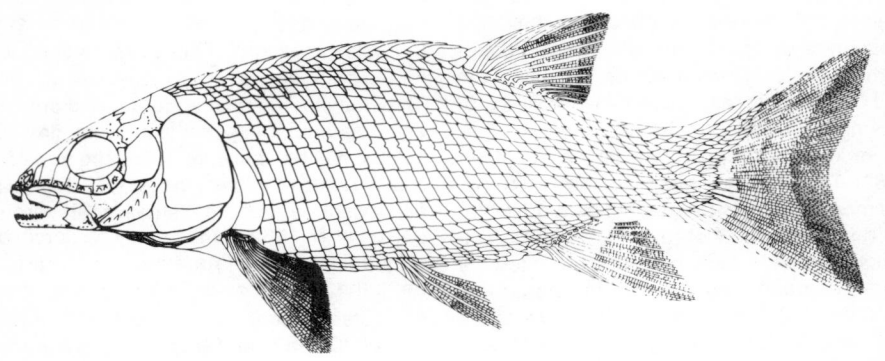

Figure 4. Reconstruction of semionotid fish from the Early Jurassic Boonton Formation of the Newark Basin. The figure is based on camera lucida drawings of a complete fish, both part and counterpart (YPM 6567) and taken from Olsen and McCune (ms).

only morphology to define species (e.g., Kornfield et al. 1982, Wake 1981), this practice at least communicates the diversity of form when information about reproductive behavior and genetics is lacking. Here, the inference may be partly justified because initial results seem to show that different "species" of semionotids, like different species of *Lepisosteus* exhibit distinct patterns of scale growth (Thomson, McCune 1983). Species of semionotids have been determined in this manner for only one Newark lake, lake cycle P4. The principal result was the description of 21 species of semionotids, 20 of them new (McCune 1982). In Figure 6C, each of these species is represented by an outline drawing and grouped with other species having the same dorsal ridge scale morphology.

In different lake cycles of the Newark Basin and in other basins of the Newark Supergroup different assemblages of fishes have been found (Olsen et al. 1982).

Semionotids are known from 8 basins of the Newark Supergroup, including the Fundy Basin in Nova Scotia, Deerfield Basin in Massachusetts, Hartford Basin in Connecticut, Newark Basin in New Jersey, Culpepper Basin in Virginia, and Dan River Basin in North Carolina (see Figure 1). Of the 45 million years spanned by Newark sediments, semionotids have been found through a period of 40 million years.

Although semionotid faunas from most Newark lakes are mainly undescribed, we have a general sense of overall diversity for a few other lakes. We illustrate several of these faunas from lakes that filled the Newark basin at various times (Figure 6), but we can not yet assign taxonomic status to any of the fish illustrated here except those from the Early Jurassic Towaco Formation, lake cycle P4.

Each of the four faunas in Figure 6 (A,B,C,D) comes from a discrete sedimentary cycle, separated from one another by millions of years and many thousands of feet of sediment. For stratigraphy of the Newark Basin see Olsen (1980a). The faunas are arranged in the stratigraphic order (A - D) of decreasing age. Each outline drawing in Figure 6C represents a distinct species (McCune 1982). Drawings in 6A, B and D probably represent species, but are pictured here only to give an approximation of diversity.

Of the Late Triassic Newark semionotids, we have only illustrated the fauna in one lake cycle from the Lockatong Formation. However, our sample comes from a large excavation. Results from that excavation, with our prospecting in another 50 Lockatong lake cycles, and the sample from an earlier excavation in North Carolina (Olsen et al. 1978) show that the diversity of semionotids in the Newark lake system throughout the Late Triassic was relatively low. In contrast, semionotids from

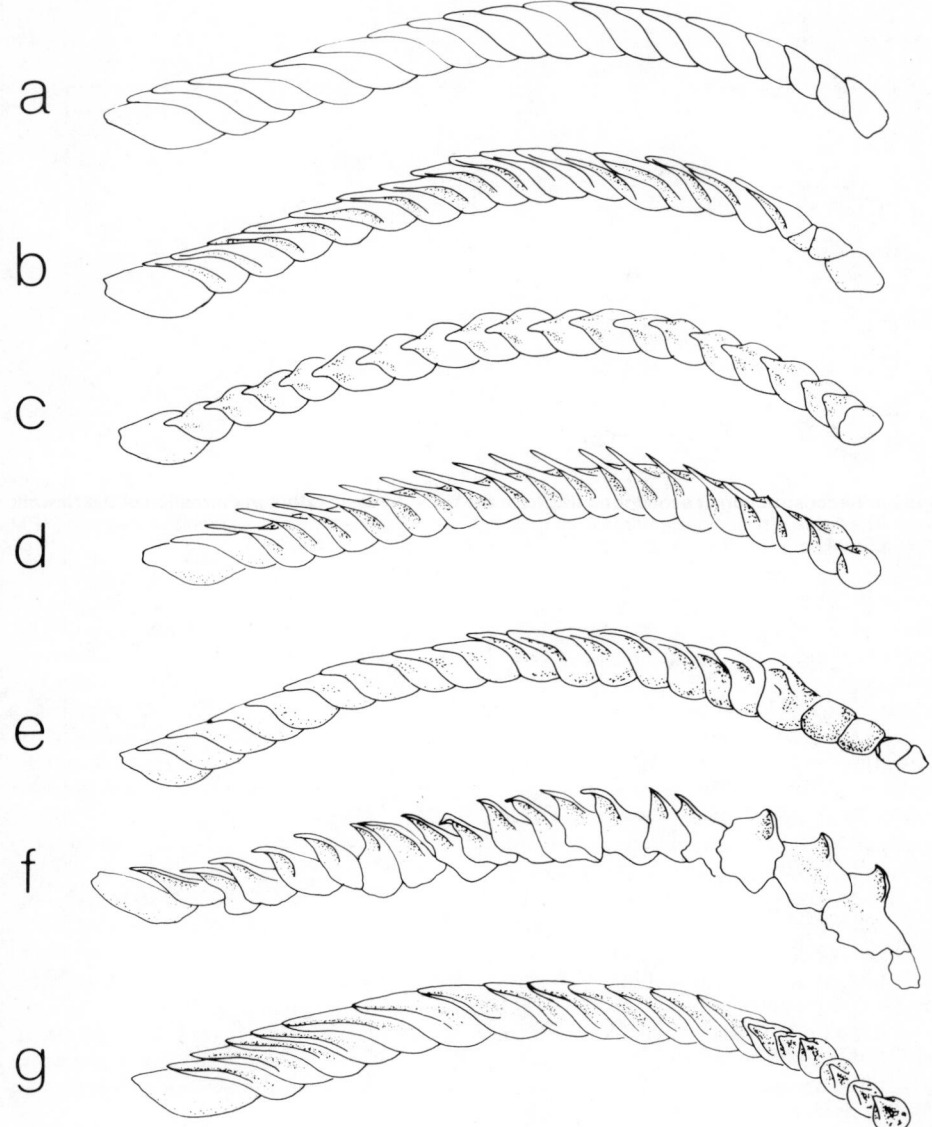

a

b

c

d

e

f

g

Figure 5. Morphological diversity of the dorsal ridge scales of seminotid fishes from cycle P4, Towaco Formation, Newark Basin. Scales rim the dorsal midline between the nape and the origin of the dorsal fin. The anterior direction is towards the right in each series pictured; spines are directed posteriorly. A: "simple" scales correspond to those of the "simple scale group" (Olsen et al. 1982). B: "modified simple" scales. C: "small" scales, corresponding to those of the "small scale group" (Olsen et al. 1982). D: "thin-spined" scales, corresponding to those of the "*S. Micropterus* group" (Olsen et al. 1982). E: "globular" scales. F: "robust" scales, corresponding to those of the "*S. tenuiceps* group" (Olsen et al. 1982). G: "concave" scales, corresponding to those of the *S. elegans* group" (Olsen et al. 1982).

Early Jurassic lakes from the Feltville, Towaco and Boonton formations in the Newark Basin were much more diverse (refer to Figure 6, B,C,D). The degree to which these faunas were endemic remains to be seen, but the fact that all semionotids from Boonton share the

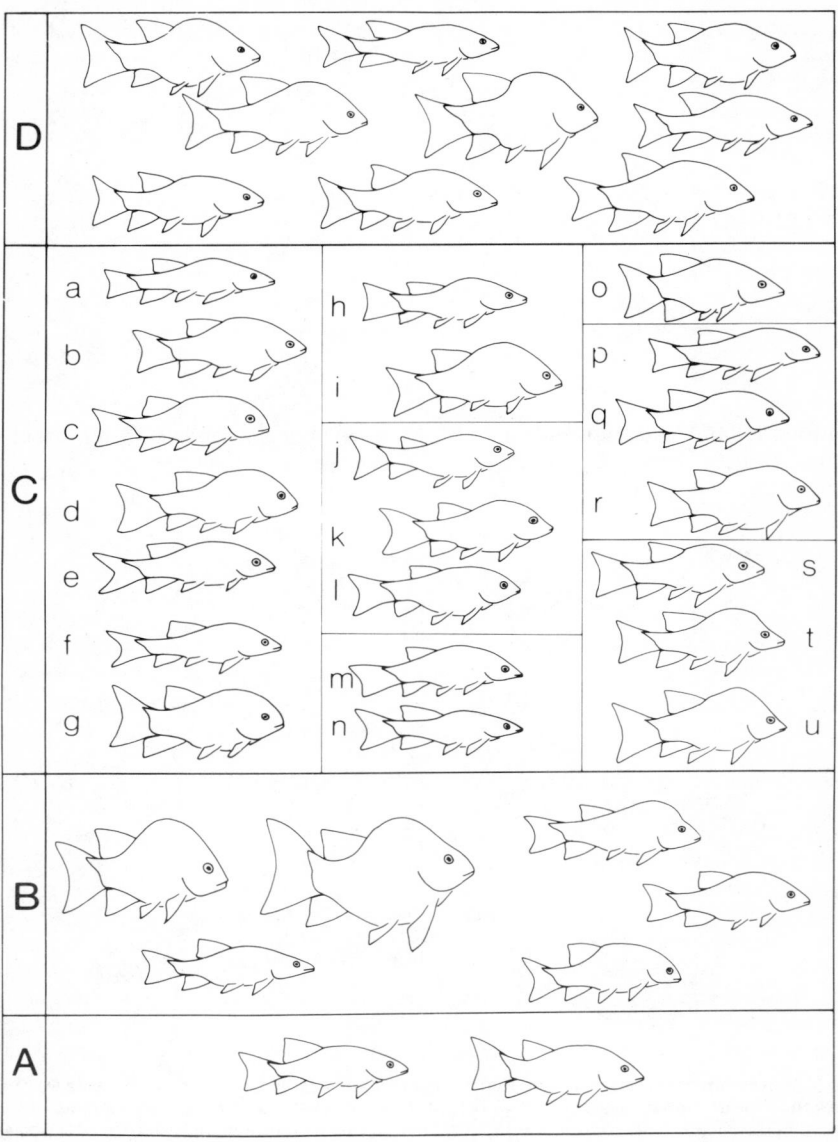

Figure 6. Array of semionotid fishes known from lakes in the Newark Basin, New Jersey. Faunas in stratigraphic order, A,B,C,D, from oldest to youngest. All drawings are based on complete specimens. Names of species are not given because descriptions have not yet been published. Species a - u correspond to those in Figure 7. A. Semionotid fauna from lake cycle W6 of the late Triassic Lockatong Formation. B. Semionotid fauna from Early Jurassic Feltville Formation. C. Semionotid fauna from cycle P4 Early Jurassic Towaca Formation. Species are grouped by dorsal ridge scale as follows: a - g, simple scales; h, i, modified simple; j, k, l, thin-spined; m, n, concave; o, small; p, q, r, globular; s, t, u, robust.

same dorsal ridge scale morphololgy suggests a high probability of endemism.

Preliminary diversity data suggest an inverse relationship between the number of semionotid taxa and the number of non-semionotid fish taxa in any given lake cycle. Where the diversity of semionotids is highest, such as Towaco Formation lake cycle P4, there are no non-semionotid fishes. Conversely, there are only one or two semionotids in most Lockatong cycles, but there are minimally three to four other genera of fishes. Our ability to deduce the ecological relationships of these fishes is limited, but we are intrigued by the high diversity of semionotids in the absence of other fishes. Interestingly, the diversity of cichlid and noncichlid fishes is also inversely related in several African lakes (Greenwood 1984). The ephemeral nature of individual Newark lakes and associated drainage patterns may have allowed chance to play an important role in excluding potential non-semionotid predators and competitors in some lakes but not in others. While predation and competition may help maintain diversity in ecological time, the absence of an associated fauna may have been one factor which allowed the evolution of semionotid diversity.

Relationships, Polytomies and Speciation

To understand the evolution of semionotids in the Newark lakes we must elucidate their relationships and distributions in space and time. We initially concentrated our taxonomic studies on the semionotids from Towaco lake cycle P4. Figure 7 shows tentative relationships of 21 species of semionotids from lake cycle P4 and previously described European species. By presenting this cladogram we do not suggest a definitive pattern of relationship for all semionotids. We mean only to organize the information we have so far amassed to facilitate discussion and future work.

The details of relationships within the Semionotidae are still ambiguous and we have not included many undescribed Newark semionotids here. The principal hindrance to working out relationships among semionotids is finding enough synapomorphies to define subgroups. A large portion of the derived characters in the cladogram are autapomorphic (33) rather than synapomorphic (10).

One of the most striking features of the cladogram is the number of unresolved polytomies. Of the 8 nodes defined by one or more synapomorphies, 5 are polytomies — nodes that include 3 to 9 branches. While there is nothing intrinsically wrong with polytomies, they are often thought to arise from "partial ignorance" (Nelson 1980), owing to shortages of specimens, taxa, or characters, particularly in paleontological studies (Patterson, Rosen 1977). However, in Newark semionotids, the kinds of characters and the completeness and high quality of specimens require us to accept the present character distributions until contrary evidence accrues. Further, there are many reasons to believe that polytomies in cladograms of species, especially diverse groups like species swarms, are appropriate to represent true character distributions, and not simply incomplete analyses.

The appeal of fully dichotomous cladograms over those that contain polytomies is their higher information content (Nelson 1980). But higher information content is a formal statement about a dichotomous versus a polytomous branching pattern in the abstract. It is not an evaluation of the degree to which a particular branching pattern reflects the relationships within a group of taxa.

Both Nelson (1980) and Wiley (1981) discuss special circumstances, such as multiple speciation, hybridization, and inclusion of ancestors, when an unresolved polytomy would reflect true character state distributions. Wiley reasoned that if a number of peripheral isolates around the range of a single ancestral species formed new species, they would share only primitive characters with other peripheral isolates and the ancestral species. The cladogram of a group with this character-state distribution would be a polytomy with as many branches as there were taxa. Assuming the same character distribution, if all the descendant species arose from a single common ancestral species that did not change through time, the cladistic pattern would still be a polytomy even if speciation were neither simultaneous nor allopatric. Operationally, the branching sequence of this non-simultaneous case could not be determined from the character distributions. Unless the actual sequence of speciation "events" could be

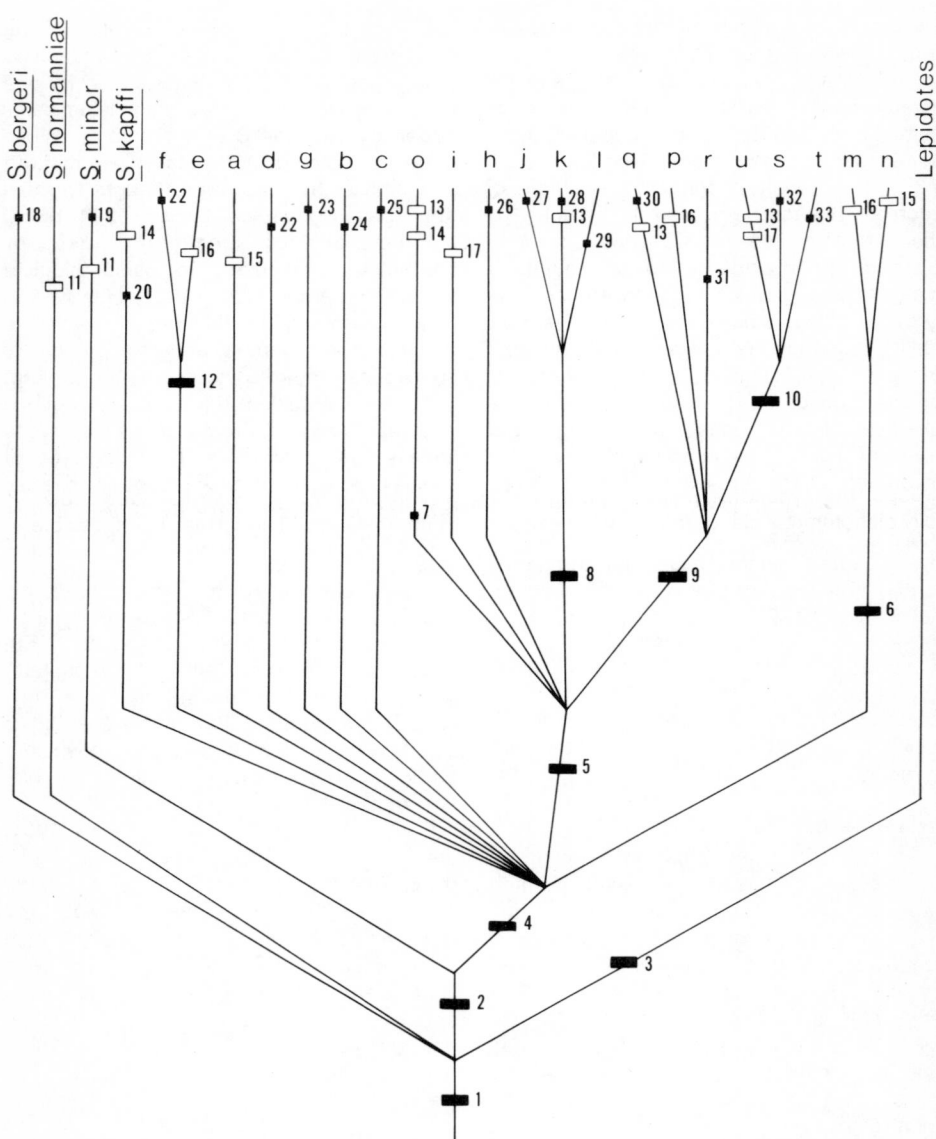

Figure 7. Tentative cladistic relationships for semionotids from P4 and European semionotids. (McCune 1982; McCune ms) Names for species a through u (also in Figure 6C) are not given because descriptions have not yet been published. Large black bars are synapomorphies. White bars represent homoplasies. Primitive character states are not shown. Key to derived character-states is as follows: 1. Dorsal ridge scales present, premaxillae with long rostral process, epiotic with posteriorly directed process, lacrymal serially subdivided. 2. Fringing fulcra reduced in number. 3. Suborbitals number greater than one. 4. Basal fin fulcra reduced in number. 5. Dorsal ridge scales laterally undercut. 6. Dorsal ridge scales concave. 7. Dorsal ridge scales with short spines, dorsal ridge scales small relative to flank scales. 8. Dorsal ridge scale spines very narrow and separated distally from base. 9. Anterior dorsal ridge scales enlarged. 10. Anterior dorsal ridge scales even larger. 11. Flank scales serrated. 12. Caudal fin forked. 13. Horizontal flank scale rows reduced. 14. Vertical flank scale rows reduced. 15. Shape "1". 16. Shape "2". 17. Shape "3". 18-23. Autapomorphic shapes.

determined, the relationships of these species must be represented as a polytomy.

The extreme case of multiple speciation is deducible simply from the biological species concept. Reproductive isolation, or sharing a mate recognition system in the sense of Paterson (1978) is an autapomorphy (Rosen 1979). If speciation is marked by little genetic or morphological divergence, there may be no synapomorphies that relate species to each other (Avise et al. 1975; Dobzhansky 1970; Kornfield 1978; Kornfield et al. 1982; Mayr 1963; Wake 1981). The cladistic relationships of sibling species must be represented by a polytomy unless there are nonmorphological synapomorphies. One may need to rely on biochemical measures of overall similarity.

At the species level, especially in diverse groups, the synapomorphies required for a fully dichotomous cladogram may not exist in some groups of organisms. A synapomorphy-poor distribution can arise if: 1) all states of a multistate character are derived directly from the ancestral condition, or 2) for one or more characters, only one species within a group shows the derived condition (an autapomorphy), or 3) little genetic or morphological differentiation is associated with reproductive isolation. All three situations are biologically tenable. Having synapomorphies to relate species depends on 1) descendants of a species faithfully inheriting all or most previous synapomorphies characterizing a lineage (few character reversals), and 2) the pattern of speciation events within a group being essentially linear (A ► B ► C) or dichotomous: (A ► B & C; B ► D & E) rather than radial: (A ► B & C & D & E). Neither assumption has been properly tested. Therefore we have no reason to believe that the shortage of synapomorphies in our cladogram for semionotids or the cladogram implied for cichlids by Greenwood (1979:269; 1980:3) is not real.

Relationships, Polarity and Parallelism

One problem with working out relationships of semionotids or other groups that have few synapomorphies available is that the cladogram will never be highly corroborated. Thus, we have to rely heavily on our decisions about polarity for the characters we do have. At low taxonomic levels, determinations about polarity may present a particular problem. For

Newark semionotids, the two most useful characters, body shape and dorsal ridge scale morphology, have many more than two character states, with 22 for body shape and 7 for dorsal ridge scales.

A scheme of character transformation for different shapes is particularly complex. Shape is variable within so many groups of fishes that outgroup comparisons are unreliable. How can the primitive state of shape in semionotids be determined? Are all shapes derived independently from the ancestral type? Can the character, shape, be broken down legitimately into a series of ratios and individuals clustered on the basis of overall similarity? Perhaps the most desirable solution would be to understand the transformation from one shape to another developmentally, or to describe it mathematically. The former is impossible in fossils and the latter analysis is beyond the scope of completed work, but a solution on lines discussed by Bookstein (1978), Humphries et al. (1981) or Raup (1966) is possible. However, a mathematical description would not necessarily agree with a developmental or evolutionary description (Raff, Kauffman 1982).

For semionotids, identity in shape was recognized by the shape grouping criteria described by McCune (1982). Different shapes were treated as autapomorphic states of the character, or character complex, shape. If shape were analyzed developmentally or mathematically, the relationships within the Semionotidae might be resolved. But resolution of a polytomy imposed by a transformation series (which is a scenario) of the states of a single character, rather than by congruence of several characters, sensu Patterson (1982), effectively reduces the level of objectivity in the cladogram to that of a tree, no better than the transformational scenario itself.

Devising a transformation series for dorsal ridge scales of semionotids appears more tractable than for body shape. One might want to consider each type of dorsal ridge scale series as derived from the primitive type independently. Alternatively, the dorsal ridge scales could be described by a series of two-state characters, the primitive state of each character being that shown in the primitive scale type (Figure 4a). The cladogram (Figure 7) was based on the latter analysis of dorsal ridge scale morphology. But both schemes of character transformation are only scenarios. Without corroborating characters or understanding of the genetic-developmental control of these morphological transformations, we can not really choose between them. Furthermore, the existence of morphological anomalies suggests that most individuals share the potential to express many or all variants.

Anomalies, Parallelism, and Speciation

In about 2 percent of all Newark semionotids examined, a dorsal ridge scale series of one type will include one or a few "sports" of another type in a single individual (Figure 8). For example, a concave scale may be mixed with simple scales; or simple concave and thin-spined or short-spined scales may occur together; or thin-spined and small scales may appear in a globular series (Figure 8 a - d). Another type of variant involves doubling or tripling of scales in the dorsal ridge series. Several variants involve doublings of only one or two scales (Figure 8 e,f), although in a few instances, a large portion of the series is doubled or tripled (Figure 8 g,h).

Flank scale doubling is concentrated in the anterior epaxial region near the dorsal ridge scales. Interestingly, the anterior flank is the last region to develop scales in *Lepisosteus* (Suttkus 1963), a variety of paleoniscids (D. Bardack, pers. comm.), and probably semionotids. Variability in scale number and shape may be more tolerable in this region because these scales develop late in ontogeny (Goldschmidt 1940:268; deBeer 1930). The occasional complete doubling of scales in the dorsal ridge scale series is particularly interesting since there is a tendency for semionotids and their sister group, the macrosemiids to develop intercalary scale rows (Bartram 1977; McCune 1982; Olsen, McCune ms). The capacity to vary scale shape and number, or tolerance of this variation, appears to be primitive for semionotids and macrosemiids.

What are the implications of this variability for the systematics and evolution of semionotids? Because anomalies occur in such low frequencies, they do not diminish the conspicuous taxonomic value of dorsal ridge scale characters. But the occurrences of supernumerary spines and dorsal ridge scale "sports" misplaced in series of differing morphologies suggest that all or many semionotid species had the genetic/developmental potential to generate any of the observed dorsal ridge scale patterns, though the potential is rarely expressed. If so, the likelihood of parallelism in closely related species may be very high (Rachootin, Thomson 1981). The same morphology could be less indicative of immediate common ancestry than of shared genetic-developmental potentials (Roth 1984) This would be due to common ancestry on a more general, higher taxonomic, level.

An analogy may be drawn with cichlid trophic polymorphism. It has been suggested that much of the "adaptive radiation" in African cichlids may have been achieved through trophic polymorphism rather than by speciation, because two morphological "species" of cichlids from Cuatro Cienegas belong to a single biological species (Sage, Selander 1975; Kornfield et al. 1982). While some trophic morphs of African cichlids may have been recognized as species, one can hardly dismiss the African cichlid radiation as taxonomic artifact. It is more interesting that the same complex of characters exhibiting intraspecific variability in the Central American *Cichlasoma* can be used to diagnose species of cichlids from the African lakes. Perhaps this intrinsic variability in trophic structures should be considered pleisiomorphic for cichlids, and viewed as the material for the morphological divergence traditionally associated with speciation. Given appropriate extrinsic conditions, such as large lakes or lack of predators, morphs of ancestral polymorphic species from Africa may have been fixed, through selection or chance, by or along with changes in breeding structure. Instead of asking whether African cichlids are oversplit, we might shift some attention to the potential range of variability in tooth morphology that all cichlids share, and the mechanisms by which particular variants are fixed.

ENVIRONMENTAL CYCLES AND EVOLUTION

Cyclic formation and evaporation of lakes which are permanent on an ecological time

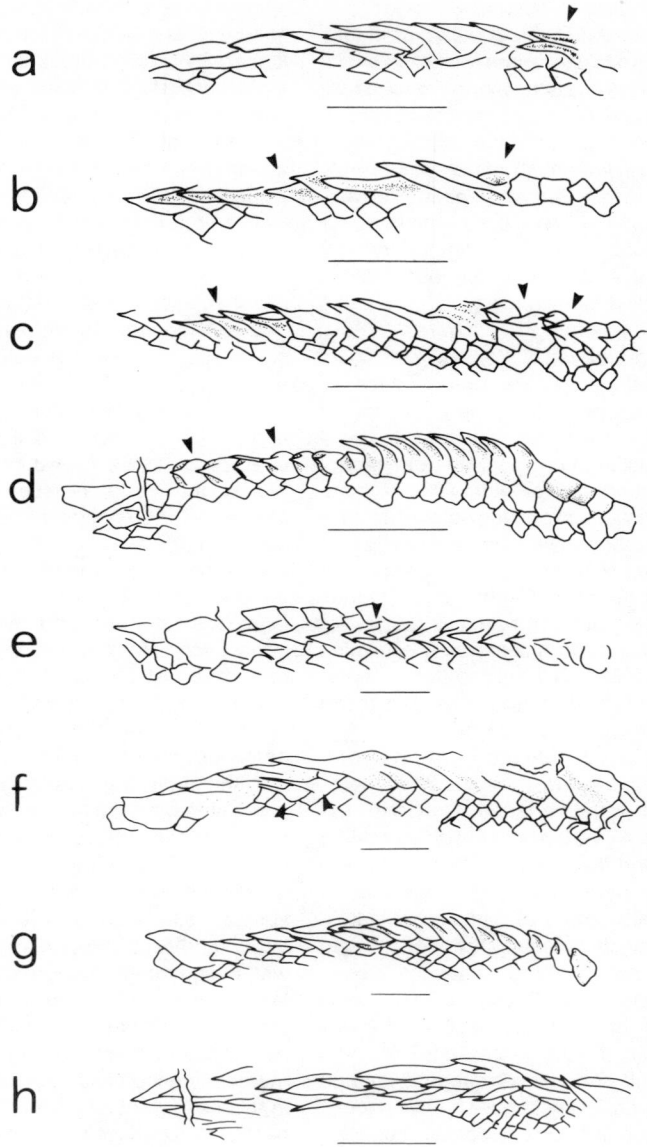

Figure 8. Anomalous variation in dorsal ridge scales from camera lucida drawings of individuals from cycle 4 Towaco Formation, Newark Basin. Specimen numbers refer to Yale Peabody Museum (YPM) specimens. Anterior portion on right dorsal ridge scales point posteriorly. Scale bars = 5 cm. a. Arrow marks two concave scales in a predominantly simple complex series (YPM 8932). b. Anterior arrow marks "thin spined" scales mixed with "simple" scales anteriorly and "concave" scales posteriorly (YPM 8869). c. Central portion comprises "simple" scales while arrows mark "small", "thin-spined" and concave scales anterior to posterior (YPM 8894). d. In this "globular" series, anterior arrow marks thin-spined scale; posterior arrow marks short-spined scale (YPM 8932). e. Arrow marks single scale doubling (YPM 8861). f. Arrows mark scale doublings (YPM 8719). g. Posterior scales are doubled (YPM 8847). h. Entire dorsal ridge scale series tripled or more (YPM 8848).

scale, are not usually discussed in relation to the evolution of fishes. Studies of fishes living in modern lakes cannot reveal the evolutionary effect of such longterm cyclicity because, from an ecological and populational perspective, the rate of climatic change is so slow as to be imperceptible to successive generations of fish. Our retrospective advantage allows us to examine the longterm evolutionary consequences of regular habitat destruction (see also Schindel, 1982), in this case periodic drying, on the species living within a lake episode.

Periodic evaporation of Newark lakes could have extinguished entire communities of fishes. This regular wholesale extinction of communities through geological time deserves consideration separate from natural selection and species selection. While communitiy extinction could be reduced to processes of natural selection, knowledge of cyclic environmental change allows a general prediction of the fate of entire assemblages without consideration of selective factors for individual species. Both community extinction and species selection predict patterns of persistence through geological time, but these predictions are completely different.

In Newark lakes, periodic drying was probably insensitive to most species differences, let alone individual variation. Only fish that could have survived in small isolated saline ponds or found their way up shrunken rivers might have survived a dry phase. If a riverine form was likely to survive better when a lake dried up, then the persistent lineage would have been the one that maintained the riverine lifestyle. Evaporation of a given lake probably would have eradicated most or all of the fishes living in that lake. In regions of high endemism, this sort of environmental change would amount to extinction for many species.

Community extinction and species selection would have occurred on about the same time scale. How do they differ? Stanley (1975:646) suggested that species selection determines the direction of transpecific evolution. Species selection does not invoke a mechanism other than ordinary natural selection. Stanley's point is that evolutionary persistence may be attained by a lineage through a member species that leaves many daughter species, either by a high rate of speciation or by persistent

speciation at a lower rate over a long period of time. This argument is in part probabilistic. If there is a large stochastic component to extinction, then the more descendants a species leaves, the better the chance of survival for that lineage. There may also be an adaptive component to Stanley's proposition. If species adapt by speciating, then a lineage has a better chance of cornering the right direction of adaptation if it speciates frequently. In the case of periodic drying, neither prolific speciation nor adaptation by speciation is relevant. Community extinction would not be stochastic, but decidedly deterministic against lake fishes.

In the case of Newark lake sequences, the hypothesis of the persistence of riverine species could be tested by comparing the faunas of two or more lakes in historical succession. If the species composition was the same in both lakes, then the dry period would appear not to have made any difference to the evolution of the group. If the species in the two lakes are different, and the species in the upper, younger lake are derived from just a few older (riverine?) forms, it would be reasonable to presume that local extinction restarted evolution in the younger lake from persistent riverine forms.

Within individual lake histories, population processes will mold future generations of lake fishes. There probably were both "Red Queen" fish competitively adapting to local conditions (Van Valen 1973), and prolific speciators living in Newark lakes. But as lake habitats disappeared every 21,000 years, whatever evolution occurred within individual lakes may have been largely irrelevant to the longterm pattern of evolution. Stanley's hypothesis of species selection would predict the persistence of species-rich lineages. But in this case, probably only the riverine forms persisted, regardless of their speciation potential. If the effect of periodically drying lakes in the Newark was indeed to select for riverine species, we are forced to see the tremendous diversity of semionotids in various Newark lakes as a kind of noise over evolutionary time.

CLIMATIC CYCLES AND AFRICAN LAKES

What is the relevance of the cyclic expansion and contraction of Newark lakes to other lake systems? How general is periodic climatic change? Is there any evidence that similar phenomena may have played a role in the history of cichlids in African rift lakes?

Periodic climatic change that produces cyclicity in environmental and sedimentary history appears to be a general phenomenon throughout the Phanerozoic record, and not just an anomaly of the Late Triassic to Early Jurassic Newark record. At tropical latitudes, the laws of celestial mechanics suggest that the precession and eccentricity cycles should produce regular variation of insolation over periods of roughly 21,000 years and 100,000 years. In the Newark Supergroup system, the principal lake level cycle of 21,000 years is modulated by a 100,000 year cycle. Bradley (1929) described a 21,000 year periodicity in lake level in the Eocene Green River Formation of Wyoming, and similar sedimentary lacustrine cycles have been described in the Devonian Old Red Sandstone of Scotland (see Olsen 1983; Donovan 1980).

In the Quaternary, periodic changes in climate —cycles of 21,000, 42,000, and 100,000 years— have been inferred from oxygen isotope ratios in deep sea cores (Berger 1978; Kukla 1977), and ascribed to astronomical causes (Imbrie, Imbrie 1979; Berger 1978; Hays et al. 1976; Imbrie 1982). In the high latitudes, these dramatic, globally synchronous changes in climate were expressed as changes in temperature and ice volume during the Great Ice Ages. In the middle and low latitudes, these climatic fluctuations were expressed as changes in precipitation, influencing the levels of even the largest lakes in East Africa (Livingstone 1975; Hamilton 1982).

Thus far our knowledge of the history of lakes Victoria, Tanganyika, and Malawi extends back through only the last major glacial maximum. For this interval, the sedimentary record indicates that fluctuations in lake level were not equally important in all lakes. Therefore, the degree to which cichlid evolution may have been influenced by lake level changes was probably different in different lakes.

Climatic changes have been most influential in Lake Victoria. Bishop and Trendall (1967) dated the oldest lacustrine sediments associated with Lake Victoria as mid-Pleistocene, about 750,000 years old. But some evidence suggests that only 14,000 years ago, Lake Victoria was at least 75 m lower than its present maximum depth of about 90 m (Kendall 1969; Livingstone 1975), or perhaps completely dry (Livingstone 1980). If the Lake Victoria basin was dry at that time, the evolution of endemic Lake Victoria cichlids may have occurred in as little as 14,000 years. Such a young age for Lake Victoria is consistent with the electrophoretic data interpreted to mean that speciation in the *Haplochromis* from Lake Victoria has occurred very recently (Sage et al. 1984).

Lake Tanganyika also experienced a major lake level drop at about the same time as Lake Victoria. Hecky and Degens (1973) argue that 14,000 years ago, Lake Tanganyika was at least 600 m lower than it is now, although Livingstone (1975) believes this drop was closer to 300 m. An increase in salinity associated with a decrease in water volume might not affect some cichlids directly, given that many species are euryhaline (Fryer, Iles 1972). However, the decrease in lake area might severely reduce the extent and diversity of habitat, thereby lowering the numbers of individuals and species that could inhabit the lake.

Less is known about Lake Malawi than about any other African Great Lake. Livingstone (pers. comm.) has not yet fully analyzed his cores from Lake Malawi, but he has seen no obvious evidence for massive changes in lake level through the last 20,000 years. Presumably, the climatic changes producing lake level changes in the other lakes had little effect on the cichlids of Malawi.

The record of earlier climatic change in African Great Lakes has not been investigated because of the tremendous logistic problems of taking long, continuous lake cores. Without longer, older cores, it is not possible to observe directly the sedimentological record of periodic fluctuation in lake levels. However, the coincidence of the last great lowering of lakes Victoria and Tanganyika with the most recent glacial advance suggests a 100,000 year periodicity. On the other hand, increased precipitation 9,000 years ago (Hecky, Degens 1973; Hamilton 1982) suggests a precipitation cycle of roughly 20,000 years. Too little is known of the relation between precipitation, solar insolation, and continental glacier formation to deduce the period of lake level oscillation in any of the African lakes. Yet it seems clear that the African Great Lake levels are under climate control and that this climatic change is periodic.

If climatic changes throughout the Quaternary resulted in very large fluctuations in levels of African lakes, our perceptions of the history and future of the African cichlid radiations may need to be modified. Irrespective of whether Lake Victoria actually dried up completely, we know that both lakes Victoria and Tanganyika experienced severe drying some 14,000 years ago. The historical evidence for this dry episode combined with the evidence for global periodic climatic change throughout the Phanerozoic, emphasizes the ephemeral nature of even the largest lakes. How might these longterm fluctuations influence the persistence of cichlid lineages? Without a longterm perspective we might expect that lineages specialized to lacustrine life would be most likely to persist. And according to the species selection argument (Stanley 1975), we might predict the persistence of *good speciator* lineages. But if African lakes, like Newark lakes, are ephemeral, we might predict that the cichlid lineages most likely to persist would not be the lake-adapted forms nor the prolific speciators, but those with a propensity to migrate or disperse and the ability to survive in small stagnant ponds.

SUMMARY

American semionotid fishes were extremely abundant and diverse in the Late Triassic and Early Jurassic. During that time they dominated the great lakes that filled a series of rift valley basins in eastern North America. The fossil record of this complex of closely related species is particularly conducive to direct historical studies of evolution: 1) Within each of several basins, large, deep, stratified lakes repeatedly formed and evaporated every 21,000 years. Thus, in a given basin there is a geological "stack" of discrete lake histories. Many of these individual lakes, which filled the basin at different times, were dominated by semionotids. 2) The paleontological detail within several individual lake histories is exceptional. Complete articulated fishes are preserved in great abundance. The micro-

layers of sediments, deposited in the anerobic hypolimnion of deep stratified lakes have preserved a yearly chronology of the fish fauna.

The morphological diversity of semionotid fishes from Newark lakes is concentrated in two complex characters —overall body shape, and the morphology of the series of dorsal ridge scales. From a single lake history in the Early Jurassic Towaco Formation of the Newark Basin, 21 species (20 new) have been described. Although most species from Newark have not yet been described, two patterns are apparent. 1) Species diversity is low in the Late Triassic and increases in the Early Jurassic. 2) Semionotid diversity within lake histories is inversely related to the diversity of non-semionotid fish taxa.

Much of the cladogram of relationships of Newark semionotids remains unresolved. Polytomies originate from a shortage of synapomorphies which may, in turn, arise from inadequacy of material or analysis. However, more importantly and more interestingly, a predominance of autapomorphous, not synapomorphous, characters at low taxonomic levels are likely to reflect actual patterns of speciation and real (complete) character distributions, especially in closely related, diverse, and rapidly speciating groups.

Most synapomorphies defining subgroups of semionotids describe features of the dorsal ridge scale series, with about 2 percent showing anomalous deviation from the 7 general types of dorsal ridge scales. The deviations do not diminish the taxonomic value of dorsal ridge scale characters, but they do suggest a pervasive potential for all or most semionotids to produce different dorsal ridge scale morphologies rather easily, increasing the probability of parallelism in these characteristics.

The periodic expansion and contraction of Newark lakes implies a level of selection, extinction of entire communites, that could not be predicted from population processes operating on an ecological time scale. Over the long term, say 22,000 years or more, the evaporation of individual Newark lakes probably "selected for" riverine fishes, and "selected against" lake fishes, irrespective of their propensity to speciate or their adaptations to local conditions.

The influences of periodic climatic change are not only reflected in the record of Newark lakes, but throughout the Phanerozoic to the present. Similar lake level fluctuations of the modern African rift valley lakes are also under climatic control, although the period of fluctuation is not known. Different lakes appear to have been variously affected by climatic change. Present evidence indicates that Lake Victoria may be as young as 14,000 years, and that Lake Tanganyika experienced as much as a 300-600 meter drop in lake level in the same period. Lake Malawi appears to be relatively stable. The rise and fall of African lakes may have been as important in the evolution of cichlid fishes as the evanescence of Newark lakes was to the evolution of semionotid fishes.

ACKNOWLEDGMENTS

Fellowship support for ARM was provided by the Miller Institute for Basic Research in Science, University of California, Berkeley. Grant support came from the National Science Foundation (DEB-7921746 to KST), National Geographic Society, Theodore Roosevelt Memorial Fund, and Sigma Xi (to ARM). We thank R Boardman for technical assistance, W K Sacco for taking the photographs, Bobb Schaeffer and John Maisey, American Museum of Natural History, and F A Jenkens, MCZ of Harvard University, for loans of specimens, and K Padian, R D Sage, S L Wing, and D W Winkler for reviewing the manuscript.

REFERENCES

Agassiz L 1833-44 *Recherches sur les Poisson Fossiles.* Neuchatel: Imprimerie de Petitpierre

Avise J C, J J Smith, F J Ayala 1975 Adaptive differentiation with little genetic change between two native California minnows. *Evolution* 29:411-426

Bartram A W H 1977 The Macrosemiidae, a Mesozoic family of holostean fish. *Bull Brit Mus (Geol)* 29:137-234

de Beer Sir G R 1930 *Embryology and Evolution.* Clarendon. Oxford

Berger A L 1978 Longterm variations of caloric insolation resulting from the earth's orbital elements. *Quart Res* 9:139-167

Bishop W W, A F Trendall 1967 Erosion surface, tectonics, and volcanic activity in Uganda. *Quart J Geol Soc Lond* 122:385-420

Bookstein F L 1978 The measurement of biological shape and shape change. *Lecture Notes in Biomathematics* 24 Springer Verlag. New York

Bradley W 1929 The varves and climate of the Green River Epoch. *U S Geol Surv Prof Papers* 158-E

Davies G R, S D Ludlam 1973 Origin of laminated and graded sediments, Middle Devonian of Western Canada. *Geol Soc Amer Bull* 84:3527-3546

Dingus L, P M Sadler 1982 The effects of stratigraphic completeness on estimates of evolutionary rates. *Syst Zool* 31:400-412

Dobzhansky T 1970 *The Genetics of the Evolutionary Process.* Columbia Univ Press. New York

Donovan R N 1980 Lacustrine cycles, fish ecology and stratigraphic zonation in the Middle Devonian of Caithness. *Scott J Geol* 16:35-50

Fryer G, T D Iles 1972 *The Cichlid Fishes of the Great Lakes of Africa.* T F H Publications. Neptune City NJ

Goldschmidt R 1940 *The Material Basis of Evolution.* Yale Univ Press. New Haven

Greenwood P H 1979 Towards a phyletic classification of the 'genus' *Haplochromis* (Pisces:Cichlidae) and related taxa, Part I. *Bull Brit Mus Nat Hist (Zool)* 35:265-322

_____ 1980 Towards a phyletic classification of the 'genus' *Haplochromis* (Pisces:Cichlidae) and related taxa, Part II: The species from Lakes Victoria, Nabugabo, Edward, George, and Kivu. *Bull Brit Mus Nat Hist (Zool)* 39:1-101

_____ 1981 *The Haplochromine Fishes of the East African Lakes.* Kraus Thomson Organization GmbH. Munich

_____ 1984 African cichlids and evolutionary theories. 141-154 A A Echelle, I Kornfield eds. *Evolution of Fish Species Flocks.* Univ Maine Press at Orono

Hamilton A C 1982 *Environmental History of East Africa: A Study of the Quaternary.* Academic Press. New York

Hays J D, J Imbrie, N J Shackleton 1976 Variations in the earth's orbit: pacemaker of the ice ages. *Science* 194:1121-1132

Hecky R E, E T Degens 1973 Late PLeistocene-Holocene chemical stratigraphy in the rift valley lakes of Central Africa. *Woods Hole Oceanogr Inst Tech Rpt.* WHOI 73-28

Hitchcock E 1819 Remarks on the geology and mineralogy of a section in Massachusetts on the Connecticut River, with a part of New Hampshire and Vermont. *Amer J Sci* 1:105-116

Humphries J M, F L Bookstein, B Chernoff, G R Smith, R L Elder, S G Poss 1981 Multivariate discrimination by shape in relation to size. *Syst Zool* 30:291-308

Imbrie J 1982 Astronomical theory of the Pleistocene Ice Ages: A brief historical review. *Icarus* 50:408-422

Imbrie J, K P Imbrie 1979 *Ice Ages: Solving the Mystery.* Enslow. Hillside NJ

Kendall R L 1969 An ecological history of the Lake Victoria basin. *Ecol Monogr* 39:121-176

Kornfield I L 1978 Evidence for rapid speciation in African cichlid fishes. *Experientia* 34:335-336

Kornfield I L, D C Smith, P S Gagnon, J N Taylor 1982 The cichlid fish of Cuarto Cienegas, Mexico: Direct evidence of conspecificity among distinct trophic morphs. *Evolution* 36:658-664

Kukla G L 1977 Pleistocene land-sea correlations. 1. Europe. *Earth Sci Rev* 13:307-374

Livingstone D A 1965 Sedimentation and the history of water level change in Lake Tanganyika. *Limnol Oceanogr* 10:607-610

———— 1975 Late Quaternary climate change in Africa. *Ann Rev Ecol Syst* 6:249-280

———— 1980 Environmental change in the Nile headwaters. 339-359 M A J Williams, H Faure eds. *Sahara and the Nile.* Blakema. Rotterdam

Manspeizer W, H L Cousminer, J H Puffer 1978 Separation of Morocco and eastern North America: A Triassic-Liassic stratigraphic record. *Geol Soc Amer Bull* 89:901-920

Mayr E 1963 *Animal Species and Evolution.* Harvard Univ Press. Cambridge

McCune A R 1982 *Early Jurassic Semionotidae (Pisces) from the Newark Supergroup: Systematics and Evolution of a Fossil Species Flock.* PhD Thesis. Yale Univ

———— ms A revision of *Semionotus* from the Late Triassic and Jurassic of Europe.

McDonald N G 1983 History of paleoichthyology in the Newark Supergroup Basins, eastern North America. *Geol Soc Amer Abstr* 15:121

Nelson G 1980 Multiple branching in cladograms: two interpretations. *Syst Zool* 29:86-91

Olsen P E 1978 On the use of the term Newark for Triassic and Early Jurassic rocks in eastern North America. *Newsl Stratigr* 7:90-95

———— 1980a The latest Triassic and early Jurassic Formations of the Newark Basin (eastern North American Newark Supergroup): Stratigraphy, structure and correlation. *New Jersey Acad Sci Bull* 25:25-51

———— 1980b Fossil great lakes of the Newark Supergroup in New Jersey. 352-398 M Manspeizer ed. *Field Studies of New Jersey Geology and Guide to Field Trips.* 52nd Ann Mtg, New York State Geol Assoc. Rutgers Univ. Newark NJ

———— 1983 Periodicity of lake-level cycles in the Late Triassic Lockatong Formation of the Newark Basin. (Newark Supergroup, New Jersey and Pennsylvania.) in press D Reidel ed. *Proceedings, NATO ASI Symposium on Milankovitch and Climate.* Dordrecht.

Olsen P E, C L Remington, B Cornet, K S Thomson 1978 Cyclic changes in Late Triassic Lacustrine Communities. *Science* 201:729-733

Olsen P E, A R McCune, K S Thomson 1982 Correlation of the early Mesozoic Newark Supergroup by vertebrates, principally fishes. *Amer J Sci* 282:1-44

Paterson H 1978 More evidence against speciation by reinforcement. *S African J Sci* 74:369-371

Patterson C 1982 Morphological characters and homology. 21-74 K A Joysey, A E Friday eds. *Problems of Phylogenetic Reconstruction, Systematics Association Spec Vol No 21.* Academic Press. London

Patterson C, D E Rosen 1977 Review of ichthyodectiform and other Mesozoic teleost fishes and the theory and practice of classifying fossils. *Bull Amer Mus Nat Hist* 158:83-172

Rachootin S P, K S Thomson 1981 Epigenetics, paleontology and evolution. 181-193 G G E Scudder, J L Reveal eds. *Evolution Today* Proc 2nd Internat Congr Syst Evol Biol. Hunt Inst Bot Documentation, Carnegie-Mellon Univ. Pittsburgh

Raff R A, T C Kauffman 1982 *Embryos, Genes, and Ancestors.* Macmillan. New York

Raup D M 1966 Geometric analysis of shell coiling: general problems. *J Paleon* 40:1178-1190

Rixon A E 1976 *Fossil Animal Remains: Their Preparation and Conservation.* Atholone. London

Rosen D E 1979 Fishes from the uplands and intermontane basins of Guatemala: revisionary studies and comparative biogeography. *Bull Amer Mus Nat Hist* 162:267-376

Roth V L 1984 On homology. *Biol J Linn Soc* In press.

Sage R D, P V Loiselle, P Basasibwake, A C Wilson 1984 Molecular versus morphological change among cichlid fishes (Pisces: Cichlidae) of Lake Victoria. 185-202 A A Echelle, I Kornfield eds. *Evolution of Fish Species Flocks.* Univ Maine Press at Orono

Sage R D, R K Selander 1975 Trophic radiation through polymorphism in cichlid fishes. *Proc Natl Acad Sci USA* 72:4669-4673

Schaeffer B 1967 Late Triassic fishes from the western United States. *Bull Amer Mus Nat Hist* 153:287-342

Schindel D E 1982 Resolution analysis: a new approach to the gaps in the fossil record. *Paleobiol* 8:315-353

Stanley S 1975 A theory of evolution above the species level. *Proc Nat Acad Sci USA* 72:646-650

Stebbins G L 1982 Perspectives on macroevolution. *Evolution* 36:1109-1118

Suttkus R D 1963 Order Lepisostei. *Fishes of the Western North Atlantic.* Memoir Sears Found for Marine Research. No 1 Pt 3. New Haven

Thomson K S, A R McCune 1983 Scale structure as evidence of growth patterns in fossil semionotid fishes. *J Vert Paleo* In press.

Van Couvering J A H 1982 Fossil cichlid fish of Africa. *Spec Pap Paleon 29.* Paleont. Assn. London

Van Houten F B 1969 Late Triassic Newark Group, north central New Jersey and adjacent Pennsylvania and New York. 314-347 S Subitzki ed. *Geology of Selected Areas in New Jersey and Eastern Pennsylvania.* Rutgers Univ Press. New Brunswick

Van Valen L 1973 A new evolutionary law. *Evol Theor* 1:1-30

Wake D 1981 Application of allozyme evidence to problems in evolution of morphololgy. 257-270 G G E Scudder, J L Reveal eds. *Evolution Today* Proc 2nd Internat Congr Syst Evol Biol. Hunt Inst Bot Documentation, Carnegie-Mellon Univ. Pittsburgh

Wiley E O 1981 *Phylogenetics: The Theory and Practice of Phylogenetic Systematics.* Academic Press. New York

Williamson P G 1981 Palaeontological documentation of speciation in Cenozoic molluscs from Turkana Basin. *Nature* 293:437-443

EVOLUTION OF SPECIES FLOCKS OF FISHES IN NORTH TEMPERATE LAKES

G R Smith and T N Todd

INTRODUCTION

To understand the origin of species flocks in lakes we must study the causes of ecological separation and reproductive isolation among populations. In allopatric speciation, a theory advanced by Mayr (1942, 1976) and Brooks (1950) and now accepted by most students of species in lakes, the origin of species that later become sympatric occurs in geographic isolation. For instance, the apparent differentiation of five cichlid species in Lake Nabugabo during a geologically brief low-water stage in Lake Victoria (Greenwood 1974) is evidence for the *fluctuation hypothesis* of allopatric speciation in lakes. A second hypothesis, which does not require fluctuation of lake size or shape, proposes *immigration* of species into a lake from allopatric sites of origin. Finally, the geography and ecology of a lake may support instead a third hypothesis: *intralacustrine* reproductive isolation and ecological differentiation of species (Fryer 1977). This hypothesis includes allopatric, parapatric, and sympatric divergence.

We will explore the possibility that some partly sympatric fish populations attain premating reproductive isolation by behavioral preference for differences in time and location of spawning: the *reproductive allopatry and allochrony* model. In suggesting this we must try to explain how reproductive isolation that is initially incomplete can permit divergence to proceed against the force of gene flow and genetic dilution of differences. We will propose that genes favoring specific locations and times of spawning act as assortative mating modifiers in the sense of Moore (1979) and Endler (1977), in models for parapatric and sympatric divergence. We will show that some populations being isolated by reproductive allopatry or allochrony may be otherwise ecologically sympatric and in competition for food resources, permitting disruptive selection as proposed by the competititve speciaton model of Rosenzweig (1978).

Definitions. We define species as genetically isolated populations or groups of populations with some demonstrable ecological and historical individuality, as for several thousands of years. Since these characteristics are not measured directly, we infer them from morphological divergence and covariation in the context of ecology, geography, and geologic history. It is fundamental to arguments that follow, to recognize that hybridization is common among many freshwater fish species, and that even lineages that have diverged for several millions of years may be incompletely isolated (Miller, Smith 1981). Here, we do not solve the problem of tolerable gene flow among "good species."

A *lacustrine species flock* is a monophyletic group of species living in one lake or originally derived in one lake. The test of monophyly is critical to the test of origin. In the two strictly allopatric hypotheses—immigration and fluctuation—the monophyletic group included members outside the lake or in different lakes at some time during the speciation process. The third hypothesis requires a model for evolution of a monophyletic species flock within one continuous lacustrine habitat.

Our ideas have been influenced by the study of what we believe are examples of *incipient* species flocks, monophyletic groups of stocks of ciscoes (*Coregonus*), which are not quite separated at the level of species. These are incompletely isolated, but partially differentiated populations. From overlapping ecological and morphological characteristics of such populations we infer that geographically adjacent (parapatric), or overlapping (sympatric), or temporally different (allochronic) spawning populations are experiencing partial divergence and limited gene exchange. We suggest that the initial isolation of such populations is a natural consequence of lacustrine ecology and that under certain circumstances, such populations may become species.

Characteristics of lakes. Ecological segregation in lakes is abetted by certain unique conditions. The critical physical features — temperature, light, pressure, density, viscosity, substrate texture, and seasonality — are correlated with depth, as are oxygen, food, and predators. In concert, these result in steeper, more powerful ecological gradients than are found in most surface terrestrial and fluvial systems. Mountains are analogous to lakes in gradients of temperature, substrate, and seasonality, and have been compared in speciation models. In both mountains and lakes, most of the species diversity occurs around the edges — at the bases of mountains, and in the shallower part of lakes.

Lacustrine species usually prefer specific temperatures, forage, light etc. (perceived in sampling as depth zones), and substrates (e.g., Kozhov 1963), though these preferences may change seasonally in response to reproductive cycles, overturn, zooplankton cycles, or other food migrations. Benthic lacustrine species are restricted laterally by substrate variations arising in different sediment sources from rivers and eroding shores. Finally, the characteristic pattern of alternating bays and headlands, especially in tectonic lakes, increases the disruption of the seemingly homogeneous habitat into zones that further select for specializations in physiology, behavior, and morphology.

We contend that these lacustrine gradients and barriers promote ecological segregation of populations through their effects on food zones, refuges, spawning sites, and nurseries. The addition of seasonal fluctuations to this system causes the existence of discontinuities between ecological niches.

If these zonations contribute to speciation in lakes, lacustrine populations should show spatial and morphological patterns without analogs in fluvial systems. Populations differentiating in headwaters of fluvial systems are normally isolated from each other by watershed divides and long reaches of downstream habitat. Populations isolated in this way provide the most frequent and familiar examples of geographic differentiation among fishes, but they are differentiating in similar habitats, under similar selective regimes, compared to species diverging by homing to different depths within a lake. Inhabitants of large streams are connected over hundreds of miles by more or less continuous habitat, and show less differentiation than other fishes (Smith 1981b). With few exceptions, fluvial fishes seem to conform to the traditional vicariance model of speciation.

We propose that the speciation process in lakes differs fundamentally in that it often involves some ecological sympatry and competition among populations that are each other's closest relatives. We expect ecological character displacement to contribute more immediately to lacustrine species differences. The difference in character divergence of fishes in lakes and rivers is related to the difference in the temporal sequence in which reproductive isolation and competitive character displacement occur. Traditional allopatric speciation models propose ecological character displacement following the initiation of reproductive isolation. The ecological model for intralacustrine speciation predicts divergence of food- and depth-related structures and behaviors concurrent with reduction of gene flow.

Finally, it is generally true that fluvial fishes living in lakes rarely undergo significant differentiation, perhaps because they are behaviorally inclined more toward dispersal and colonization (Fryer, Iles 1969). The fishes that normally undergo intralacustrine differentiation in lakes are usually restricted to those lakes

ALTERNATIVE HYPOTHESES

There are three alternative hypotheses concerning the isolation and evolution of fish populations in lakes. Each is phrased in terms of inferred historical speciation processes. Corollaries of each hypothesis describe expected evidence, such as expected geographic distribution of sister species in the taxon including the supposed species flock. Mechanisms for the evolution described by hypotheses 1 and 2 are universally accepted by systematists. Hypothesis 3 is more difficult to accept.

Hypothesis 1. Immigration. At least one member of each species pair in the supposed species flock arose outside the lake and immigrated into it after evolution of reproductive isolation. For lacustrine speciation, this is a null hypothesis.

Corollary 1. Each species in the assemblage should have a closely-related sister species living, or represented by fossils, outside the lake.

Hypothesis 2. Fluctuation. Each species pair evolved the potential for reproductive isolation while its members were allopatric in lakes separated by a low-water stage of a larger parent lake. Sympatry was achieved when high waters united the daughter lakes and their species. This hypothesis is favored by most workers, and it is important to show how many examples support it.

Corollary 2. Lake systems at low-water stages, such as lakes Nabugabo and Victoria, will contain populations that are diverging in ways that might permit them to coexist in later, connected stages of the system. In a basin with a history of prior fluctuations, lakes in low stage should show sympatric species evolved in previous low stages, if multiplication of species is to be realized. Geologic evidence must document barriers, fluctuations, or tectonic events sufficient to bring about the vicariance and recolonization events. Relatives outside the lake should be secondary dispersants. There should be cladistic evidence that relatives outside the lake are terminal rather than basal branches in the group's cladistic history.

Hypothesis 3. Intralacustrine reproductive allopatry and allochrony. Sister species in the flock diverged in isolation that resulted from population-specific differences in time or place of spawning within one lake. Homing and timing errors among spawners cause gene flow that might or might not be overcome by any selection for changes in morphology or behavior. Such hypotheses are usually rejected because of theoretical difficulties with the proposed mechanisms. It is important to establish criteria for evaluation based on data.

Corollary 3. The flock should consititute a monophyletic group. There should be no

relatives outside the lake, or they should be obvious migrants from the lake, or should be a sister group of the whole flock. Representatives of incipient species pairs will occasionally be sampled together, sympatric with regard to food and space, but must be allopatric or allochronic in reproduction. In the case of small, localized populations of fishes with low vagility in large lakes, classic allopatric speciation is a possibility. Evidence for this mode takes the form of complete absence of strays sympatric with the sister form.

Hypotheses 1-3 lead to important questions about the relationships between reproductive isolation, ecological divergence, and species differences in morphology. The relationships are contrasted in the hypotheses to follow.

Hypothesis 4. Evolution in allopatry permits divergence of relatively neutral characters. In allopatry, for example among diverging populations in separate lakes or river drainages, divergent characters are likely to be associated with growth, efficiency, or some other aspect of within-population competition. Some potential for reproductive isolation is evolved as a pleiotropic byproduct of selected responses to habitat and community differences. Ecological sympatry of the divergent forms occurs later, bringing with it greater potential for competition among close relatives and possible character displacement.

Corollary 4. Most-recently diverged sister species that evolved in strict allopatry will differ primarily in meristic, morphometric, or biochemical characters less likely to be related to competition for shared resources. To the extent that prior reproductive isolation is postulated, gene flow would be rare. However, a common variant of this scenario involves incomplete reproductive isolation and gene flow early in the history of sympatry, since morphological divergence (especially of fishes) usually proceeds well in advance of the accumulation of mechanisms for reproductive isolation (Dobzhansky 1951). Hybridization should broaden the distribution and overlap of the diverged characters, but not necessarily homogenize them (Smith et al. 1983), especially if there are ecological differences. In either case, the basic prediction is that differences evolved in allopatry are less likely to involve structures directly related to competition with close relatives. Hybridization could obscure the features predicted to distinguish this hypothesis from Hypothesis 5.

Hypothesis 5. Morphological consequences of divergence in sympatry. Competition forces divergence of subpopulations to different utilization of resources such as food and breeding sites. Characteristics related to resource utilization diverge as a result of ecological character displacement. Reproductive separation is the consequence of a simple difference in time or place of spawning, so that gene flow occurs in proportion to strays and dilutes any tendencies for divergence that are not strongly selected for.

Corollary 5. Sister species or incipient sister species differ primarily in morphological or life history characters functionally related to competition for resources. Effects of gene flow will be manifest in the absence of biochemical differences among species. Potential for gene flow may be assessed by analysis of proportion of homing and timing strays among spawners. Morphological variability and overlap may be high, though mean differences in ecologically significant characters may be large. Similarly, nearly-neutral allozymes should show few differences among sister species, but any strongly related to resource utilization should show divergence.

For those examples (or parts) for which Hypotheses 1, 2, and 4 can be rejected, the theoretically more challenging problem remains. Which of the diverse models for allopatric, parapatric, and sympatric speciation might be applicable to intralacustrine differentiation in fishes? In the examples to follow we attempt to make choices between the general hypotheses and offer suggestions on the last question.

EXAMPLES

The two major examples of fish species flocks in north temperate lakes are the 36 kinds of sculpins, in or related to the family Cottidae, in Lake Baikal (Taliev 1955), and the 8 kinds of ciscoes (Salmonidae: Coregoninae) which were once common in the Laurentian Great Lakes (Koelz 1929). Many other northern species flocks also involve coregonine fishes: three species of *Prosopium* in Bear Lake of Utah and Idaho (Snyder 1919), many sympatric pairs of *Coregonus* and of

Prosopium in Canada (McPhail, Lindsey 1970), of *Coregonus clupeaformis* in Canada and the northeastern United States (Fenderson 1964 Lindsey et al. 1970; Kirkpatrick, Selander 1979), and populations of *Coregonus* in northern European lakes (Svardson 1979; Bergstrand 1982). In Pliocene Lake Idaho, at least 8 kinds of sculpins in three genera make up a related assemblage (Smith et al. 1983). Incipient differentiation can be seen in lake trout of the Laurentian Great Lakes (Goodier 1981) and walleye of Lake Erie (Deason 1936). Four kinds of trout are differentiating in Lake Ohrid, Yugoslavia and Albania (Stancovic 1960). Two forms of landlocked salmon, *Oncorhynchus nerka*, have diverged in Lake Kronotskiy in the Soviet Union (Kurenkov 1977). Lake Biwa, Japan, shows differentiation among catfish (*Parasilurus*) and crucian carp (*Carassius*) (Tomoda 1962, 1965; Taniguchi, Ishiwatari 1972).

From the similarities and differences among some of these examples we will extract some general hypotheses about the origin of reproductive isolation in lake populations. Among whitefish and trout, most examples involve relatively youthful lakes 10^4 to 10^5 years old. The sculpins, living in lakes that existed for 10^6 or more years have diverged more. Several of the flocks among our examples have few relatives outside their lakes. Preliminary cladistic analysis of these provides some evidence bearing on the manner of origin.

The greatest degree of morphological differentiation is found in trophic and other ecologically significant features of microphagous sculpins and of whitefish with benthic or benthopelagic habits specialized for lacustrine life. Life history differences in time and place of spawning, but not trophic characters, have diverged in the predaceous trout and walleyes.

All of the species of interest show significant localization of spawning sites and restriction of spawning time. The degree of reproductive isolation appears to be correlated with the age of the lakes.

BEAR LAKE WHITEFISHES

Bear Lake, a small (36x11 km) lake on the Utah-Idaho border, has three species of whitefish, *Prosopium spilonotum, abyssicola,* and *gemmiferum,* that would seem to constitute a classic, though miniature flock. The three other species in the genus are *cylindraceum* of northeast Asia and northern North America, *williamsoni* of western North America, and *coulteri* of northwestern North America, and Lake Superior (Table 1). The Bear Lake species are thought to be endemic, but two of them or their vicariants are found in sediments of Lake Bonneville, Utah, to which, Bear Lake was tributary through the Bear River during the past 30,000 years (Smith et al. 1968). Therefore the group is endemic to a system involving several basins, supporting the immigration or fluctuation hypotheses.

Relationships outside the lake offer added reasons for thinking that the evolution of these whitefishes might involve more than Bear Lake. *Prosopium cylindraceum, williamsoni,* and *spilonotum* are similar in size, osteology, meristics, and spawning season (Table 1), but these may be plesiomorphic characters (see abyssicola below); *williamsoni* might be the most specialized of the trio in its adaptation to life in streams. Regarding the other two species, Eddy's suggestion (1957) that *abyssicola* is a subspecies of *coulteri* is negated by numerous differences (Table 1, Figure 1), so we cannot regard these as sister groups in support of the immigration hypothesis. *C. abyssicola* is probably the sister group of *spilonotum.*

The distinctiveness of Bear Lake species in diverse characteristics, including non-ecological meristics, argues against their recognition as a monophyletic flock in Bear Lake, though they could be a monophyletic flock of the Bonneville Basin. They are certainly divergent from each other in spawning time and depth (Table 1), and in food, suggesting considerable ecological displacement. *C. gemmiferum* eats mostly zooplankton, *abyssicola* eats mostly ostracods, and *spilonotum* eats mostly insects (Sigler, Miller 1963). Nevertheless, the weight of the evidence suggests that they originated elsewhere and immigrated into Bear Lake.

If two species of Bear Lake whitefishes evolved outside the lake, the importance of checking two aspects of other examples of possible intralacustrine speciation is strongly indicated. First, without the fossil relatives of Bear Lake forms outside the basin, we would probably be satisfied to regard the fishes as a relatively old species flock. Second, the ecological partitioning inside the lake is a criterion we would otherwise offer as evidence for differentiation having occurred in the lake. Obviously, ecological partitioning can evolve among sympatric species without regard to their manner of origin. The differences in non-ecology-related meristic characters, however, support the idea that most of the Bear Lake whitefishes, unlike the example to follow, either originated in allopatry or diverged earlier in the Pleistocene.

GREAT BASIN FLUVIAL LAKES: A NULL EXPERIMENT

Three great lake systems, Bonneville, Lahontan, and Death Valley, of the Pleistocene in Utah, Nevada, and California, illustrate the low-water conditions postulated by the fluctuation model of lacustrine speciation. Each is currently at one of many temporary lows in its history (Hubbs, Miller 1948; Miller 1948; Smith 1978, 1981a,b). If the fluctuation hypothesis is valid, the relict lakes

TABLE 1. Characteristics of the species of *Prosopium*
(Data: Sigler, Miller 1963; Scott, Crossman 1974; original measurements)

Species	Range	mm adult length	gill rakers	lateral line scales	dorsal rays	anal rays	maxilla	spawning time	depth
cylindraceum	NE Asia N. America	400+	15-20	74-106	11-13	8-12	medium	Oct-Nov	
williamsoni	W Nor Amer	400+	19-26	76-89	12-13	11-13	medium	Nov	
spilonotum	Bear Lake	400+	18-22	76-86	10-12	9-11	medium	Nov Dec	shallow
abyssicola	Bear Lake	250	19-23	67-78	10-12	9-11	short	Jan Feb	15-30 m
gemmiferum	Bear Lake	190	37-45	69-76	10-11	9-12	long	Jan	0.5-2 m
coulteri	NW N Amer L Superior	150	13-20	54-70	10-14	11-15	medium-long	Nov Jan	

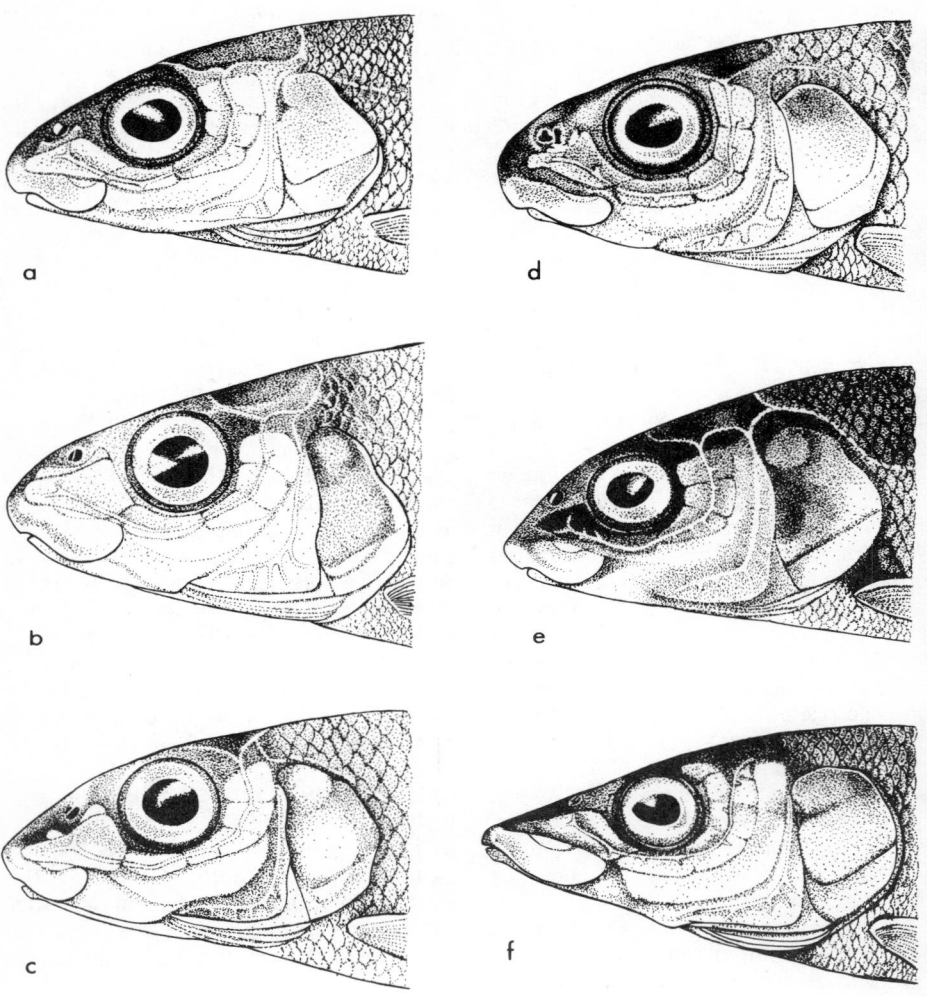

Figure 1. Heads of six species of *Prosopium*, of the Bear Lake (Utah and Idaho) whitefishes and their relatives. Illustrations augmented with certain bone shapes after clearing and staining. The Bear Lake group consists of b) *spilonotum*, c) *abyssicola*, and f) *gemmiferum*. *P. williamsoni* (e) lives in tributaries to Bear Lake and in medium-elevation streams and lakes in intermountain United States north to SW Canada; a) *cylindraceum* is found in NE Asia, N North America S to the Great Lakes, and New England; d) *coulteri* lives in Lake Superior and NW North America. Drawn by S. Jacobshagen.

should therefore contain species flocks of minnows, suckers, trouts, and sculpins brought together by previous high-water connections of isolated basins. Further, currently disconnected fragments of the systems should contain the isolated incipient species to be brought together in the next phase of high-water sympatry.

Only among the Lake Lahontan minnows,

Gila bicolor and its relatives, is there evidence for such a process. *Gila bicolor* has two sympatric forms, distinguished by high and low numbers of gill rakers. But these forms are equally likely to be remnants of the previous pluvial phase, if the suggested connection between Hypotheses 3 and 5 is valid. The subspecies with high-raker numbers is a large, lake fish. *Eremichthys acros*, endemic to

Soldier Meadows, Nevada, could be a dwarfed former isolate of *Gila*, but even if it is, the case does not demonstrate extensive speciation in allopatry because only two cases at most can be cited, despite fluctuations between nine separate sub-basins and the maximum lake area, covering more than 20,000 km ^{2}.

The Bonneville drainage shows possible sympatric pairs in sculpins (Smith et al. 1968), but these could have evolved just as well in the pluvial part of the cycle. There is little evidence for differentiation of species in the dry part of the cycle, although the allopatric sculpins *C. extensus* and *echinatus* of Bear Lake and Utah Lake are a possible example. *Gila atraria* shows considerable geographic variation, including forms with early maturation and small size. *Iotichthys phlegethontis* is a small relative of *Gila atraria* that might have originated in this way. These examples are disappointingly few, considering the vast size of the lake at high water levels, over 40,000 km ^{2}, and the many isolated daughter drainages containing populations of these fishes at low water levels.

Only in the Death Valley pupfishes (*Cyprinodon nevadensis, salinus, diabolis*) do we see an incipient flock that appears to be differentiating since the previous pluvial stage (Miller 1948). The *Cyprinodon* of southwestern North America are highly interfertile and rarely sympatric (Liu 1969; Turner, Liu 1977). That they differ modestly in ecologically insignificant characters (Miller 1948) suggests that they are products of the current allopatric phase of the cycle, and that they have differentiated to the level of semispecies (Liu 1969). By contrast, the ecologically important morphological differences predicted by Hypothesis 5 to occur among sympatric species are seen in the *Cyprinodon* of Lake Chichancanab Yucatan (Humphries 1984).

In Death Valley, as in the rest of the Great Basin, the fossil evidence shows no more diversity than that seen now (Smith et al. 1968; Smith 1981a; Taylor, Smith 1981). It is doubtful that the *Cyprinodon* of this system have been or are becoming a sympatric species flock. Divergence only to the level of semispecies in about 12,000 years of postpluvial isolation argues for caution in regard to supported examples of rapid speciation, which are popular in discussions of species flocks. The example illustrates only modest divergence in allopatry. Our next example suggests that in sympatry, ecologically significant differences have occurred in the same time span, although the products are, similarly, semispecies.

We conclude that cycles of extreme fluctuation, as seen here, would usually cause extinction due to reduced habitat size, or genetic homogenization of different isolates, rather than extensive speciation (Smith 1978, 1981a,b). More moderate fluctuations, producing geographic barriers without severely reducing habitat size, are necessary if new species are to be produced while the established diversity is protected.

CISCOES OF THE LAURENTIAN GREAT LAKES

Because of the history of semi-isolated fluctuating lakes with a limited geological time range, the eight nominal kinds of *Coregonus* in lakes Nipigon, Superior, Michigan, Huron, Erie, and Ontario provide evidence relative to the above hypotheses. Many of the forms live in Lake Nipigon, which has been separated from Lake Superior for over 8000 years. These facts suggest that much of the differentiation among the seven or eight kinds occurred in the first half of the lakes' history (Bailey, Smith 1981). It is doubtful that all of the forms separately colonized the basins following deglaciation, a scenario that would require their persistence in rapidly changing glacial lakes. Evidence instead for their time- and place-dependent reproductive isolation makes this unlikely. The monophyletic species flock definition is probably most applicable to two assemblages, taken separately: the *artedii-nipigon* group and its possible sister group, *hoyi* and *kiyi*, and the *zenithicus-alpenae-johannae-reighardi-nigripinnis* group. In other words, we suspect two colonizations of the lakes following the last glaciation. Here we follow the taxonomy of Koelz (1929) with modifications (Scott, Crossman 1973; Todd, Smith 1980). Koelz' classification depended especially on length and number of gill rakers which show broad overlap among species (Figures 2-6). The lack of non-overlapping characters (Koelz 1929 Table 6) raises doubts about the reproductive isolation and species level recognition of the taxa (Bailey, Smith

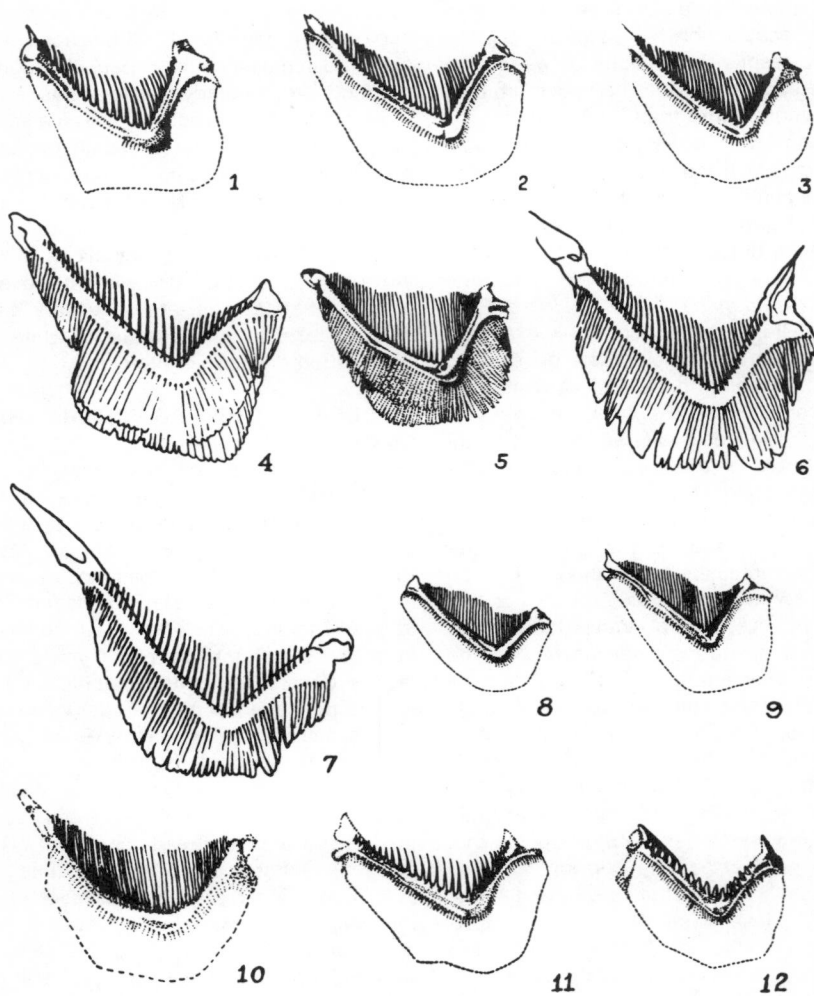

Figure 2. Variation in lengths of gill rakers among Great Lakes coregonids. Copied from Koelz 1929, figure 13. Left branchial arches (rakers up) of 1) *C. johannae,* 2) *C. alpenae,* 3) *C. zenithicus,* 4) *C. reighardi,* 5) *C. nigripinnis,* 6) *C.* [*zenithicus*] *cyanopterus,* 7) *C. kiyi,* 8) *C. hoyi,* 9) *C. artedii,* 10) *C.* [*artedii*] *nipigon,* 11) *C. clupeaformis,* 12) *Prosopium cylindraceum.*

1981). Todd (1981) showed minimal isozyme differences among the forms in Lake Superior. We suggest that none of the species is more than extrinsically reproductively isolated from all, or perhaps any, of the others. If so, we are justified in examining them as an incipient species flock.

According to the three hypotheses for flock origin we would expect: (1, immigration) pairs of sister species with no more than one inside and at least one from outside the basin; (2, fluctuation) allopatry of incipient sister species paired in different lakes inside the basin; or (3, Intralacustrine reproductive allopatry and allochrony) sympatric sister species or races within each lake. Evaluation of these expectations requires that we be able to recognize sister groups as well as secondary sympatry

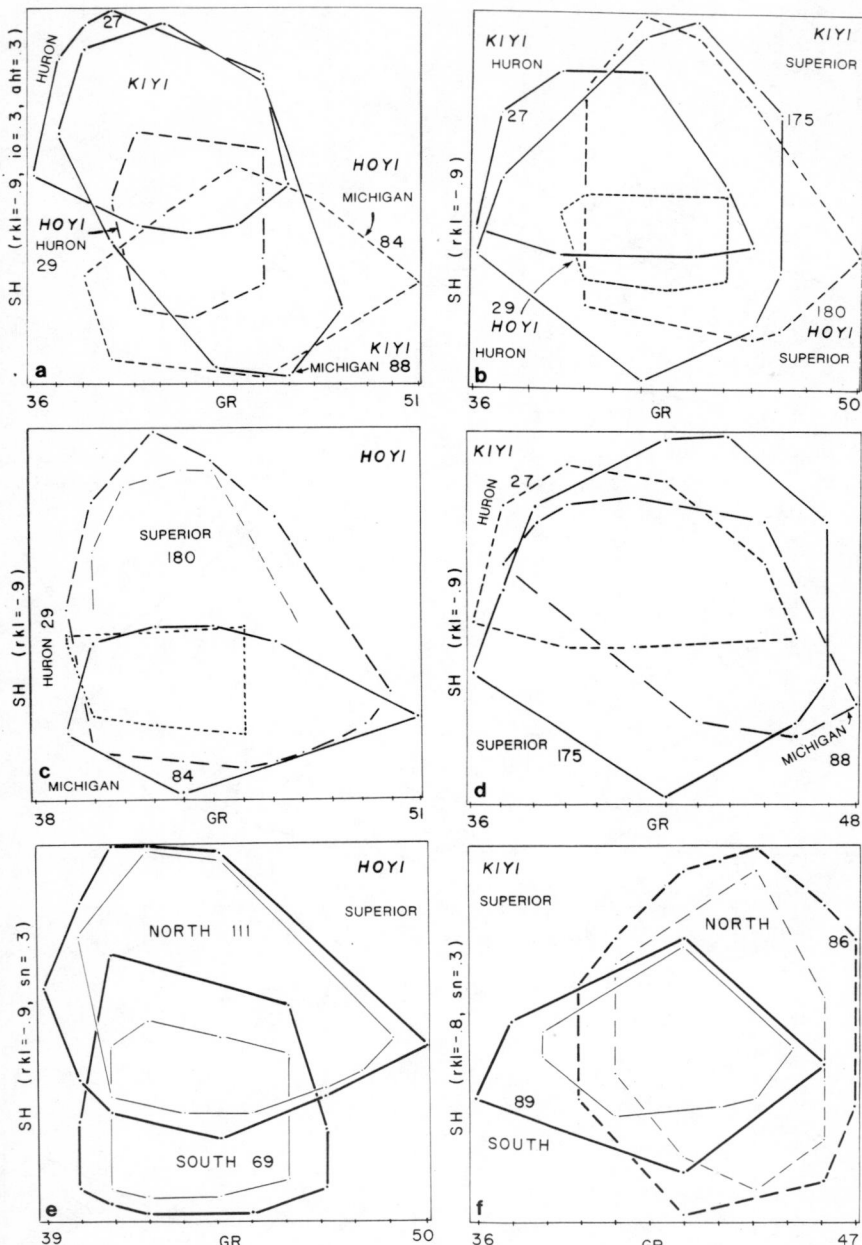

Figure 3. Cluster outlines of populations of *Coregonus hoyi* and *kiyi* from the upper Great Lakes.
Vertical axes are sheared (size-free, Humphries 1981) principal components. Variable coefficients above
0.3 in parentheses. Horizontal axes are numbers of gill rakers. Cluster outlines include peripheral individuals , and, in thinner lines, the next most peripheral individuals. The two convex polygons are plotted where presence or absence of outliers is emphasized. Numbers in scatter plots are sample sizes.
All samples are from Data Set B of Table 2 (collected by the U. S. Fish and Wildlife Service, 1950's &
60's). Abbreviations of variable names explained in Table 2.

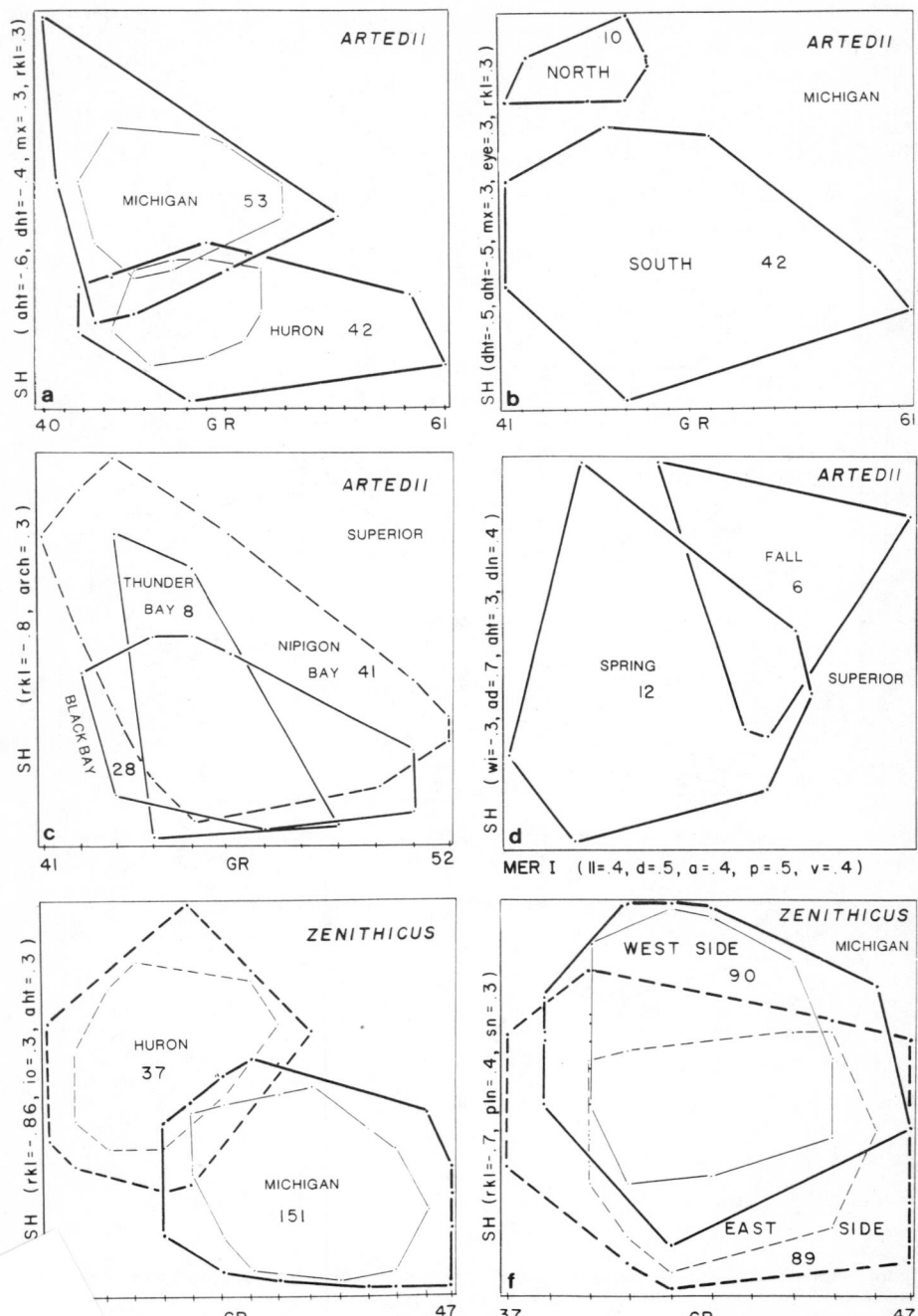

ster outlines of populations of *C. artedii* and *C. zenithicus* of the Great Lakes. See
igure 3. Comparison of spring and fall spawning *artedii* is based on Data Set A, Table 2.
is first principal component of the six meristic variables. The coefficients are highly cor-
tions of variable names explained in Table 2.

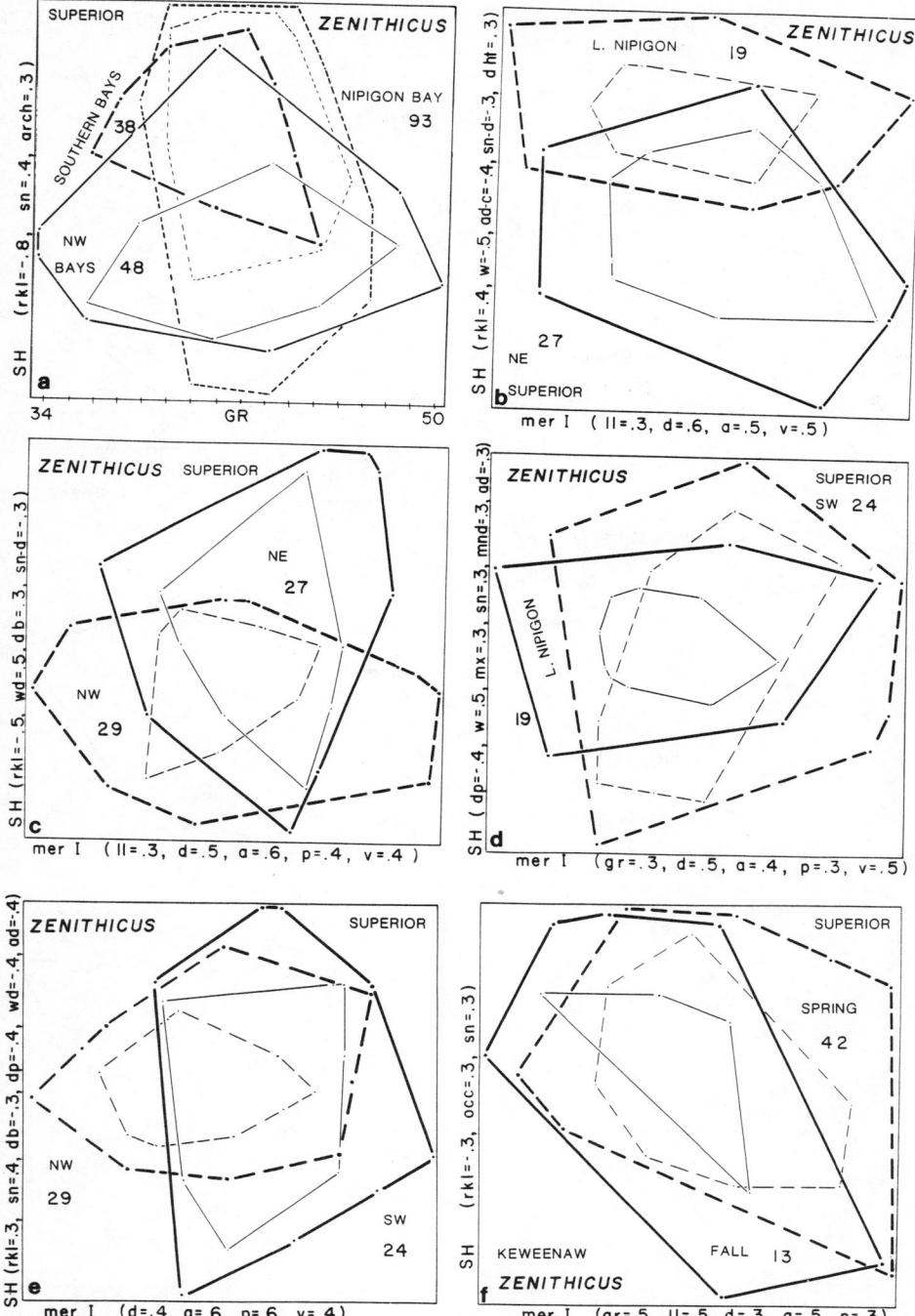

Figure 5. Cluster outlines of populations of *C. zenithicus* of the Great Lakes. Axes explained in legends of Figures 3 and 4; abbreviations explained in Table 2.

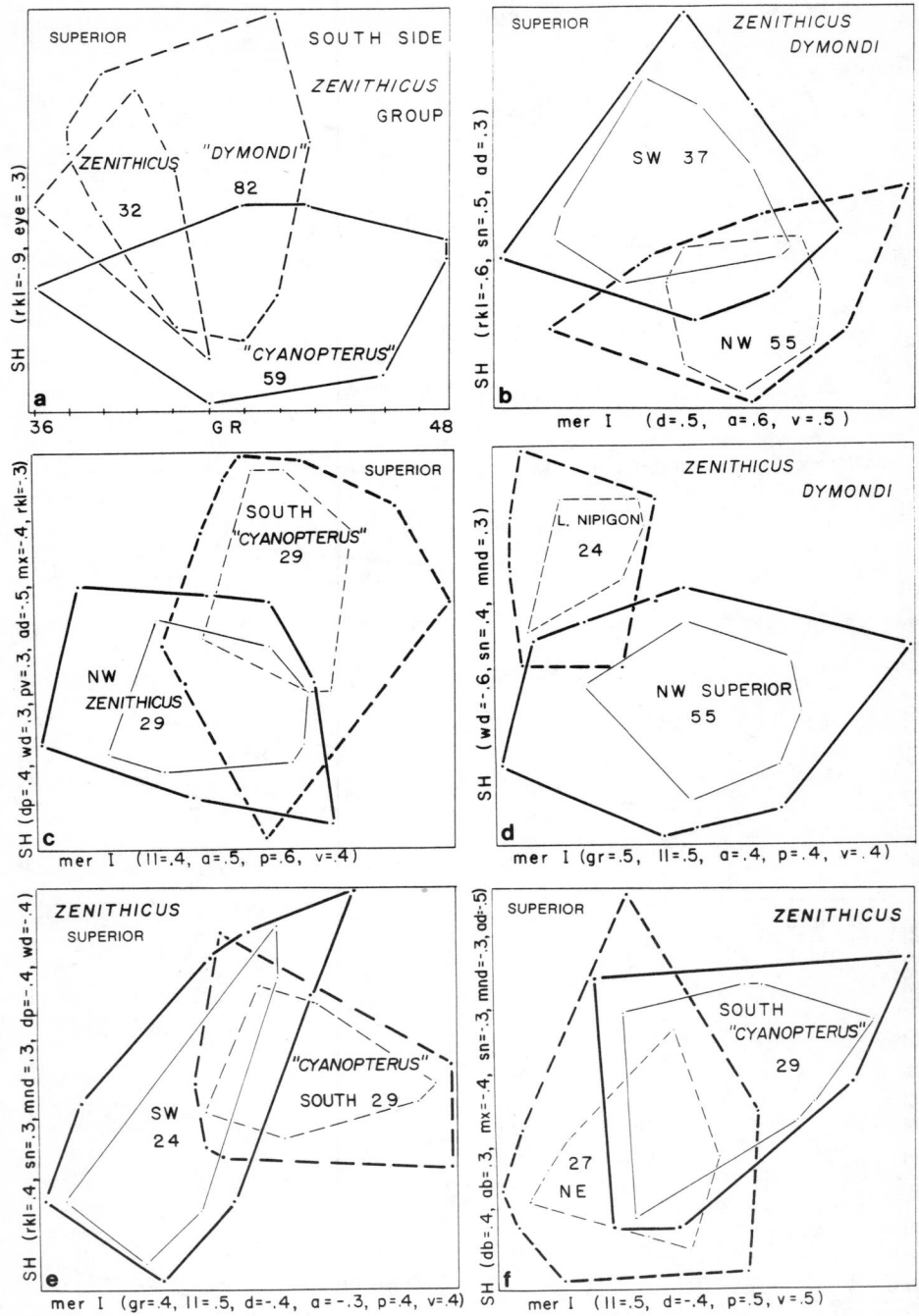

Figure 6. Cluster outlines of populations related to *C. zenithicus* of Lake Superior. Axes explained in legends of Figures 3, 4; abbreviations explained in Table 2.

and allopatry resulting from fluctuations and dispersal.

The species pairs are usually parapatric or partially sympatric during most of the year and allopatric or allochronic during their spawning seasons (Figure 7). For example, *kiyi* is usually sympatric and parapatric by depth with *hoyi*, its sister species, in all lakes except Nipigon, where *kiyi* does not occur. Similarly, *johannae*, now extinct, was parapatric by depth with *zenithicus*, its sister species, in lakes Michigan and Huron. In our analyses, *alpenae* and *zenithicus* are indistinguishable, and have been combined. *Coregonus nipigon* of Lake Nipigon is sympatric with its sister form, *artedii*, from which it is probably not specifically distinct (Scott, Crossman 1973). Considerable differentiation among sympatric and parapatric forms of *zenithicus* in Lake Superior constitute some of the most interesting examples of reproductive allopatry (Todd, Smith 1980).

From the presence of *zenithicus* outside the basin (Clarke 1973; Scott, Crossman 1973), we infer that it was an initial colonist following glaciation (Bailey, Smith 1981). *Coregonus artedii*, which is also widely distributed outside the basin, shows extensive allopatric differentiation and some sympatric differentiation. It is also a likely original colonist. The other species may have been derived in the basin in post-glacial time. We will explore the possible mechanisms for their evolution by examining the pattern of population differences within and between lakes.

The semi-isolation of the six lakes and the history of fluctuations in the depth and breadth of the connections between them provide the necessary conditions for the fluctuation hypothesis. Weak biological evidence for the fluctuation hypothesis is seen in slight interlake differentiation in *hoyi* between lakes Michigan, Huron, and Superior (Figure 3a,b,c), and in *kiyi* of lakes Superior, Michigan, and Huron (Koelz 1929; Figure 3a,b,d). Stronger evidence is provided by *C. kiyi orientalis* of Lake Ontario, which averages about five more gill rakers on the first arch than other *kiyi*. Substantial differentiation of *artedii* and *zenithicus* between lakes Michigan and Huron (Figure 4a,e), *zenithicus* between lakes Superior and Nipigon (Figures 5b,6d) and some other examples given by Koelz (1929)

also suggest differentiation in separate lakes. The maximum separation of lakes and sub-basins occurred between 10,600 and 8,000 years ago (Prest 1970). The eastern and western basins of Lake Superior might have been separated briefly; Lake Huron might have been divided in three; and separate lakes occupied the northern and southern basins of Lake Michigan.

Not all of the observed divergence is amenable to this explanation. There are several examples of within-lake differentiation: *hoyi* in Lake Superior (Figure 3e), *kiyi* from Lake Superior (Figure 3f), *artedii* in Lake Superior (Figure 4d; see also Koelz 1929), *artedii* in Lake Michigan (Figure 4b), *zenithicus* in Lake Michigan (Figure 4f) and *zenithicus* in Lake Superior (Figures 5a,c,6a,b,c,e,f). In many of the examples, divergence is allopatric or parapatric, but in some we see sympatric races. Sympatric populations of Lake Superior *zenithicus* separated by spawning time are shown by Figure 5f and of Lake Superior *artedii* by Figure 4f. Other possible examples are indicated by the forms of Lake Superior *zenithicus* (*dymondi* and *cyanopterus*) (Figure 6a,e) and are diagrammed in Figure 7. (Several examples of *artedii* and *zenithicus* with different spawning times, but no observed morphological differentiation, are not illustrated). At least two of the examples of differentiation between lakes (*kiyi* Figure 3a; *hoyi* Figure 3c) are partly attributable to within-lake variation, as indicated in the broader scatter in Figure 3, and thus might support the intralacustrine model as well.

Ciscoes are pelagic as young and epibenthic as adults, and feed on zooplankton and crustaceans. With regard to hypotheses about the kinds of characters that have diverged, the different examples provide ambiguous results. There is evidence for strong divergence of ecological characters in all comparisons — allopatric and sympatric (Table 2), perhaps because all are sympatric with other members of the superspecies, even when allopatric with respect to their sister groups. Unfortunately, the number of available comparisons of sympatric races is not large enough to allow statistical tests of significance. But there is evident importance among morphometric characters associated with the gill rakers and

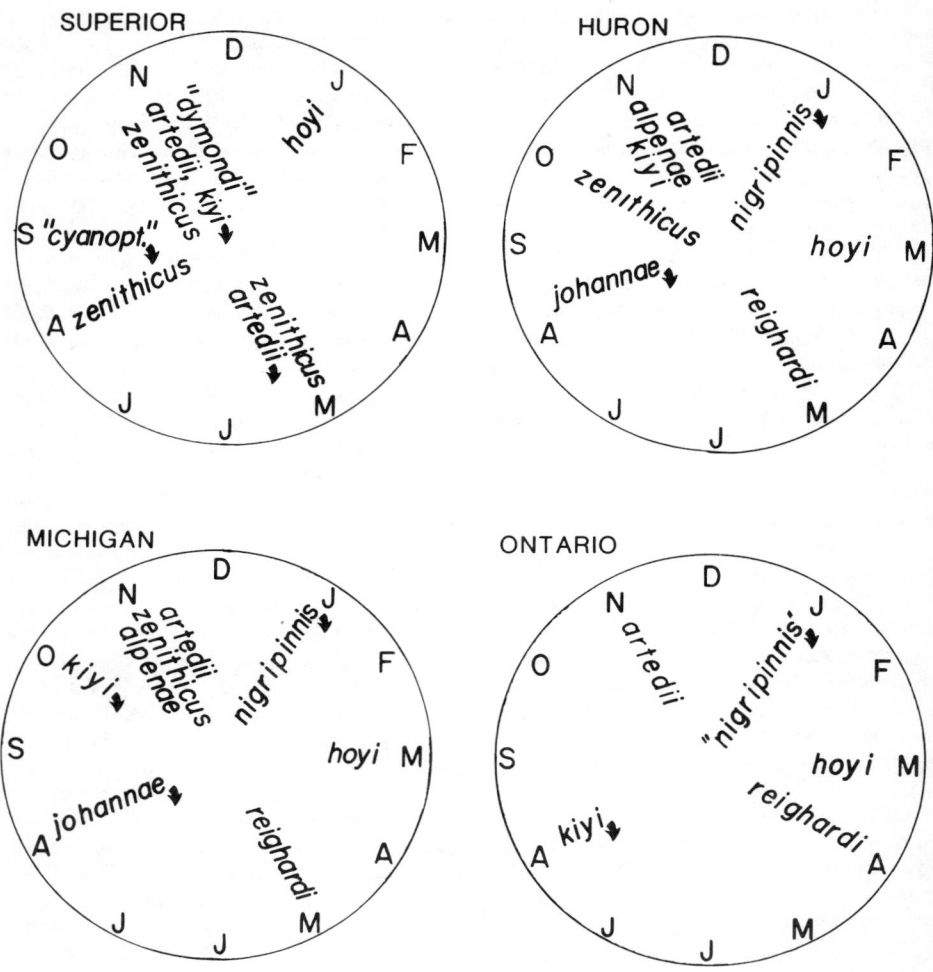

Figure 7. Modal spawning times for ciscoes in four Great Lakes. Abbrevations for months; arrows indicate deep water (120-179 m) spawners. Data from Koelz (1929) except fall-spawning *zenithicus* from Lake Superior (Todd, Smith 1980) and spring-spawning *artedii* from Lake Superior collected at 130-140 m by Todd at Copper Harbor.

mouth, indicating support for Hypothesis 5. The length and number (spacing) of gill rakers is directly related to food capture, and differences in these features have been shown by Bergstrand (1982) to be associated with partitioning of food resources among sympatric *Coregonus*. There is evidence for the importance of gill rakers and their responsiveness to ecological character displacement (Lindsey 1981; Dunham et al. 1979).

Inexplicably, within-population coefficients are as high as between-population coefficients in Table 2. Gill rakers are partly heritable (Svardson 1952; Loch 1974; Todd et al. 1981), although environmental effects do occur. The possibility that polymorphism in raker length is fundamental to the speciation process (West-Eberhard 1983) deserves further study.

Among the meristic variables, gill raker number is not as prominant as expected, given that intergroup and interpopulation

Table 2. Relative importance of variables in principal component analyses of cisco populations of the Great Lakes
(Principle components calculated on correlation matrices of meristic variables and covariance matrices of logs of morphometric variables. Average values of coefficients on principle components are shown.)

Variables	LAKE SUPERIOR: 22 morph,. 6 mer.				ALL LAKES: 14 morph.			
	within pop	between lake	within–lake parapatric	sympatric	within pop	between lake	within–lake parapatric	sympatric
morphometrics								
snout length (sn)	.1	.2	.3	.2	.2	.1	.3	.2
mandible length (mnd)	.1	.2	.2	.2	.1	.1	.2	.1
maxilla length (mx)	.0	.2	.2	.1	.1	.1	.2	.2
eye diameter (eye)	.0	.1	.1	.1	.1	.1	.1	.1
interorbital width (io)	–	–	–	–	.1	.1	.2	.1
head depth (occ)	.2	.1	.2	.2	.1	.0	.1	.1
head length (hl)	.1	.1	.1	.1	.1	.1	.1	.1
head & body length (hl)	.1	.1	.1	.1	.1	.1	.1	.1
branchial arch l. (arch)	–	–	–	–	.1	.1	.2	.1
mid gill raker length (rkl)	.4	.3	.4	.5	.7	.8	.7	.8
max dorsal ray l. (dht)	.2	.2	.1	.1	.2	.2	.1	.1
max anal ray length (aht)	.2	.1	.1	.1	.2	.3	.2	.2
pectoral fin length (pln)	–	–	–	–	.1	.0	.1	.1
pelvic fin length (vln)	.2	.2	.1	.1	.1	.1	.1	.0
max dorsal fin l. (dln)	.1	.2	.1	.2				
adipose fin 1. (ad)	.3	.1	.3	.3				
dorsal fin base (db)	.1	.2	.2	.0				
anal fin base (ab)	.1	.1	.1	.1				
snout to dorsal (sn-d)	.1	.2	.1	.0				
snout to adipose (sn-a)	.0	.0	.0	.0				
dorsal to adipose (d-a)	.1	.2	.1	.0				
pectoral to pelvic (pv)	.1	.1	.1	.0				
pelvic to anal (av)	.2	.1	.1	.2				
adipose to caudal (ad-d)	.2	.2	.1	.1				
body width (w) or (wd)	.3	.5	.3	.1				
body depth (dp)	.2	.2	.3	.2				
N	20	4	9	3	30	7	8	5
meristics								
gill rakers	.4	.2	.3	.4				
lateral line scales	.3	.2	.4	.5				
dorsal rays	.3	.4	.4	.4				
anal rays	.3	.5	.5	.5				
pectoral rays	.3	.3	.4	.3				
pelvic rays	.4	.4	.4	.2				
N	16	4	9	3				

The Lake Superior data are taken from 425 specimens examined by T. N. Todd. Most samples were taken in the 1920's by Walter Koelz. The data set for all lakes is from 1126 specimens collected under supervision of LaRue Wells, measured under supervision of S H Smith, Great Lakes Laboratory, U. S. Fish and Wildlife Service. Specimens deposited in the Museum of Zoology.

differences in scale counts and fin-ray numbers are less than in those in gill raker numbers among these fishes (Koelz 1929 Tables 7, 17-80). Strong covariance among the meristic variables is implied by the values in Table 2.

In summary, the incipient species flocks of Great Lakes ciscoes provide examples supporting intra-lake and inter-lake differentiation. The intra-lake examples include some for which geographic separation might be the agent of isolation and others which support the hypothesis of reproductive allopatry and allochrony. Within lakes the cause of reduced gene flow between populations is apparently differentiation of time and place of spawning, as suggested by Koelz (1929) and Svardson (1965). In these cases reproductive isolation among the incipient species seems dependent on migration path and maturation time, both of which are proximally related to temperature and depth preference, and are ultimately related to seasonal timing of food resources available, especially to early life history stages. The schooling behavior of these fishes is an important factor in their differentiation (Svardson 1965). Population differences show an imbalance toward food capture structures, suggesting ecological character displacement (Table 2).

SCULPINS OF LAKE BAIKAL

Taliev (1955) recognized 36 species and subspecies of sculpins in Lake Baikal. This is probably the most ancient lacustrine species flock among fishes. Its differentiation has proceeded to the family level over a time span covering much of the late Cenozoic. Russian workers separate *Comephorus baicalensis* and *dybowskii* highly-derived pelagic, ovoviviparous predators, as a family separate from their sister genera, *Batrachocottus* and *Cottocomephorus*, which remain classified in the Cottidae with the remainder of the Baikal sculpins. Taliev's subfamilies, Cottomephorinae (*Paracottus, Procottus, Metacottus, Batrachocottus,* and *Cottocomephorus*) and Abyssocottinae (*Cottinella, Asprocottus, Abyssocottus*), comprise the rest of the group (Figure 8). The species and subspecies, with ecologic data, are listed in Table 3. A cladistic and more conservative taxonomy would recognize one family in the lake, Cottidae, with *Paracottus* assigned to the genus *Cottus* (Sideleva 1979; Koryakov, Sidelev 1976). The other genera, possibly excepting *Metacottus*, are clearly differentiated groups of species held together with good synapomorphies. There are repeated trends toward loss of preopercular armament and exposure of lateralis canals on the bones in conjunction with development of specialized sensory structures.

Relative to the immigration hypotheses, we are interested in the examples of Baikal species outside the lake. Differentiation of ecologically significant characters within sister species in the lake are of interest with respect to the hypothesis of intralacustrine evolution. There is some evidence for the immigration hypothesis. *Cottus (Paracottus) kessleri bauntovi* of lake Baunt differs from *kessleri* in Lake Baikal by dorsal and anal fin-ray counts. *C. kessleri* may represent one of two primitive immigrants from which the sculpin flock might be derived (Taliev 1955 Fig 169). On the other hand, *Asprocottus kozovi* of Lake Baunt is related to specialized Baikal species.

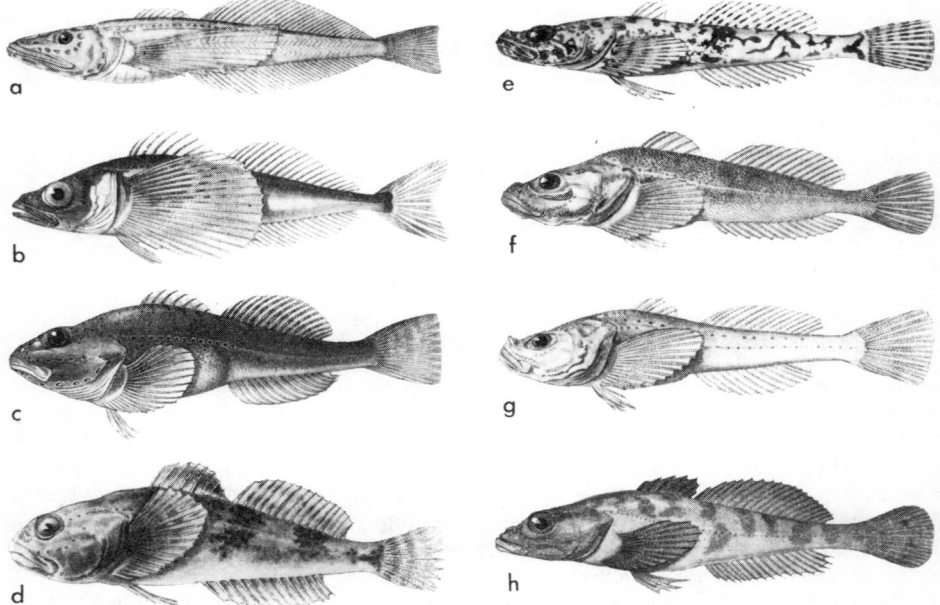

Figure 8. Illustration of Baikal sculpins (Taliev 1955). a) *Comephorus baikalensis,* b) *Cottocomephorus* inermis, c) *Batrachocottus nikolskii,* d) *Cottus kneri,* e) Abyssocottus godlewskii, f) Asprocottus herzensteini, g) *Cottinella boulengeri,* h) *Procottus jeittelesi minor*

Table 3. Ecological characteristics of 36 species and subspecies of Baikal sculpins
Data source: Taliev 1955; Kozhov 1963
(Parenthetic values = pelagic fishes. Depth Zone: x = more common occurrence; − = less common.)

Species	rock	rock & silt	silt	silt & sand	sand	sand& rock	0–5	5–100	100–300	300–500	500+	spawning months
Paracottus kneri	50	10	6	14	16	5	x	−				MJJ
P. insularis	100						x					
P. kessleri	5	6	6	24	46	13	x					
P. k. lubricus				26	63	11	x	−				
P. pelagicus					67	33	x					
Batrachocottus baicalensis	42	15	7	4	7	25	−	x	−			FMA
B. b. pachytus			100					x	−			FM
B. uschkani	25					75		x	−			
B. multiradiatus		33	43	22		2		x	−			MA
B. nikolskii	10	29	49	10	3			−	−	x		
Cottocomephorus grewingki (pel.)	(10)	(10)	(17)	(24)	(23)	(17)	x			x		MAMJJ
C. g. alexandrae	(6)	(8)	(38)	(31)	(10)	(6)		x	−	−		FM
C. inermis	(8)	(8)	(33)	(25)	(13)	(12)	x	−	−	−	−	F
Metacottus gurwici						1 ind.	x	−				
Procottus jeittelesi	11	22	36	21	3	7	x	−	−	−		JFM
P. j. minor		73	27					−	x			
P. j. major			18	54	18	9		−	x	−	−	JF
Asprocottus herzensteini	6	19	40	21	8	6		−	x	−	−	SONDJ-MJ?
A. h. parmiferus	11	57	7		4	22		−	x			
A. h. platycephalus				70	20	10		−	x	−	−	
A. h. intermedius			22	59	15	4		−	x	−		
A. h. abyssalis		42	58						−	x		
A. pulcher			33	67				−	x	−		
A. megalops		19	50	31				−	x	−		
A. m. eurystomus		21	56	16	3	3		−	x	−	−	MAM
A. gibbosus		29	59	12					−	x	x	JF
Cottinella boulengeri		44	56								x	DJ
C. werestschagini		14	86								x	O-?
Abyssocottus korotneffi		24	69	7					−	x	−	DJF
A. godlewskii	3	65	14	15	2			−	x	−		SONDJFM
A. g. griseus	12	31	28	21	7			−	x	−	DJ
A. pallidus	3	30	57	9	1			−	−	x	−	
A. bergianus		66	24	10				−	x	−	MAM
A. elochini	100							x				
Comephorus baicalensis (pelagic)								x	−	−		JA
C. dybowskii (pelagic)								x		−		SN

It differs from *A. megalops* and *A. herzensteini* of Lake Baikal in the length and pore size of the lateral line and the number of pelvic rays (Taliev 1955). Lake Baunt is in one of a series of connected valleys in the Tsipa tectonic depression, tributary to the Vitim River in the Lena drainage, 500 km west of the north end of Baikal, which is in the Angara-Yenesey drainage. *Cottus kneri*, an unspecialized Baikal species, is also in the Vitim drainage (Kozhov 1963 226). It also might be an immigrant and a basal species in the sculpin flock (Taliev 1955 fig 169). The Baunt and Vitim records indicate that Lake Baikal and its fauna are part of an older lake system with connections to the northwest, different from the connections seen today. The fluctuation hypothesis is supported (but not exclusively) by these three occurrences. A third sculpin said to be related to the Baikal fauna is known from Lake Agata on the Putoran Plateau (Koryakov, Sidelev 1976). It differs from

C. (P.) kneri by small size, elongate body, large lateral line pores, corneal overgrowth, and number of pectoral rays. This record is almost 4,000 km northwest, in the Yenesey drainiage. It further supports the idea that *Cottus (P.) kneri* is a sister group to the other sculpins in Lake Baikal. These distributional facts provide reasonable examples of expectations from the immigration hypothesis.

Many distributional and morphological characteristics of the Baikal sculpins support the reproductive allopatry hypothesis. All but the least specialized species are restricted to waters that are moderately cold (4-12 C; below 5 m) or cold and abyssal (Taliev 1955; Kozhov 1963), where they show broad but often distinctive preferences for depth and substrate (Table 3). They also show some diversification of spawning times (Table 3), though less than shown by ciscoes. Examination of the characteristics separating 16 sister pairs reveals an interesting distribution of kinds of characters. Differences in the size, distribution, and accessories to lateralis sensory pores are most prominent (8 examples), followed by size and shape of paired fins (7), eye size (6), color and pattern (6), position or shape of prickles (5), development of preopercular spines (3), head width or depth (3), specializations of fin membranes (3), number of dorsal or anal rays (3), rostrum shape (1), anal papillae (1), and separation of the dorsal fins (1). Objective evaluation of ecological significance is difficult, but it appears that the sensory and locomotor differences between these species differ from those that separate the two dozen species of American *Cottus*. Species of American *Cottus* are separated from each other by small differences in numbers of dorsal and anal rays, preopercular armature, reduction of lateralis pores, reduction of palatine teeth, pelvic and pectoral ray number, and relative head size.

Reproductive isolation seems patent among most Baikal species, however the divergence of intralacustrine subspecies and the evidence for hybridization between *C. kneri* and *kessleri* (Taliev 1955) suggest that in Lake Baikal, initial divergence may occur in the absence of complete reproductive isolation.

Differentiation of several other groups of fishes within Lake Baikal also provides evidence for the hypothesis of reproductive allopatry and allochrony. Four stocks of *Coregonus autumnalis* spawn in different tributaries to the lake. The intralacustrine subspecies of *Coregonus lavaretus* differ in spawning localities within the lake and in its tributaries. Two forms of *Thymallus arcticus* in Lake Baikal differ in color, fatness, habitat, and substrate preference. Two forms of *Lota lota* differ in their choice of coastal versus tributary habitats in the lake (Kozhov 1963).

In summary, the fluctuation hypothesis finds some support among the few sculpins found in adjacent basins as well as in Baikal. The distribution of *Cottus kneri* and *kessleri* and their relatives provides evidence for the immigration hypothesis. The broader distributional and morphological evidence, however, provides support for depth-related ecological divergence of parapatric populations mostly in the intermediate (5-500 m) depth zone (Kozhov 1963; Table 3). Behavioral preferences for depth and depth-related conditions such as substrate texture, especially during reproduction, provide the most likely causes of assortative mating. Because of the tremendous size of the lake (length 636 km; depth 1620m) and the benthic habits of much of the diversity, geographic separation is probably a significant barrier to gene flow among the localized populations. This is also supported by differentiation of some fishes, molluscs, and gammarids in the northern and southern basins within Baikal (Brooks 1950; Kozhov 1963). Where interpopulation gene flow occurs, however, ecological differences probably have contributed to divergence through mechanisms discussed below.

SCULPINS OF LAKE IDAHO

The North American counterpart to the sculpins of Lake Baikal is found among sculpins in the Pliocene stage of Lake Idaho (Smith et al. 1983). Lake Idaho was a rift lake about the size of Lake Ontario that occupied the western Snake River plain for about two million years in the Pliocene. Six species of sculpins have been described in three genera, *Myoxocephalus, Kerocottus,* and *Cottus*, and two additional species have been discovered, but are undescribed. A ninth species (*Cottus*) lived in tributaries to the lake. Except for the latter, Lake Idaho sculpins were, so far as known, endemic to that lake. A fair sampling of fish fossils, including sculpins, from surrounding Pliocene lakes is documented (Smith 1981a), but no relatives of the Lake Idaho sculpins have been found, except among recent northern *Myoxocephalus*, which Taliev (1955) suggests is a relative of part of the baikal assemblage.

Like Baikal sculpins, those of Lake Idaho had specializations of the lateralis sensory system. In contrast to Baikal sculpins, those of Lake Idaho are characterized by elaborations of strong spines on the preopercles and the suborbital bones (Smith 1975) presumably because predators such as trout, salmon, squawfish, bass-size sunfish and catfish, which are large enough to consume adult sculpins, were more diverse and more widespread in the shallower Lake Idaho. In Lake Baikal, three species of trout are rare, being restricted to river mouths, while pike, perch, and burbot are restricted to bays and shore habitats (Kozhov 1963). The Baikal seal is an important predator on sculpins; otters were common in Lake Idaho. In both lakes there are or were many fish species predatory on young and deepwater sculpins. The most important are whitefish, *Coregonus autumnalis*, in Lake Baikal and, as evidenced by sedimentary association, *Prosopium prolixus* in Lake Idaho.

The most derived Lake Idaho sculpins are predominantly found (and are the dominant forms) in relatively deepwater sediments, though not as deep as in Baikal. The stratigraphic record shows the shallow-water species, *Kerocottus divaricatus* and *Cottus calcatus*, to be alone in sediments of the older half of the lake's history (Smith et al. 1983) with the other species added later in time and in deeper habitats, as postulated for the sculpins of Lake Baikal by Kozhov (1963).

DISCUSSION

The origin of closely related fish species in north temperate lakes is here examined with respect to three hypotheses: accumulation by 1) immigration, 2) speciation in allopatry caused by fluctuation of water level, and 3) intralacustrine isolation by reproductive allopatry and allochrony. Five lacustrine examples have been discussed. Three species of whitefishes in Bear Lake seem to illustrate Hypothesis 1; i.e., they comprise a pseudo-

flock largely accumulated by immigration from surrounding drainages. Evidence for this conclusion is the existence, outside of Bear Lake, of sister forms to the three species. Relative to Hypothesis 2, pluvial lakes of the Great Basin in western United States demonstrate that large-scale fluctuations caused extinction, not species flock formation.

Ciscoes of the Laurentian Great Lakes, examined in more detail, demonstrate an active role for all three processes in the origin of eight semispecies and their numerous subpopulations. The sculpins of Lake Baikal, though much older, with more trenchant divergence, are characterized by distributions, relationships, and specializations that suggest a role for each of the three modes of origin. The sculpins of Pliocene Lake Idaho provide evidence for progressive occupation of deeper waters by new, more specialized species, as postulated by Kozhov (1963) for Lake Baikal.

In the examples purporting to show reproductive allopatry and allochrony, the causes of initial isolation and divergence are unkown. We try to infer them from distributional and morphological facts in the context of ecological and evolutionary models, but the arguments are still drawn piecemeal from different examples, and the models remain speculative. Nevertheless, it now seems possible to narrow the field of possible explanations to a small set of basically familiar hypotheses that are mutually consistent.

In our north temperate examples of flocks, it has been observed that the fishes have diverged especially in depths, times, and/or substrates used for reproduction. Koelz (1929) said that however the cisco species originated, they were maintaining segregation by time and place of spawning. Fryer (1977) noted general agreement that lacustrine fishes show their most restrictive substrate preferences during reproduction. Therefore, we must look to aspects of reproductive behavior that not only contribute to assortative mating, but seem to involve an adaptive landscape (Wright 1931) with peaks of high fitness separated by areas of low fitness. Fitness in this sense would be estimated by growth rates associated with different combinations of phenotypes and resources. Because the dominant mortality occurs among young, probably under density-dependent conditions

(Cushing 1974), the key ecological factor probably is abundance of food available to young life-history stages. This is therefore a period that could select for timing of the entire life history.

Integration of these observations into a cohesive theory of population divergence imposes three requirements: 1) ecological evidence for discrete differences in spawning time and place; 2) evidence for distinctness in distribution of available zooplankton; 3) a model for disruptive selection leading to divergence of two populations from one.

First, it is clear that spatial gradients in depth, temperature, substrate, food, light, etc. exist in lakes. Each of these can be traced more or less clinally through space with local steepness at the thermocline, at substrate ecotones etc. Patchy distribution of substrate and food present few significant barriers to mobile searchers like ciscoes or pelagic sculpins, but probably act as stronger barriers to negatively bouyant sculpins. When temporal variation is added to the system, however, the total possible niche space presents major low-fitness areas even to the pelagic fishes; being at the right place at the right time for abundant food, favorable metabolic and maturation temperature, and successful reproduction requires relatively unerring reaction to patchy resources. The areas of low fitness are those combinations of spawning times and places in which the fish have reduced growth because of inadequate food or suboptimal temperature, or mortality by predation. Successful adaptation to the system invloves schooling behavior (Svardson 1965), homing behavior (Horrall 1981), and probably, recognition of one's own population by olfactory cues (Nordeng 1977; Svardson 1965). The experimental and biochemical evidence for discreteness of spawning populations was summarized by Todd and Smith (1980).

The functionally demanding connections between spawning time, depth, temperature, food, and survival are most easily documented for coregonines because of the rich fish-cultural literature for this group. Eggs of *C. artedii*, for example, hatch in 236 days at 0.5 C, but 92 days at 5.6 C (Colby, Brooke 1970). Therefore, production of offspring that are timed to exploit a peak of zooplankton availability

requires spawning at a location and a time that will provide the critical temperature. The role of incubation temperature in regulating the time of hatching to coincide with peak plankton availability has been demonstrated (Ammann 1949). Russian workers have shown that in Lake Baikal, which retains ice until well into spring, hatchery fry must be timed to hatch and be planted before breakup of the ice, because at that time, plankton are concentrated near the surface where the phototaxic fry feed (Chernyayev 1973). Coregonine fry are sight feeders requiring light, and they capture prey one at a time (Janssen 1978). Fry feed in the first several days after hatching, before absorption of the yolk sac; failure to feed leads to mortality (Bogdanova 1972), with allowable starvation time negatively correlated with temperature (John, Hasler 1956). Early prey include rotifers, *Cyclops, Bosmina,* small *Daphnia,* (found to be essential in some studies) and possibly ostracods (Fluchter 1980; Brooke and Todd unpubl.). Different species of *Coregonus* may have different preferences (Marciak 1979). Similarly, food availability is critical at the time of metamorphosis (Kokova 1978).

In the Great Lakes, each species of zooplankton shows a single annual abundance peak in the summer, with a rapid spring rise, a rapid decrease in the autumn, and low winter numbers. Species peaks are roughly coincident, but maximum abundances are at different depths, and there are early and late species at different localities and depths (Wells 1960; Watson 1974). In Lake Ontario, 23 species show peaks at depths ranging from 15-83 m, which vary seasonally and diurnally over tens of meters (Wilson, Roff 1973). Horizontal variation is considerable. Many cladoceran species are confined to bays, inshore waters, and near the surface, and there are north-south gradients. Lakewide sampling of Lake Ontario in all seasons showed variation from January to August of 10-157 mg/m^3 in the upper 50 m, and in Saginaw Bay, Lake Huron, concentrations of individuals were as high as 324,000/m^3 (Watson, Carpenter 1974). Biomass may vary by a factor of 2 to 4 during a month, being higher in the bays and littoral zone, and low in the pelagic zone and at greater depths.

Maximizing the utilization of this resource

with optimal hatching time would enable a tremendous increase in survival of offspring. But year-to-year variations may be large (Watson 1974). The resulting average reproductive success would nevertheless select strongly for traits leading to migration to the right place at the optimal time (Horrall 1981). Such timing also requires an accurately regulated metabolic and maturation response to temperature. Migratory or timing errors would lead to low fitness, except when they coincide with unexploited zooplankton availability. Such errors are the basis of colonization and gene flow among populations. The most successful schedulings of reproduction, hatching, and growth are those in which optimal sequences of food resources and ambient temperatures are tracked through yearly migration cycles. Viewed in this way, the nature of low fitness regions of niche space becomes more apparent.

Several models outline mechanisms for differentiation in spatial and ecological gradients. Endler's (1977) model for differentiation along clines is applicable to allopatric, parapatric, and possibly sympatric populations. Rosenzweig's (1978) model for competitive speciation proposes a mechanism for differentiation of sympatric populations and predicts differentiation in ecologically relevant characters.

The potential for competitive speciation depends on the existence of phenotypes adapting to density-dependent variation in fitness (adaptive peaks) over nichespace, with each phenotype exploiting other subniches to some extent. Assuming food to be a critical limiting resource for Great Lakes cicsoes, density-dependent variation in fitness may be estimated by changes in growth rate. A phenomenon replicated several times in the exploitation of Great Lakes ciscoes is an increase in individual growth rate in response to a reduction in numbers. In Lake Michigan from 1963 to 1970, as numbers of C. hoyi declined, weight nearly doubled, initially at least, because of increased growth rate (Brown 1970). A similar phenomenon evidently occurred in Lake Huron in the late 1960's (Wells, McLain 1973). These observations provide evidence for density-dependent growth in this species. Density-dependent survivorship has not been measured but may be inferred for conditions prior to the advent of commercial fishing, lamprey depredations, and alewife competition.

Rosenzweig's model postulates colonization from an occupied to an unoccupied adaptive peak, such as must have occurred many times in the postglacial history of the lakes, following the original colonization. Resulting competitive pressure emanates from both peaks to the intervening trough, i.e., from density-dependent population pressures at the edges of the optimal combinations of feeding times and localities. If fitness at the trough is pushed below zero by competition from the sides, the decreased survival of phenotypes exploiting that time and locality creates a disruptive gap, and gene flow involving the straying or intermediate phenotype is depressed. In the model, fitness is defined as the *rate of change* of the logarithm of the density of the ith phenotype. The relevant phenotypes are tendencies to forage at certain localities and depths at certain times and the associated food-capture structures. Food-capture structures are selected as a result of the kinds of food available in the area searched, rather than by local choices (MacArthur, Pianka 1966).

The model provides a mechanism for late- and early-spawning populations to competitively depress the fitness of individuals with intermediate times. It can be extended to habitats or foraging locations at different depths if the food availability has the necessary peaks at those depths, or if combinations of food, predator avoidance, and growth are discontinuous. The predominance of morphological traits related to food capture among the phenotypic differences, along with the behavioral-physiological differences in spawning times and locations, is suggestive of the kinds of phenotypes that should be subject to the competitive speciation model.

The alternative model (Endler 1977; Moore 1981) explains divergence of the premating isolating mechanisms — spawning differences — as a result of selection for traits that promote spawning at the natal grounds with related individuals. The genes promoting such homing and recognition behavior act as assortative mating modifiers, as modeled by Moore (1979,1981), leading to differentiation of parapatric populations. This model does not provide a mechanism for divergence of ecological traits as immediately as the Rosenzweig model, but it is simpler with regard to postulation of the origin of premating isolation.

SUMMARY

The sculpins of Lake Baikal and the ciscoes of the Laurentian Great Lakes are the best examples of north-temperate species flocks of fishes. Each comprises, at most, one or two monophyletic groups that evidently differentiated wholly within the lakes, following the immigration of one or two ancestral kinds. Evidence for initial immigration is seen in the existence of a few generalized sister groups outside the lakes.

A prevalent hypothesis for lacustrine speciation is the fluctuation model, which proposes that low-water division of the lacustrine habitat into daughter lakes permits allopatric speciation of the fishes. Sympatric assemblages are created by reuniting the lakes and their inhabitants at high-water stages. Among Baikal sculpins and Great Lakes ciscoes, there are a few examples that might be explainable by this model. Baikal was once part of a larger system of lakes, and the Laurentian Great Lakes have gone through a low-water stage with more isolation among and within lakes than now exists. A significant part of the diversity is not explainable in this way, however. Also, the pluvial lakes of the Great Basin of western North America provide only a few small examples possibly attributable to this kind of speciation, despite a history of fluctuations that theoretically should have caused abundant speciation.

Intralacustrine variation among incipient species and other populations of Great Lakes ciscoes and among species and subspecies of Baikal sculpins suggests that most of the divergence has occurred within the lakes. Though some of the intralacustrine divergence might be strictly allopatric, most has probably occurred despite some gene flow. It is postulated that differentiation begins with reproduction at different localities and times — reproductive allopatry and allochrony. Physiological interaction between temperature (depth) and maturation time, and behavioral tendencies to home to the spawning grounds and mate with related individuals are the basis of premating reproductive

isolation. Genes for these traits serve as the "assortative mating modifiers" permitting divergence (Endler 1977; Moore 1981).

Among the phenotypic differences between populations, morphological traits related to food capture are prominent. These suggest that ecological character displacement has been important in the divergence. Rosenzweig's (1978) competitive speciation model may explain initial divergence of populations in competition for zooplankton.

ACKNOWLEDGMENTS

J Humphries helped with sheared principal component analysis. W Dominey, E H Brown, A R Templeton, F L Boodstein, and R L Elder read the manuscript and made important contributions. L Wells of the U.S. Fish & Wildlife Service and the 1982 Species Flock Seminar, Museum of Zoology, provided data and ideas. Partial support by NSF Grant DEB-8011562 to R L Strauss & G R Smith. Contribution 613 of the Great Lakes Fishery Laboratory, USFWS.

REFERENCES

Ammann E 1949 Erfahrungen und Erfolg der Kaltwassererbrutung bei Felchen. Int Assoc Theor Appl Limn 10:38-42

Bailey R M, G R Smith 1981 Origin and geography of the fish fauna of the Laurentian Great Lakes Basin. Can J Fish Aqua Sci 38:1539-1561

Bergstrand E 1982 The diet of four sympatric whitefish species in Lake Parkijaure. Rept Inst Freshwater Res Drottningholm 60:5-14

Bogdanova L S 1972 The transition to exogenous feeding in the larvae of Coregonus lavaretus ludoga. Poljakov. J Ichth 12:531-537

Brooks J L 1950 Speciation in ancient lakes. Quart Rev Biol 25:30-60, 131-176

Brown E H 1970 Extreme female predominance in the bloater (Coregonus hoyi) in Lake Michigan in the 1960's. 501-514 C C Lindsey, C S Woods eds. Biology of Coregonid Fishes. Univ of Manitoba Press. Winnipeg

Chernyayev Z A 1973 Reproduction and development of the Lake Baikal whitefish (Coregonus lavaretus baicalensis Dyb.) in relation to problems of artificial propagation. J Ichth 13:216-230

Clarke R McV 1973 The systematics of ciscoes (Coregonidae) in Central Canada. PhD Thesis Univ Manitoba. Winnipeg

Colby P J, L T Brooke 1970 Survival and development of lake herring (Coregonus artedii) eggs at various incubation temperatures. 417-428 C C Lindsey, C S Woods eds. Biology of Coregonid Fishes. Univ Manitoba Press. Winnipeg

Cushing D H 1974 The possible density dependence of larval mortality and adult mortality in fishes. 101-111 J H S Blaxter ed. The Early Life History of fish. Springer-Verlag. Heidelberg

Deason H J 1936 *Morphometric and life history studies of the pike-perches (Stizostedion)* of Lake Erie. PhD thesis. Univ Michigan. Ann Arbor

Dobzhansky T 1951 *Genetics and the Origin of Species.* Columbia Univ Press. NY

Dunham A E, G R Smith, J N Taylor 1979 Evidence for ecololgical character displacement in western American catostomid fishes. *Evolution* 33:877-896

Eddy S 1957 *How to Know the Freshwater Fishes.* Brown. Dubuque Iowa

Endler J 1977 *Geographical Variation, Speciation and Clines.* Princeton Univ Press. Princeton NJ

Fenderson O C 1964 Evidence of subpopulations of lake whitefish, *Coregonus clupeaformis,* involving a dwarfed form. *Trans Am Fish Soc* 93:77-94

Fluchter J 1980 Review of the present knowledge of rearing whitefish (Coregonidae) larvae. *Aquacult* 19:191-208

Fryer G 1977 Evolution of species flocks of cichlid fishes in African lakes. *Z Zool Syst Evol* 15:141-165

Fryer G, T D Iles 1969 Alternative routes to evolutionary success as exhibited by African cichlid fishes of the genus *Tilapia* and the species flocks of the Great Lakes. *Evolution* 23:359-369

Goodier J L 1981 Native lake trout (*Salvelinus namaycush*) stocks in Canadian waters of Lake Superior prior to 1955. *Can J Fish Aqua Sci* 38:1724-1737

Greenwood P H 1974 The cichlid fishes of Lake Victoria, East Africa: The biology and evolution of a species flock. *Bull Brit Mus (Nat Hist)* Suppl 6:1-134

Horral R M 1981 Behavioral stock-isolating mechanisms in Great Lakes fishes with special reference to homing and site imprinting. *Can J Fish Aqua Sci* 38:1481-1496

Hubbs C L, R R Miller 1948 Correlation between fish distribution and hydrographic history in the desert basins of western United States. 17-166 *The Great Basin with Emphasis on Glacial and Postglacial Times.* Univ Utah Bull 38 Biol Ser 10:1-191

Humphries J M 1984 Genetics of speciation in pupfish from Laguna Chichancanab, Mexico. 129-140 A A Echelle, I Kornfield eds *Evolution of Fish Species Flocks.* Univ Maine Press at Orono

Humphries J M, F L Bookstein, B Chernoff, G R Smith, R L Elder, S G Poss 1981 Multivariate discrimination by shape in relation to size. *Syst Zool* 30:291-308

Janssen J 1978 Feeding-behavior repertoire of the alewife, *Alosa pseudoharengus* and the ciscoes *Coregonus hoyi* and *C. artedii. J Fish Res Bd Can* 35:249-253

John K R, A D Hasler 1956 Observations on some factors affecting the hatching of eggs and the survival of young shallow-water cisco, *Leucichthys artedii* LeSueur, in Lake Mendota, Wisconsin. *Limn Ocean* 1:176-194

Kirkpatrick M, R K Selander 1979 Genetics of speciation in lake whitefishes in the Allegash basin. *Evolution* 33:478-485

Koelz W 1929 Coregonid fishes of the Great Lakes. *Bull U.S. Bur Fish* 43:297-643

Kokova V Y 1978 The feeding of larvae of the whitefish, *Coregonus lavaretus ludoga,* the Baikal omul, *Coregonus autumnalis migratorius* and the 'peled', *Coregonus peled. J Ichth* 18:953-960

Koryakov Y A, G N Sidelev 1976 The sculpins (Cottidae) of Lake Agata on the Putoran Plateau. *J Ichth* 16:498-501

Kozhov M 1963 *Lake Baikal and its Life.* W Junk. The Hague

Kurenkov S I 1977 Two reproductively isolated groups of kokanee salmon *Oncorhynchus nerka kennerlyi* from Lake Kronotskiy. *J Ichth* 17:526-533

Lindsey C C 1981 Stocks are chameleons: plasticity in gill rakers of coregonid fishes. *Can J Fish Aquat Sci* 38:1497-1506

Lindsey C C, J W Clayton, W G Franzin 1970 Zoogeographic problems and protein variation in the *Coregonus clupeaformis* whitefish species complex. 127-146 C C Lindsey, C W Woods eds. *Biology of Coregonid Fishes.* Univ Manitoba Press. Winnipeg

Liu R K 1969 The comparative behavior of allopatric species (Teleostei, Cyprinodontidae: *Cyprinodon*) PhD thesis. Univ California Los Angeles.

Loch 1974 Phenotypic variation in the lake whitefish *Coregonus clupeaformis* induced by introduction into a new environment. *J Fish Res Bd Can* 31:55-62

MacArthur R H, E Pianka 1966 On optimal use of a patchy environment. *Amer Natur* 100:319-327

Marciak Z 1979 Food preferences of juveniles of three coregonid species reared in cages. *Sp Publ European Maricul Soc* 4:127-137

Mayr E 1942 *Systematics and the Origin of Species.* Columbia Univ Press. New York.

_____ *Evolution and the Diversity of Life.* Belknap, Harvard Univ Press. Cambridge Mass

McPhail J D, C C Lindsey 1970 Freshwater fishes of northwestern Canada and Alaska. *Fish Res Bd Can Bull* 173:1-381

Miller R R 1948 The cyprinodont fishes of the Death Valley system of eastern California and southwestern Nevada. *Misc Publ Univ Mich Mus Zool* 42:1-80

Miller R R, G R Smith 1981 Distribution and evolution of *Chasmistes* (Pisces: Catostomidae) in western North America. *Occ Pap Mus Zool Univ Mich* 696:1-46

Moore W S 1979 A single-locus mass-action model of assortative mating, with comments on the process of speciation. *Heredity* 42:173-186

_____ Assortative mating genes selected along a gradient. *Heredity* 46:191-195

Nordeng H 1977 A pheromone hypothesis for homeward migration in anadromous salmonids. *Oikos* 23:155-159

Prest V K 1970 Quaternary geology of Canada. 676-764 R J W Douglas ed. *Geology and Economic Minerals of Canada.* Geol Surv Canada. Ottawa

Rosenzweig M L 1978 Competitive speciation. *Biol J Linn Soc* 10:275-289

Scott W B, E J Crossman 1973 Freshwater fishes of Canada. *Fish Res Bd Can Bull* 184:1-966

Sideleva V G 1979 Morphology of the lateral line system of Eurasian sculpins of the subfamily Cottinae (Cottidae). *J Ichth* 19:434-446

Sigler W F, R R Miller 1963 *Fishes of Utah.* Utah Dept Fish Game. Salt Lake City

Smith G R 1975 Fishes of the Pliocene Glenns Ferry Formation, southwest Idaho. *Univ Mich Mus Paleo., Pap Paleo.* 14:1-68

_____ 1978 Biogeography of intermountain fishes. 17-42 K Harper ed. *Intermountain Biogeography.* Great Basin Natur Mem 2

_____ 1981a Late Cenozoic freshwater fishes of North America, *Ann Rev Ecol Syst* 12:163-193

_____ 1981b Effects of habitat size on species richness and adult body size of desert fishes. 125-171 R J Naiman, D J Soltz eds. *Fishes in North American Deserts.* Wiley. New York

Smith G R, J G Hall, R K Koehn, D J Innes 1983 Taxonomic relationships of the Zuni Mountain Sucker. *Catostomus discobolus yarrowi. Copeia* 1983:37-48

Smith G R, W L Stokes, K F Horn 1968 Some late Pleistocene fishes of Lake Bonneville. *Copeia* 1968:807-816

Smith G R, K Swirydczuk, P J Kimmel, B H Wilkinson 1983 Fish biostratigraphy of Late Miocene to Pliocene sediments of the western Snake River Plain, Idaho. *Idaho Bur of Mines Geol.* In press.

Snyder J O 1919 Three new whitefishes from Bear Lake, Idaho and Utah. *Bull U.S. Bur Fish* 36(1917-18):1-9

Stancovic S 1960 *The Balkan Lake Ohrid and its Living World.* W Junk. The Hague

Svardson G 1952 The coregonid problem. IV. The significance of scales and gill rakers. *Rep Inst Freshwater Res. Drottningholm* 33:204-232

_____ 1965 The coregonid problem. VII. The isolating mechanisms in sympatric species. *Rep Inst Freshwater Res. Drottningholm* 46:95-123

_____ 1979 Speciation of Scandinavian *Coregonus. Rep Inst Freshwater Res. Drottningholm* 57:1-95

Taliev D N 1955 Sculpins of Baikal. (Cottoidei). Acad Sci USSR. Moscow

Taniguchi N, T Ishiwatari 1972 Inter- and intraspecific variations of muscle proteins in the Japanese Crucian carp I. Cellulose-acetate electrophoretic pattern. *Japanese J Ichth* 19:217-222

Taylor D W, G R Smith 1981 Pliocene molluscs and fishes from northwest Nevada and southwest California. *Contr Univ Mich Mus Paleontol* 25:339-413

Todd T N 1981 Allelic variability in species and stocks of Lake Superior ciscoes (Coregoninae). *Can J Fish Aquat Sci* 38:1808-1813

Todd T N, G R Smith 1980 Differentiation in *Coregonus zenithicus* in Lake Superior. *Can J Fish Aquat Sci* 37:2228-2235

Todd T N, G R Smith, L E Cable 1981 Invironmental and genetic contributions to morphological differentiation in ciscoes (Coregoninae) of the Great Lakes. *Can J Fish Aquat Sci* 38:59-67

Tomoda Y 1962 Studies of the fishes of Lake Biwako — I. Morphological study of the three species of catfish of the genus *Parasilurus* from Lake Biwako, with reference to their life. *Japanese J Ichth* 8:126-146

_____ 1965 Morphological studies on the development of the Crucian carp in Lake Biwa — I. Ontogenetic development of Nigoro-buna, and some remarks on the differentiation of Gengorobuna and Nigoro-buna. *Bull Osaka Mus Nat Hist* 18:3-30

Turner, B J, R K Liu 1977 Extensive interspecific genetic compatibility in the New World killifish genus *Cyprinodon. Copeia* 1977:259-269

Watson N H F 1974 Zooplankton of the St Lawrence Great Lakes — species composition, distribution, and abundance. *J Fish Res Bd Can* 31:783-794

Watson N H F, G F Carpenter 1974 Seasonal abundance of crustacean zooplankton and net plankton biomass of lakes Huron, Erie, and Ontario. *J Fish Res Bd Can* 31:309-317

Wells L 1960 Seasonal abundance and vertical movements of planktonic Crustacea in Lake Michigan. *Fishery Bull* 60:343-369

_____ 1966 seasonal and depth distribution of bloaters (*Coregonus hoyi*) in southeast Lake Michigan. *Trans Amer Fish Soc* 95:388-396

Wells L, A L McLain 1973 Lake Michigan: Man's effects on native fish stocks and other biota. *Great Lakes Fish Comm Tech Rept* 20. Ann Arbor Michigan

West-Eberhard M J 1983 Polymorphism, speciation and phylogeny. ms.

Wilson J B, J C Roff 1973 Seasonal vertical distributions and diurnal migration patterns of Lake Ontario crustacean zooplankton. *Proc 16th Conf Great Lakes Res* 1973:190-203

Wright S 1931 Evolution in Mendelian populations. *Genetics* 16:97-159

CYPRINIDS OF LAKE LANAO, PHILIPPINES: TAXONOMIC VALIDITY, EVOLUTIONARY RATES AND SPECIATION SCENARIOS

IRV KORNFIELD AND KENT E CARPENTER

Let one of the "new" biologists leave his laboratory and apply his methods to the fishes of Lake Lanao; perhaps he might then make a real contribution to the study of evolution.   A W C T Herre, 1933

INTRODUCTION

The endemic cyprinid fauna of Lake Lanao,[1] Mindanao has interested evolutionary biologists since Herre first brought it to the attention of the scientific community in 1924. The Lanao cyprinids have played a central role in the development of the species flock concept (Greenwood 1984) have been widely cited in discussions of evolutionary rates (Beadle 1974), and are germane to the current dialogue on macroevolution (e.g. Humphries, Miller 1981; Reid 1980)

Taxonomic background. Herre (1933) wrote that in Lanao there exist 17 (later 18) species of Cyprinidae, all descended from a single and still extant parental taxon, *Puntius binotatus.* This variable riverine species, broadly distributed over much of Southeast Asia, is abundant on Mindanao, but absent from Lake Lanao and the Lanao plateau. The unusual degree of morphological radiation characterizing the Lanao cyprinids prompted creation of four endemic genera. Of the included fishes, one (*Mandibularca resinus*) was so divergent that an even higher taxonomic rank was informally proposed. The morphological change seen in these endemic genera suggested to Myers (1960) a process by which higher taxonomic categories originated. He coined the term *supralimital specializations* to capture the idea that some of the Lanao endemics transcended the *familial* limits of all other cyprinids combined.

Despite the phenetic distinctiveness of some taxa and recognition of all endemic species by the indigenous Maranao people,

Reid (1980) suggested that species assignments might be in error; the diversity of cyprinids in Lanao might be a taxonomic artifact. Based on recent studies of goodeids (Turner, Gosse 1980) and cichlids (Kornfield, Taylor 1983), the additional possibility exists that some of the endemic forms might actually represent discrete morphological variants of a single biological species.

A variety of hypotheses have been advanced to explain the origin and diversity of the Lanao cyprinids. A number of scientists believed that the species complex arose within the lake (Herre 1933; Myers 1960; Beadle 1974), but others have suggested an allopatric origin (Woltereck 1941; Kosswig, Villwock 1964, 1974; Wood 1966). Conventional morphological studies have been unable to distinguish between these two alternatives (Wahl 1972, 1976). Related to this question is the possibility that multiple colonization or extensive hybridization could have increased diversity in the fauna (Kosswig, Villwock 1964).

Geological History. The age and origin of Lanao, and hence its fauna, is speculative (Reid 1980). Located on a plateau approximately 700 m above sea level, Lanao (357 km^2) is a relatively deep (max. depth 112 m) monomictic lake with a simple bottom topography (Frey 1969). It is situated in an isolated drainage system (1680 km^2) with a single high gradient river (including a waterfall in excess of 60 m) as its only outlet. The limnology, community structure and community energetics of Lanao have been well characterized (Lewis 1979; Frey 1969).

In Maranao legend, the lake was formed when Archangel Diabarial removed the then-mighty sultanate of Mantapoli to the center of the earth, creating a huge basin on the plateau of Mindanao (Zaide, Saber 1975). A less romantic scenario (Smith 1924; Myers 1960) is that the lake formed by volcanic impoundment when lava dammed southwesterly flowing plateau streams. However, Macawaris (1971) states the the Lanao depression is a distinctive and isolated topographic feature of the region, and he views the damming scenario as improbable; instead, he suggested, "... that the [Lanao] depression has been caused by a large collapsed basin of structural origin ... modified by constructive volcanism." In partial agreement with this, Woltereck (1941) suggested that the deep trough running parallel to the south shore of the lake probably corresponds to an old volcanic crater.

There can be little doubt that volcanism and tectonic activity were important in the formation of Lanao. Subsurface blocks of volcanic debris extend along the northern and eastern shores of the basin. The region is inherently unstable and a major fault follows the long axis of the lake (Washburn 1977). Local tectonic activity is not infrequent. In 1955, an earthquake of intensity 7-8 centered in the middle of the basin caused a large portion of the shoreline near Togaya to sink into the lake (Rabor, Rabor 1972)

The age of Lake Lanao is in dispute. Quoting Willis, Myers (1951, 1960) stated that the lava damming event occurred about 10,000 years ago. Based on a brief examination of fossils from the lake, Wood and Wood (1963) said that the lake originated 8-11,000 years ago. Subsequently, the idea of a very recent origin was accepted by Beadle (1974), Kosswig and Villwock (1974) and Lowe-McConnell (1969). Alternatively, Frey (1969) suggested that the lake was formed in the Tertiary, and Lewis (1978), quoting a personal communication from Frey, stated that rocks from the lava dam near the south end of the lake were dated at 3.6 and 5.5 x 10^6 years (mid-Pliocene) by potassium-argon methods. Herre (1933) also suggested that the lake was not of recent origin. However, Lanao has never been cored and no critical geological studies have been undertaken.

Equally speculative is the date at which Cyprinidae first colonized the Philippines. Several independent studies show that both the aquatic and the terrestrial vertebrate faunas of the Philippines were derived from Malaysia via Borneo to the west (Brown, Alcala 1970; Darlington 1957; Heaney 1978, 1983). Cyprinids from Borneo (Sabah) passed along two now partially submerged land bridges; a southern Sulu bridge to Mindanao, and a northern bridge to the Island of Palawan (Figure 1). A few cyprinids subsequently spread to the islands of Busuanga and Mindoro, but they did not colonize Bohol, Cebu, Dinagat, Leyte, Luzon, Negros, Panay, or Samar, the other major islands of the Philippine Archipelago (Herre 1953). We note that the absence of cyprinids from islands adjacent to Mindanao, particularly Dinagat, Samar and Leyte, and the low diversity of cyprinids in the Philippines, compared with Borneo, are not consistent with vicariance interpretations.

The effective magnitude of sea level (SL) lowering during glacial episodes, viz. the emergence of dry land bridges, is not known with certainty for the area around the Sulu Sea. Examination of the submarine relief along the probable Bornean migration routes reveals three major depressions which would need to have been exposed for fluvial colonization by cyprinids to occur: a 144 m deep along the northern route and 210 and 274 m deeps along the Sulu route (Figure 1). Recent stratigraphic studies indicate that the SL lowerings sufficient to connect the Philippine Islands with Borneo may have occurred twice during the Quaternary (Gascoyne et al. 1979; Jongsma 1970). However, since estimates of eustatic variation may be significantly discordant over space (Emery et al. 1971), hypotheses about the timing of faunal exchange must be approached cautiously. Heaney (1983) suggests that a land bridge may have existed between Borneo and Palawan during the middle Pleistocene, about 175,000 years ago. However, since the probable southern Sulu migration route contains depressions almost twice as deep as those to Palawan, colonization of Mindanao could not have occurred then, but must have been earlier. Myers (1960) believed that freshwater fishes did not arrive on Mindanao earlier than the Pliocene.

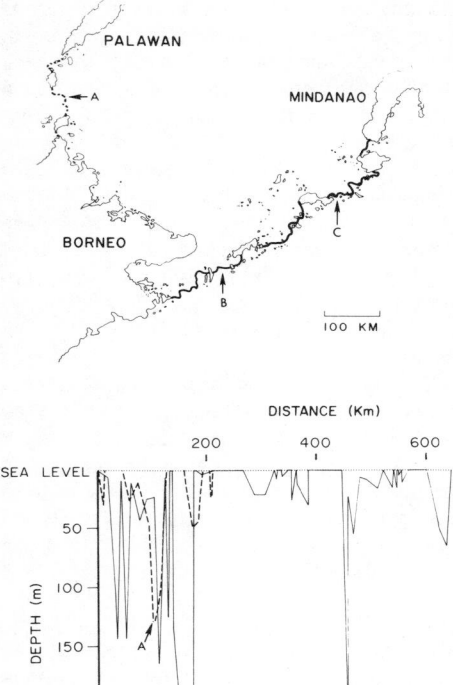

Figure 1. Projected landbridges: Sulu Sea.
Above: Probable migration routes: Dashed line; from Borneo to Palawan. Solid line; from Borneo to Mindanao.
Below: Maximum submarine elevations along migration routes. Major submarine depressions are indicated along the Palawan migration route (A: Balabac Strait, 144m) and the Mindanao migration route (B: Sibatu Passage, 274m; C: Capual Channel, 210m).

Data on more recent SL lowerings are contradictory. Though it has been generally accepted that a global eustatic SL minimum of 130 m occurred approximately 15,000 yrs ago (Chappell, Thom 1977; Emery et al. 1971; Orchiston 1979), stratigraphic studies off Australia in the Arafura Sea and the Coral Sea have suggested that SL may have been 175 m lower than at present, approximately 16,000 years ago (Jongsma 1970; Veeh, Veevers 1970). Thus, the geological data do not

exclude a recent Borneo-Palawan land connection.

Unfortunately, a major caveat complicates any simple assessment of cyprinid colonization of the Philippines: the fish may have been able to traverse saltwater bridges. This was explicitly suggested to explain the unusual island distribution of *Puntius binotatus* (Darlington 1957; cf. Myers 1960). If this and other species were able to tolerate seawater, our concern about eustatic variation would be irrelevant and reconstruction of zoogeography would be difficult.

Status of Lanao Fauna. Aside from Herre's taxonomic work, the morphological studies of Wahl (1972, 1976) and ecological characterization of *P. lindog* (Escudero et al. 1980; Sanguila et al. 1975), the Lanao cyprinids have been subjected to much speculation but no critical study. Unfortunately, time for direct examination has probably expired. Faunal changes in Lanao over the last two decades have been dramatic. Both the diversity and the abundance of the commercial catch have declined by more than an order of magnitude. Despite initial optimism about current diversity (Kornfield 1982), we now believe that no more than three of the 18 original endemic cyprinids are extant. Representatives of all endemic genera apparently have gone extinct. Of the remaining taxa, we were able to obtain live specimens of only two endemic species: *Puntius lindog* and *P. tumba*.

Using limited material from the lake and samples of congeners collected from Mindanao and elsewhere in the region, we critically examined several controversies surrounding the unique fauna of Lanao. Based primarily on comparative electrophoresis, we present here our systematic and zoogeographic conclusions and offer our speculations on some of the evolutionary problems that may be irresolvable because of various faunal perturbations.

MATERIALS AND METHODS

All specimens were collected live, and stored either frozen on dry ice (Borneo, Palawan, Thailand) or for less than two weeks in a conventional freezer (Mindanao). Fish were transported to the United States on dry ice and transferred to an ultracold freezer, at

– 80° C, for storage up to six months.

Specimens and collection localities were the following: *Puntius altus* from Macklong River (Bachiralongorn Dam Reservoir) at Kanchanaburi Provincial Fisheries Station, Thailand; *P. binotatus* from three localities: 1) in Borneo from Tintap Stream, a tributary of Moyong River, at Babagon Fish Breeding Station, Penam Pang, Sabah, Malaysia; 2) in Palawan from Balsahan River, Iwahig Penal Colony Compound, and 3) in Mindanao from Mimbalut Creek, a tributary of Linamon River near Barrio Mimbalut, Lanao del Norte; *P. tumba* was collected from the Dawaling Stream, 100 m from Lake Lanao in Barrio Dawaling, Lanao del Sur; *P. lindog* from commercial fishers in Lake Lanao off Timbangalan, 5 km from Marawi City.

Specimens were compared to preserved materials in the California Academy of Science to confirm identifications. In several cases, preliminary field identifications proved erroneous.

Preparations of specimens and horizontal starch gel electrophoresis of tissue homogenates followed standard procedures (Kornfield et al. 1982; May et al. 1979). Six buffer systems were used to resolve the products of 20 putative isozyme loci: **1)** N-(3-aminopropyl)-morpholine, pH 6.1 (Clayton, Tretiak 1972) - glyceraldehyde-phosphate dehydrogenase (*GAP-1,2*), lactate dehydrogenase (*LDH-1*), phosphoglucose isomerase (*PGI-1,2*); **2)** LiOH, ph 8.1 (Ridgway et al. 1970) - malic enzyme (*ME*) peptidase (*PEP-1,2*); **3)** Tris-borate-EDTA, pH 8.7 (Markert, Faulhaber 1965) - creatine kinase (*CPK*), phosphomannose isomerase (*PMI*); **4)** Tris-citrate, pH 6.9 (Whitt 1970) - isocitrate dehydrogenase (*IDH-1,2*); **5)** Tris-citrate, pH 8.7 (Poulik 1957) - general protein (*GP-1,2*); **6)** Tris-citrate, pH 6.3 (Selander et al. 1971) - (*LDH-2*), malate dehydrogenase (*MDH-1,2,3*), phosphoglucomutase (*PGM-1,2*). Designation of isozymes and alleles (electromorphs) follows the nomencalture suggested by Allendorf and Utter (1979). Allele mobilities for individual loci are relative to the common allele in *P. tumba*. The BIOSYS-1 program of Swofford and Selander (1981) was used to calculate Nei's (1972) estimates of genetic similarities (I) and distances (D) among sampled populations.

Salinity tolerance experiments on *P. tumba* from the Dawaling Stream were performed at Mindanao State University. In acute tolerance studies, fish were transferred directly from the running freshwater to 10 liter tanks holding aerated solutions of 5, 10, 25, and 50 percent reconstituted seawater (32 °/oo; Instant Ocean, Aquarium Systems, Eastlake OH). In gradual tolerance experiments, fish were placed in 10 liter containers of aerated freshwater to which reconstituted seawater was continually added, increasing ambient salinity at the rate of approximately 10 °/oo per 24 hours.

Submarine profiles for the Sulu Sea were obtained from navigational charts 4707, 4720 and 4722 of the Philippine Coast and Geodetic Survey, Manila. Elevations and geography around the Lanao plateau were taken from topographic maps 2539 and 2544 (PCGS) and NASA LANDSAT images E116801311601 and E3043901193D.

RESULTS

Electrophoresis. Allele frequencies for all samples are presented in Table 1. *Puntius altus* is morphologically very divergent from the other taxa examined. It has been placed in a phyletic group distinct from *P. binotatus* and its allies (Taki et al. 1978). *P. altus* was included as an outgroup to assess the discriminating power of the electrophoretic conditions employed in this study. It differed from the other congeners examined at 70 percent of the resolved loci. We are thus reasonably confident that our buffer systems were capable of detecting most of the overt electrophoretic differences among the Philippine and Bornean species of *Puntius*.

Compared to *P. binotatus* from Borneo (Sabah) and Palawan, levels of variability were noticeably higher in the taxa from Mindanao. Heterozygosity for *P. binotatus* from Mindanao (12.6%) was more than double that observed in the two samples of allopatric conspecifics, and was equivalent to that in *P. tumba* and *P. lindog* (Table 1).

Genetic differentiation among the species varied substantially (Table 2). First, as expected, *P. altus* was markedly different from its congeners; average genetic identity ($\bar{I}$) was 0.25. Second, genetic differentiation among the three allopatric samples of

Table 1. Allele frequencies and levels of isoenzyme variability in *Puntius* species in Mindanao and elsewhere in southeast Asia

Locus	allele	*Puntius binotatus* (Borneo)	*Puntius binotatus* (Palawan)	*binotatus* (Mindanao)	*Puntius tumba*	*Puntius lindog*	*Puntius altus*
CPK	0.95						
	1.00	1.00	1.00	0.44	0.63	0.33	1.00
	1.09			0.56	0.37		
	1.11					0.67	
GAP-2	1.00	1.00	1.00	1.00	1.00	1.00	1.00
	1.14						
GAP-1	0.80					0.06	
	1.00	1.00	1.00	0.84	0.59	0.02	1.00
	1.50			0.08	0.13	0.79	
	1.90			0.08	0.24	0.13	
	2.50				0.04		
GP-2	1.00	1.00	1.00	1.00	1.00	1.00	1.00
GP-1	0.91						
	0.95				0.25		
	1.00	1.00	1.00	1.00	0.75	1.00	1.00
IDH-2	0.94						
	1.00	1.00	0.25	0.38	0.77	0.50	1.00
	1.03		0.75	0.62	0.23	0.50	
IDH-1	1.00	1.00	1.00	1.00	1.00	1.00	1.00
LDH-2	1.00	1.00	1.00	1.00	1.00	1.00	1.00
	1.03						
LDH-1	*1.00*		*1.00*	0.33	0.54		
	1.17				0.15	0.07	
	1.34	*1.00*				0.18	
	1.50			0.17	0.04	0.07	
	1.60				0.04	0.46	
	1.67			0.50	0.08	0.11	
	1.83				0.15	0.11	
	2.00						*1.00*
	2.17						
MDH-3	1.00	1.00	1.00	1.00	1.00	1.00	1.00
	1.13						
MDH-2	1.00	1.00	1.00	1.00	1.00	1.00	1.00
	1.31						
MDH-1	1.00	1.00	1.00	1.00	1.00	1.00	1.00
	1.19						
ME	1.00	1.00	1.00	1.00	1.00	1.00	1.00
PEP-2	0.97						
	1.00	1.00	1.00	1.00	1.00	1.00	1.00
	1.05						
PEP-1	0.90				0.05		
	1.00	1.00	1.00	0.95	0.95	0.68	1.00
	1.04			0.05		0.32	
PGI-2	1.00	1.00	1.00	0.32	0.87	0.65	1.00
	1.05			0.68	0.13	0.35	
	1.15						
PGI-1	0.70	0.06					
	0.80	0.72	0.10				
	1.00	0.22	0.90	0.84	0.97	1.00	1.00
	1.40			0.16	0.03		
	0.00						
PGM-2	0.97		0.10				
	1.00	1.00	0.40	1.00	1.00	1.00	1.00
	1.03		0.50				
	1.10						
PGM-1	1.00	0.92	1.00	1.00	1.00	1.00	1.00
	1.05	0.08					
	1.15						
PMI	0.90						
	0.95						
	1.00	1.00	1.00	1.00	1.00	1.00	1.00
Avg. n /locus		7.3	6.8	13.5	15.5	16.5	1.0
% loci polymorph/ species		9.5	14.3	33.3	33.3	28.6	--
% Loci heterozygous / individual		2.7	5.0	12.6	14.6	14.0	--

A locus is polymorphic if the most common allele frequency ◄ 0.99

Table 2. Genetic similarities among *Puntius* species from Mindanao and elsewhere in southeast Asia.

(Above diagonal, I; below diagonal D (Nei 1972)

		PBB	PBP	PBM	PT	PL	PA
P. binotatus (Borneo)	[PBB]	--	0.781	0.668	0.657	0.647	0.241
P. binotatus (Palawan)	[PBP]	0.248	--	0.635	0.656	0.612	0.245
P. binotatus (Mindanao)	[PBM]	0.404	0.454	--	0.913	0.898	0.223
P. tumba	[PT]	0.421	0.421	0.091	--	0.892	0.294
P. lindog	[PL]	0.435	0.491	0.108	0.114	--	0.225
P. altus	[PA]	1.421	1.407	1.500	1.226	1.491	--

P. binotatus was considerable. The sample of *P. binotatus* from Borneo was more similar to conspecifics from Palawan (I = 0.78) than to conspecifics from Mindanao (I = 0.67). The genetic similarity between Palawan and Mindanao populations (I = 0.64) was the lowest of the three comparisons. *P. binotatus* from Mindanao displayed a greater number of unique, autapomorphic, alleles (11) than fishes from Palawan (4) or Borneo (2). Third, *P. tumba* from Lanao Plateau was closely related to *P. binotatus* from the coastal area of Mindanao (I = 0.91). No fixed alleles differentiated either species, but *P. tumba* displayed two unique alleles in low frequency at *LDH-1*. Fourth, *P. lindog* from Lake Lanao was equally divergent from *P. tumba* and the Mindanao sample of *P. binotatus* ($\overline{I}$ = 0.90). *P. lindog* possessed a fixed unique allele at *GP-1* (0.91) and two unique alleles in low frequency at *LDH-1*. Finally, the two Lanao endemics, *P. lindog* and *P. tumba*, were much more closely related to *P. binotatus* from Mindanao ($\overline{I}$ = 0.90) than they were to *P. binotatus*

from Palawan ($\overline{I}$ = 0.63) or from Borneo ($\overline{I}$ = 0.65). Genetic similarities from all examined taxa are summarized in a simple phenogram (Figure 2).

Salinity Tolerance. Cyprinids from Mindanao probably can tolerate only limited salinity. Acute exposure to salinities in excess of 25 percent seawater (8 °/oo) caused violent respiratory distress and rapid death (◄1 hr). Fish acutely exposed to 5 and 10 percent seawater solutions also showed respiratory distress, but survived for at least 10 hrs when experiments were unintentionally terminated. In studies where salt concentrations were gradually elevated over 36 hrs, *P. tumba* tolerated solutions of approximately 50 percent seawater (mean = 16.3 °/oo, range = 14-18 °/oo, n = 3).

DISCUSSION

Taxonomic Validity and Evolutionary Rates. *P. lindog* is the only endemic we obtained from Lanao proper, while *P. tumba* was collected from inflowing streams less than 100 m from

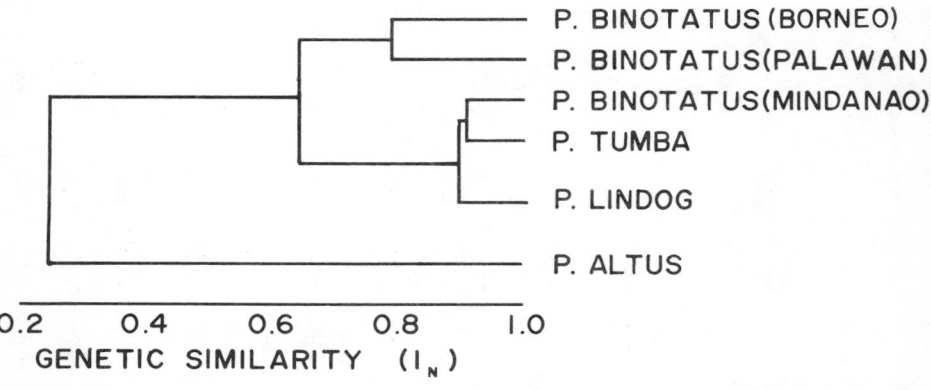

P. BINOTATUS (BORNEO)

P. BINOTATUS(PALAWAN)

P. BINOTATUS(MINDANAO)

P. TUMBA

P. LINDOG

P. ALTUS

0.2   0.4   0.6   0.8   I.0

GENETIC SIMILARITY (I_N)

Figure 2. UPGMA phenogram of relationships among *Puntius* from Mindanao and other localities in southeast Asia, based on genetic similarities, I_N. Cophenetic correlation equals 0.998.

Figure 3. Endemic cyprinids from Mindanao.
A: *Puntius binotatus* from the Linaman River, Lanao del Norte. B: *Puntius tumba* from the Cabasaran Stream, Lanao drainange. C: *Puntius lindog* from Lake Lanao.

the lake. These taxa were collected effectively in sympatry, and if they represent a single biological species, they should exhibit homogeneous gene frequencies at all polymorphic loci (Sage, Selander 1975). That is, in the absence of intense selection, gene flow getween conspecific populations should rapidly eliminate any differences. Alternately, if these forms are reproductively isolated, they generally should differ significantly in allelic frequencies (Allendorf, Phelps 1981). Allelic frequency differences between *P. lindog* and *P. tumba* unambiguously confirm their integrity as separate biological species. Frequencies were significantly different at seven of eight polymorphic loci (Table 1).

It is most important to note, regarding the taxonomic concerns of Reid (1980) that *P. lindog* and *P. tumba* are quite similar in appearance (Figure 3). From a purely morphological perspective, it would not seem unreasonable to consolidate these taxa as Banister (1973) recently did for African *Barbus*. However, demonstration of genetic independence between these two similar forms suggests that ecophenotypic variation should not be indiscriminately invoked for the Lanao fauna. Clearly, cyprinid taxonomists should augment conventional morphological data with genetic information when possible. Because of extinctions, electrophoretic comparisons with other Lanao endemics are now impossible. However, we think it reasonable to accept Herre's basic insights regarding the taxonomy of the other and *much more distinctive* endemic species. In particular, based on our finding for *P. lindog* and *P. tumba*, we feel confident in recognizing *Spratellicypris, Ospatalus, Cephalakompsus*, and *Mandibularca* as containing biological species. We are less confident regarding our decisions on the other Lanao endemics, and admit that several nominal species, such as *P. disa*, could be intraspecific variants. Regardless, based on this limited electrophoretic evidence, the high diversity described for the cyprinid complex of Lanao is apparently real.

Interpretation of genetic distances allows some preliminary statements on relationships and evolutionary chronologies of the Lanao fauna. Evidence for a molecular clock is overwhelming (Thorpe 1982), but accuracy of the clock is much debated. For sizable levels of differentiation, the relationship between time and genetic distance is nonlinear (Corruccini et al. 1980; Nei 1981). Nevertheless, by judicious use of calibration constants, an approximate *range* of absolute divergence times can be obtained for our data. In the electrophoretic literature, estimates of divergence time almost exclusively use the genetic distance (D) of Nei (1972). The relation between time and genetic distance is cD, where c represents a rate constant. For small values of D $\blacktriangleleft$1, values of c have ranged between 0.8×10^6 and 18.0×10^6 yrs (Avise, Aquadro 1982; Thorpe 1982). We use these two extreme constants, which differ by a factor of 25, conservatively to bracket estimates of divergence time.

Our electrophoretic data are incompatible with the idea that the Lanao species complex arose only 10,000 yrs BP. The magnitude of genetic distance between *P. lindog* and *P. tumba* (D = 0.114) falls into a divergence time estimate of between 91,000 and 2,060,000 yrs. While this range is too broad to indicate usefully the age of Lanao, it clearly indicates that speciation did not occur in the Holocene, as Myers (1960) and others have suggested. Two additional considerations support this conclusion. 1) The existence of autapomorphic alleles in both *P. lindog* and *P. tumba* (Table 1) point to an extended period of genetic isolation. 2) Elevated heterozygosities in these two species suggest that population bottlenecks, which could be associated with speciation, were not recently experienced. The heterozygosities of these samples, though substantially higher than generally observed in other teleosts (Nevo 1978) are within the range reported by Avise (1977a) for species of the North American cyprinid genus *Notropis*. Both heterozygosity and effective number of alleles increase slowly after severe reductions in population size (Nei et al. 1975).

Our electrophoretic information is ambiguous with regard to the taxonomic status of *P. tumba*. According to Herre, *P. binotatus*, the original colonizer of Mindanao, gave rise to *P. tumba*. The latter species, widely distributed on the Lanao plateau, is the presumed ancestor of all Lanao cyprinids (Herre 1933). Since *P. binotatus* is absent from the Lanao plateau, but occurs over the rest of Mindanao, it is possible that *P. tumba* is only of subspecific rank. Because these two taxa are allopatric, differences in gene frequencies cannot be used to determine whether they are separate biological species. The level of genetic similarity between these taxa (I = 0.913) is conventionally regarded as indicating conspecificity (Avise 1974; Thorpe 1982), but in North American cyprinids, several examples of equivalent or even higher genetic similarities have been recorded for biological species (Avise 1977b; Avise et al. 1975; Buth 1979; Buth, Burr 1978). In the absence of additional data, we provisionally retain specific rank for *P. tumba*.

The salt tolerance experiments and analysis of allopatic samples of *P. binotatus* permit us some insight into the cyprinid colonization of the Philippines. The tolerance experiments rule out the possibility, advanced by Darlington (1957), that *P. binotatus* may have been able to traverse marine waters. Our data convincingly suggest that even small gaps of normal seawater can prevent dispersal of *Puntius* species. Though it is possible that even slower acclimatization to sea water than used in our experiments might result in greater salinity tolerance (Lagler et al. 1977), studies with other cyprinids (Holliday 1969; Kornfield unpubl.) support the idea that salinity concentrations in excess of 50 percent sea water are probably lethal. We affirm the statements of Myers (1949, 1951) that the Cyprinidae should be considered primary freshwater fishes. Thus, freshwater fluvial connections were necessary for colonization of the Philippines.

The levels of genetic differentiation between samples of *P. binotatus* from Borneo and Mindanao (Table 2) suggests that fish from Mindanao should be considered distinct species (see Thorpe 1982:151). It is difficult to propose an appropriate specific epithet. *P. quinquemaculatus*, described by Seale and Bean (1907) from Mindanao near Zamboanga, is considered a probable variety of *P. binotatus* by Herre (1953). Though no electrophoretic comparisons have yet been made between samples from Lanao del Norte and Zamboanga Provinces, it may be appropriate to recognize *P. quinquemaculatus* on Mindanao as a sister species of *P. binotatus*.

Genetic differences of the Borneo-Palawan comparison (D = 0.247) relative to the Borneo-Mindanao comparison (D = 0.403) are consistent with the landbridge dispersal hypothesis championed by Herre (1933), Darlington (1957) and others. Since depth contours along the northern migration route to Palawan are closer to current SL than depths along the southern route to Mindanao (Figure 1), populations of *P. binotatus* on Palawan are thus likely to have had more recent genetic contact with fishes from Borneo than Mindanao populations had with Borneo, regardless of the magnitude of past SL fluctuations. Using the very conservative calibrations of the molecular clock, the last fluvial connection from Borneo to Palawan was between 197,000 and 4,450,000 yrs BP and for Mindanao between 332,000 and 7,260,000 yrs

BP. Holocene landbridges clearly can be excluded. Based on these temporal data, we suggest that the SL depression of 145-170 m during the middle Pleistocene (Gascoyne et al. 1979; Jongsma 1970) probably did not allow cyprinid colonization of Palawan. However, we can not exclude the possibiity that faunal exchange of terrestrial organisms may have taken place around this time. Our conservative estimates for divergence are compatible with Myers' (1960a) belief that cyprinids colonized Mindanao in the late Tertiary.

We believe, as did Herre (1933) that *P. binotatus*, or a very closely related sister taxon from Borneo, was the sole ancestor of the Lanao cyprinid complex. With the exception of *P. binotatus*, all of the many cyprinid species of North Borneo (Inger, Chin 1962; Kottelat 1982) possess multiple morphological attributes not present in any of the Lanao endemics. *P. seale*, which occasionally has been confused with *P. binotatus*, occurs in Sabah in the Labuk-Segama and Tawau watersheds, the postulated source area for Mindanao colonization. However, *P. seale* differs from *P. binotatus* and all Lanao endemics in coloration, ray morphology and scale count. Aside from species in the Lanao drainage, the cyprinid fauna of Mindanao is depauperate (Herre 1924, 1928, 1953). The other taxa present (*Nematabramis* spp. and *Rasbora* spp.) are conspicuously different from *Puntius*. To deny that *P. binotatus* or a closely related form was the sole ancestor of the Lanao species, we must postulate the existence of additional and now extinct Bornean species and explain their disappearance from both Borneo and waters of Mindanao outside Lanao. We suggest that multiple colonization and hybridization schemes (Kosswig, Villwock 1964, 1974) are not supported by any available evidence. Thus, by virtue of possessing a single common ancestor and a restricted geographic distribution, the endemic cyprinids of Lake Lanao represent a species flock sensu Greenwood (1984).

Speciation Scenarios. Part of the evolutionary appeal of the Lanao story, as with other species flocks, derives from the possibility that speciation occurred in some novel, non-allopatric mode. This idea, first mentioned by Herre (1933), was uncritically accepted in subsequent reviews by Myers (1960) and Beadle (1974). But Herre's descriptive paper (1933) was contradictory. While he clearly thought that intralacustrine speciation had occurred [2], he also noted that at least one endemic, *P. clemensi*, "... occurs in the *lakes* of the Lanao plateau, particularly in Lake Lanao ..." Since all other plateau lakes are outside the Lanao drainage basin (Figure 4), speciation therefore need not have been exclusively intralacustrine. In a detailed report on aquatic studies in the Philippines, Woltereck (1941) mentioned that the presumed endemic *Mandibularca resinus* also occurred outside Lanao proper. Though Brooks (1950) had noted Herre's findings, Kosswig and Villwock (1964) were the first to appreciate the importance of extra-Lanao distribution. Collecting two "Lanao endemics" from Lake Dapao outside the Lanao drainage basin, and acknowledging Woltereck's work, they vigorously concluded that speciation in the fauna must have been allopatric. However, their position was based on an age of 10,000 years for Lanao, a period of time they considered insufficient to permit intralacustrine radiation.

Distributional evidence in favor of the allopatric claim is marginal. Outside the lake, "Lanao endemics" have consistently been collected only from Lake Dapao (Table 3). That *P. tumba* occurs in this lake and many, if not all others on the plateau is not informative since this taxon may be, as noted before, a subspecies of *P. binotatus*. Of the other "Lanao endemics" in Lake Dapao, *P. clemensi* was collected by Herre in the 1920's and by Kosswig and Villwock during their 1963 expedition. *P. baoulan* was obtained only by Kosswig's group. The absence of *P. baoulan* from the collections of both Herre and Woltereck is significant. Laka Dapao is a moderately large (24.2 km [2]) and deep (62 m) "crater lake" situated in an isolated drainage basin directly adjacent to Lanao. We suggest that the proximity between watersheds may have facilitated a very recent transfer of fishes to Dapao.

It is widely acknowledged that freshwater fishes have been intentionally introduced into many lakes of Southeast Asia to augment local food supplies. A number of species were routinely carried about by native peoples

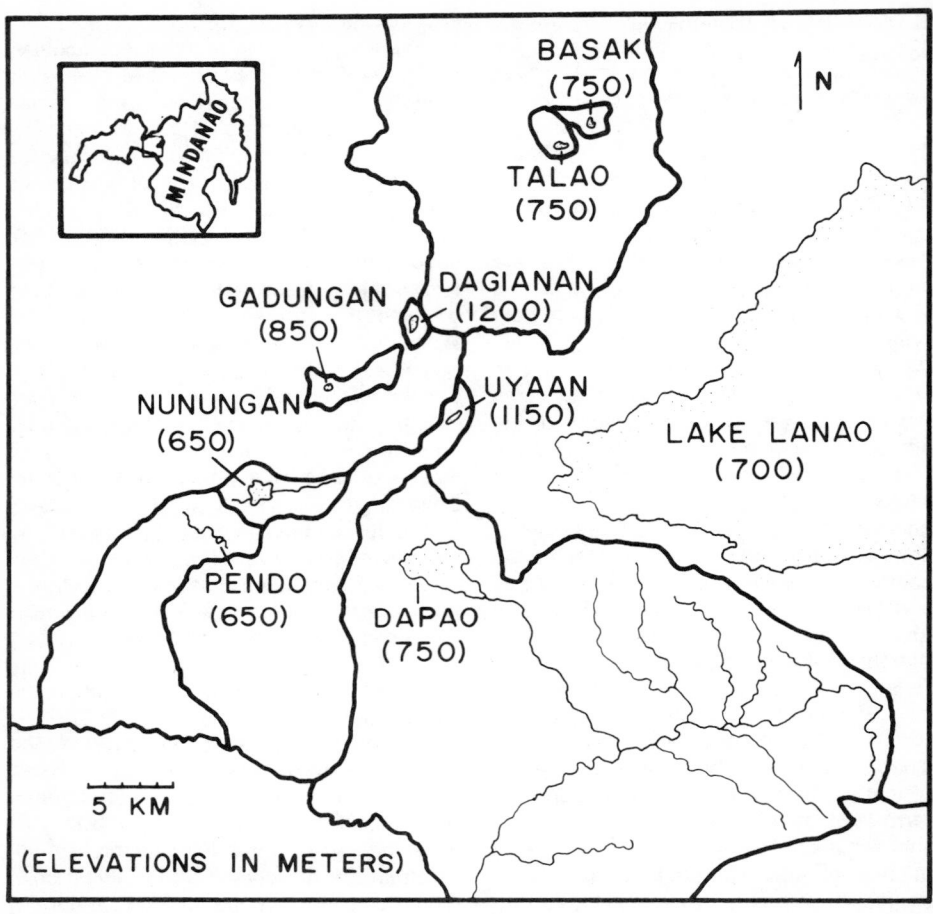

Figure 4. Satellite lakes in independent drainage basins bordering Lake Lanao. Unnamed lakes are provisionally identified after closest villages. Solid lines indicate drainage boundaries.

(Myers 1951) and on Mindanao, carp (*Cyprinus carpio*) were placed in lakes Dapao, Nunungan and Lanao early in this century (Herre 1933) [3] *P. baoulan* is the most prized food fish of the Maranao (Herre 1933) [4] This species occurs at great depths in Lanao, but is caught at the surface after violent storms. Because of its high local value, it was an ideal candidate for introduction in Lake Dapao. That the Maranao themselves transferred fish to Dapao from Lanao is evidenced by the unfortunate introduction, four years ago, of *Hypseleotris agilis* (Kornfield 1982). The apparent absence of *P. baoulan* in Dapao

prior to 1963 strongly supports the introduction argument.

It is more difficult to explain the existence of *P. clemensi* in Lake Dapao and the presence of the distinctive *Mandibularca resinus* in waters outside Lanao. Though anomalous, the extra-Lanao occurrence of the latter species was noted independently by several groups (Table 3). Thus, the observations may have been real. Why would this most specialized cyprinid, which presumably was restricted to an ecologically narrow zone in the Agus River draining Lanao, be found in small lakes? A simple natural phenomenon

Table 3. Reported occurrence of endemic cyprinids outside the Lanao drainage basin. [1]

Species	Distribution	Reference
Mandibularca resinus	Lake Uyaan,[2] small streams & lakes of Lanao plateau [3] Agus River watershed	Herre (1924) Woltereck (1941) Wood (1966)
Puntius baoulan [4]	Lake Dapao	Kosswig & Villwock (1964)
Puntius clemensi	lakes of the Lanao plateau Lake Dapao	Herre (1933) Kosswig & Villwock (1964)
Puntius manalak [5]	lakes of the Lanao plateau	Brooks (1950)
Puntius tumba	Lake Dapao	Herre (1924); Woltereck (1941) Kosswig & Villwock (1964)
	Lake Nunungan Lake Talao Lake Uyaan	Herre (1924); Woltereck (1941) Woltereck (1941) Herre (1924); Woltereck (1941)

[1] Wood and Wood (1963) reported that they collected virtually all Lanao endemics outside the lake proper, but this report is uniformly viewed with skepticism.

[2] Reported by fishers but discounted by Herre (1924)

[3] Though Woltereck (1941:150) states: "*Mandibularca* wurde auch in einigen der Kleinseen und Bache des Lanaoplateaus gefunden," he did not list this species as present in his systematic account of the fauna of Lakes Dapao, Nunungan, Talao and Uyaan, examined during the 1932 Wallacea expedition.

[4] Both Kosswig & Villwock (1964) and Wahl (1972, 1976) mention that Woltereck also found *P. baoulan* in Lake Dapao, but Woltereck (1941) does not mention it in his collections.

[5] Herre (1933) stated that this species is *only* known from Lake Lanao; Brooks (1950) clearly misread Herre's distributional account.

may be responsible. In particular, we suggest that this species, and perhaps others, including *P. clemensi*, was recently dispersed by waterspouts.

Waterspouts [5] occur irregularly, but not infrequently, on Lake Lanao (Figure 5). When they touch shore, they do only modest damage and do not travel great distances (P Gowing pers comm). We have no information on the tracks of local waterspouts. However, because of their proximity and elevations, two drainages adjoining Lanao present good targets to receive transported materials (Table 4). The lakes in these drainages, Dapao and Uyaan, are precisely the ones reported to contain "Lanao endemics."

Rains of live fish have been thoroughly documented in many parts of the world, including Asia (Gudger 1929). Though unusual, this mode of transport occasionally has been invoked to explain anomalous zoogeographic distributions (Darlington 1957). Two physical attributes of waterspouts and tornados constrain transport of fishes: pressure and temperature. Pressure inside the vortex drops to about 10 percent of normal

atmospheric pressure. The swimbladders of fish with no external openings (physoclists) could be ruptured or severely distorted under such conditions. However, cyprinids are physostomes with a pneumatic duct into the foregut (Jones, Marshall 1953; Steen 1970). Temperatures inside the vortex may be 10° C lower than ambient. We have not investigated thermal tolerances of Lanao fishes. However, preliminary experiments with a congener, *P. tetrazona* indicate it can tolerate dramatic temperature reductions over short periods of time (Kornfield, unpubl.). In sum, we do not believe that either pressure or temperature conditions associated with waterspouts would inhibit transport of live Lanao cyprinids.

Even if both human introductions and waterspout transport could be convincingly dismissed, there is no evidence against an intralacustrine origin for the Lanao species flock. Excepting Lake Dapao, the persistent occurrence of Lanao endemics elsewhere in the region has not been confirmed. In particular, none of the specialized lacustrine species occur in the many streams draining into Lanao. Though the lake region is

Table 4. Targets for waterspout transport around Lake Lanao

Lake	Surface area (km [2])	Elevation (m)	Minimum displacement from Lanao watershed vertical (m)	Minimum displacement from Lanao watershed horizontal (km)
Basak [1]	6.1	750	250	3.0
Dagianan	8.1	1200	600	1.5
Dapao	24.2	750	200	1.0
Gadungan [1]	2.0	850	400	4.5
Nunungan	12.1	650	400	4.5
Pendo	4.0	650	400	10.0
Talao	8.1	750	250	5.0
Uyaan	4.0	1150	450	1.0

[1] The lake does not have a formal name; provisionally identified with closest village.

Figure 5. A large waterspout on Lake Lanao near Marawi City, 1978. Photographed by Dr. Peter G. Gowing, Director, Dansalan Research Center.

tectonically unstable, elevations separating adjacent watersheds are considerable (Table 4), and we know of no standard mechanisms which could sequentially isolate, connect, and re-isolate the water and faunas of Lanao and these separate drainages. Further, the topography immediately around the lake is not conducive to the formation of ephemeral pools, a mechanism proposed for spatial isolation in other species flocks (Greenwood 1965; McKaye, Gray 1984). We do not deny the possibility that a portion of the Lanao fauna may have originated in some standard allopatric mode. But topographic and distributional evidence suggest that some of the endemic radiation occurred within the lake.

We have only in part responded to Herre's challenge quoted at the head of this paper. We have confirmed many of his early insights regarding the evolution of Lanao and its fauna. However, the most exciting components of the story remain unresolved.

SUMMARY

The cyprinid fauna of Lake Lanao, Mindanao presents a widely acknowledged but controversial example of adaptive radiation. In the classic account, a species complex of 18 endemic taxa arose within the lake approximately 10,000 years ago from a single common ancestor that had colonized Mindanao from Borneo. The distinctive morphology of several endemic taxa prompted the recognition of four endemic genera. However, all aspects of this story have been questioned, including: validity of the phenetically defined taxa, identity of ancestral species, age of the radiation and mode of speciation. Unfortunately, over-exploitation and exotic introductions have decimated the fauna, so that now only three or fewer endemic cyprinids are present. We collected two cyprinids from Lanao, *Puntuis lindog* and *P. tumba*, and samples of *P. binotatus*, the presumed ancestral species, from Mindanao, Borneo and Palawan. Using these materials, we were able to examine several of the controversies surrounding the unique fishes of Lanao.

Electrophoretic analysis revealed significant gene frequency differences between the two Lanao endemics and unambiguously established their status as separate biological species. The level of genetic similarity between the two endemic species suggests that speciation probably occurred recently, though not as recently as the Holocene. Salinity tolerance experiments and electrophoretic comparisons of allopatric samples of *P. binotatus* support the idea that cyprinids colonized Mindanao from Borneo (Sabah) over a Sulu landbridge before the mid-Pleistocene. By distributional evidence, we affirm that *P. binotatus*, or a now-extinct sister taxon, probably was the direct ancestor of all Lanao endemics. Because of their presumed monophyly and restricted occurrence, the cyprinids of Lake Lanao are a species flock.

Evidence for an allopatric origin of the Lanao fauna is weak. Human introductions and local transport by waterspouts may explain the anomalous occurrence of endemics outside the Lanao drainage basin. We suggest that the species flock may have originated within the lake.

NOTES

1. The name is derived from *Ranao* (the Maranao word for "lake") which was corrupted into *Lanao* (the Cebu-Bisayan term for "lake.") Thus, like Lake Nyassa (= Malawi) in Africa (Ransford 1966), the anglicized name means "Lake Lake".
2. He stated that "... changing conditions *in* the lake ... imposed new demands upon the parent stock of fishes and eventually new forms of more or less permanence or fixity appeared."
3. In 1975 Alvin Seale introduced black bass (*Micropterus salmoides*) into Lanao at the request of General Jack Pershing, then Governor of the Moro Provinces (Seale 1946; Villalaz 1966).
4. Herre stated (1926) that *P. baoulan* "... is always much thicker from side to side than any of the other cyprinids and is very fat, so that it can be fried in its own grease without adding any oil to the pan. This is a character highly prized by the Mohammedans, as they use no

lard and must import from the lowlands nearly all the coconuts used to furnish fat. The few Christian Filippinos in the Lanao plateau also value this fish greatly since their diet is nearly always deficient in fat."
5. Waterspouts are called *apos* (ancestors) or *bulawi* (spirits) by the Maranao. Waterspouts have been reported from several lakes containing species flocks including Baikal (St George 1969) and Tanganyika (Brichard 1978). In northern Laka Malawi, waterspouts are not uncommon and have been obverved to transport fishes (T D Iles pers. comm.).

ACKNOWLEDGMENTS

We thank C Atwood, C Emam, P Gowing L Heaney, W Rainboth, S Mohamad, J Siegel, and R Silang for valuable aid and information in field or laboratory. H Dewitt, C Chubs, B Glanz, P H Greenwood, J Seigel and S Tyler commented on the manuscript. M Klauser assisted with translations. J Karnasuta arranged for fieldwork in Thailand and provided laboratory space at the National Inland Fisheries Institute, Bangkhen. C Y Won arranged for collection of fishes in Sabah. E Quetuilo arranged collection and logistical support in Palawan. Collecting activities in Mindinao, labwork at Mindanao State University, and logistical support were made possible by A Macawaris and S Madid. Lanao del Sur Governor M A Dimapora arranged safe transit to Laka Dapao. Fieldwork in Sabah was authorized by the Ministry of National Unity, Government of Malaysia. Activities in the Philippines were made possible by F Gonzales, Director, Bureau of Fisheries and Aquatic Sciences, Manila.

This study was supported by the Office of the Vice President for Research and Public Service, University of Maine, the Migratory Fish Research Institute, T F H Publications, Inc, the East-West Center, the American Philosophical Society, and in part, NFS grant DEB 78-24074.

REFERENCES

Allendorf F W, S R Phelps 1981 Use of allelic frequencies to describe population structure. *Can J Fish Aqu Sci* 38:1507-1514
Allendorf F W, F M Utter 1979 Population genetics of fish. 407-454 W Hoar, D Randall, J Brett eds. *Fish Physiology* V 8. Academic Press. New York

Avise J C 1974 Systematic value of electrophoretic data. *Syst Zool* 23:465-481

———— 1977a Genic heterozygosity and rate of speciation. *Paleobiol* 3:422-432

———— 1977b Is evolution gradual or rectangular? Evidence from living fishes. *Proc Natl Acad Sci* 74:5083-5087

Avise J C, C J Aquadro 1982 A comparative summary of genetic distances in the vertebrates. 151-185 M Hecht, B Wallace, Prance eds. *Evolutionary Biology*. Plenum. New York

Avise J C, J J Smith, F J Ayala 1975 Adaptive differentiation with little genic change between two California minnows. *Evolution* 29:411-426

Banister K E 1973 A revision of the large *Barbus* (Pisces, Cyprinidae) of East and Central Africa. *Bull Br Mus Nat Hist (Zool)* 26:1-148

Beadle L C 1974 *The Inland Waters of Tropical Africa*. Longman

Brichard P 1978 *Fishes of Lake Tanganyika*. THF Publs Neptune City New Jersey

Brooks J L 1950 Speciation in ancient lakes. *Quart Rev Biol* 25:30-60, 131-176

Brown W C, A C Alcala 1970 The zoogeography of the herpetofauna of the Philippine Islands, a fringing archipelago. *Proc Calif Acad Sci* 38:105-130

Buth D G 1979 Biochemical systematics of the cyprinid genus *Notropis* - I. The subgenus *Luxilus*. *Biochem Syst Ecol* 7:69-79

Buth D G, B M Burr 1978 Isozyme variability in the cyprinid genus *Campostoma*. *Copeia* 1978:298-311

Chappell J, B G Thom 1977 Sea levels and coasts. 275-291 I J Allen, J Golson, R Jones eds. *Sunda and Sahul, Prehistoric Studies in Southeast Asia, Melanesia and Australia*. Academic Press. London

Clayton J W, D N Tretiak 1972 Amine-citrate buffers for pH control in starch gel electrophoresis. *J Fish Res Bd Can* 29:1169-1172

Corruccini R S, M Baba, M Goodman, R L Ciochon, J B Cronin 1980 Nonlinear macroevolution and the molecular clock. *Evolution* 34:1216-1219

Darlington P J 1957 *Zoogeography: The Geographical Distribution of Animals*. Wiley. New York

Emery R V, H Niino, B Sullivan 1971 Post-Pleistocene levels of the East China Sea. 381-390 K K Turekian ed. *The Late Cenozoic Glacial Ages*. Yale Univ Press. New Haven

Escudero P T, O M Gripaldo, N M Sahay 1980 Biological studies of the *Glossogobius giurus* (Hamilton and Buchanan) and the *Puntius sirang* (Herre) in Lake Lanao. *J Fish Aquaculture* 1:1-153

Frey D G 1969 A limnological reconnaisance of Lake Lanao. *Verh Int Verein Limnol* 17:1090-1102

Gascoyne M, G J Benjamin, H P Schwartz 1979 Sea-level lowering during the Illinoian glaciation: evidence from a Bahama "blue hole." *Science* 205:806-808

Greenwood P H 1965 The cichlid fishes of Lake Nabugabo, Uganda. *Bull Br Mus Nat Hist (Zool)* 12:315-357

———— 1984 What *is* a species flock? 13-20 A A Echelle, I Kornfield eds. *Evolution of Fish Species Flocks*. Univ Maine Press at Orono

Gudger E W 1929 More rains of fishes. *Ann Mag Nat Hist* 10:1-26

Heaney L R 1978 Island area and body size of insular mammals: evidence from the tri-colored squirrel (*Callosciurus prevosti*) of southeast Asia. *Evolution* 32:29-44

———— 1983 Species richness, dispersal, and vicariance of mammals in the Philippine Archipelago: A review of the problems. *Abst 63rd Ann Mtg Amer Soc Mammal* Gainesville Florida

Herre A W 1924 Distribution of the true fresh-water fishes in the Philippines. I. The Philippine Cyprinidae. *Phil J Sci* 24:249-307

———— 1926 Two new fishes from Lake Lanao. *Phil J Sci* 29:499-503

———— 1928 True fresh-water fishes of the Philippines. 242-247 R Dickerson ed. *Distribution of Life in the Philippines*. Bur of Science, Manila

———— 1933 The fishes of Lake Lanao: A problem in evolution. *Amer Natur* 67:154-162

———— 1953 Check list of Philippine fishes. *Res Rpt 20* U S Dept of Interior.

Holliday F G T 1969 The effects of salinity on the eggs and larvae of teleosts. 293-311 W S Hoar, D J Randall eds. *Fish Physiology*. V 1 Academic Press, New York

Halloway N H 1982 North Palawan block, Philippines - its relation to Asian mainland and role in evolution of South China Sea. *Amer Assn Petrol Geol Bull* 66:1355-1383

Humphries J M, R R Miller 1981 A remarkable species flock of pupfishes, genus *Cyprinodon*, from Yucatan, Mexico. *Copeia* 1981:52-64

Inger R F, P K Chin 1962 The fresh-water fishes of North Borneo. *Fieldiana (Zool)* 45:1-268

Jones F R H, N B Marshall 1953 The structure and functions of the teleostean swimbladder. *Biol Rev* 28:16-83

Jongsma P 1970 Eustatic sea level changes in the Arafura Sea. *Nature* 228:150-151

Kornfield I 1982 Report from Mindanao. *Copeia* 1982:493-495

Kornfield I, B D Sidell, P S Gagnon 1982 Stock definition in Atlantic herring (*Culpea harengus harengus*): genetic evidence for discrete fall and spring spawning populations. *Can J Fish Aqu Sci* 39:1610-1621

Kornfield I, J N Taylor 1983 A new species of polymorphic fish, *Cichlasoma minckleyi* from Cuatro Cienegas, Mexico (Teleostei: Cichlidae). *Proc Biol Soc Wash* 96:253-269

Kosswig C, W Villwock 1964 Das Problem der intralakustrischen Speziation in Titicaca und Lanaosee. *Verh dt zool Ges* 1964:95-102

———— 1974 Remarks concerning the phylogenetic status of the cyprinids of Lake Lanao (Mindanao). *Phil Scientist* 11:21-25

Kottelat M 1982 A small collection of fresh-water fishes from Kalimantan, Borneo with descriptions of one new genus and three new species of Cyprinidae. *Rev suisse Zool* 89:419-437

Lagler K, J E Bardach, R R Miller, D R M Passino 1977 *Ichthyology*. Wiley, New York

Lewis W M Jr 1978 A compositional, phytogeographical and elementary structural analysis of the phytoplankton in a tropical lake: Lake Lanao, Philippines. *J Ecol* 66:213-266

_____ 1979 *Zooplankton Community Analysis.* Springer Verlag, New York

Lowe-McConnell R H 1969 Speciation in tropical fresh water fishes. *Biol J Linn Soc* 1:51-75

Macawaris A G 1971 Hydrology of the Agus River below Lake Lanao. *Mindanao State Univ Res Ctr.* Marawi City.

Markert C L, I Faulhaber. 1965 Lactate dehydrogenase isozyme patterns of fish. *J Exp Zool* 159:319-332

May B, J E Wright, M Stoneking 1979 Joint segregation of biochemical loci in Salmonidae: results from experiments with *Salvelinus* and review of the literature on other species. *J Fish Res Bd Can* 36:1114-1128

McKaye K R, W N Gray 1984 Extrinsic barriers to gene flow in rock-dwelling cichlids of Lake Malawi: macrohabitat heterogeneity and reef colonization. 169-184 A Echelle, I Kornfield eds. *Evolution of Fish Species Flocks.* Univ Maine Press at Orono

Myers G S 1949 Salt-tolerance of fresh-water fish groups in relation to zoogeographic problems. *Bijdr Dierk* 28:315-322

_____ 1951 Fresh-water fishes and East Indian zoogeography. *Stanford Ichthy Bull* 4:11-21

_____ 1960 The endemic fish fauna of Lake Lanao, and the evolution of higher taxonomic categories. *Evolution* 14:323-333

Nei M 1972 Genetic distance between populations. *Am Nat* 106:283-292

_____ 1981 Genetic distance and molecular taxonomy. 7-22 Y P Altukhov ed. *Problems in General Genetics.* Proc 14th Internat Congr Genet. MIA Publishers. Moscow

Nei· M, T Maruyama, R Chakraborty 1975 The bottleneck effect and genetic variability in populations. *Evolution* 29:1-10

Nevo E 1978 Genetic variation in natural populations: patterns and theory. *Theor Pop Biol* 13:121-177

Orchiston D W 1979 Pleistocene sea level changes and the initial aboriginal occupation of the Tasmanian region. *Mod Quaternary Res. Southeast Asia* 5:91-103

Poulik M D 1957 Starch gel electrophoresis in a discontinuous system of buffers. *Nature* 180:1477-1479

Rabor D, L Rabor 1972 Maranao ecological source of life. *Phil Biota* 6:145-150

Ransford O 1966 *Livingston's Lake: The Drama of Nyassa.* Murray. London

Reid G McG 1980 'Explosive speciation' of carps in Lake Lanao (Philippines) -- Fact or fancy? *Syst Zool* 29:314-316

Ridgway G S, S W Sherburne, R D Lewis 1970 Polymorphism in the esterases of Atlantic herring. *Trans Am Fish Soc* 99:147-151

Sage R D, R K Selander 1975 Trophic radiation through polymorphism in cichlid fishes. *Proc Nat Acad Sci* 72:4669-4673

Sanguila W M, A Rosales, B Tabaranza Jr, S Mohamad 1975 Notes on the food habits of *Pun-*

tius sirang and *Glossogobius giurus. Mindanao J* 1:3-11

Seale A 1946 *Quest for the Golden Cloak.* Stanford Univ Press. Stanford California

Seale A, B A Bean 1907 On a collection of fishes from the Philippine Islands, made by Maj. Edward A Mearns, Surgeon U.S. Army, with descriptions of seven new species. *Proc U S Nat Mus* 33:229-248

Selander R K, M H Smith, S Y Yang, W E Johnson, J B Gentry 1971 IV. Biochemical polymorphism and systematics in the genus *Peromyscus*. I. Variation in the old field mouse (*Peromyscus polionotus*). *Studies in Genetics IV.* Univ Texas Publ7103:49-90

Smith W D 1924 Geology and mineral resources of the Philippine Islands. Publ 19, Bureau of Science, Manila

St George G 1969 *Siberia: The New Frontier.* McKay. New York

Steen J B 1970 The swim bladder as a hydrostatic organ. 414-444 W S Hoar, D J Randall eds. *Fish Physiology.* Academic Press. New York

Swofford D L, R B Selander 1981 BIOSYS-1: A computer program for analysis of allelic variation in genetics. Dept of Genetics and Development. Univ Illinois. Urbana-Champaign.

Taki Y, A Katsuyama, T Urishido 1978 Comparative morphology and interspecific relationships of the Cyprinid genus *Puntius. Japan J Ichthyol* 25:1-8

Thorpe J P 1982 The molecular clock hypothesis: biochemical evolution, genetic differentiation and systematics. *Ann Rev Syst Ecol* 13:139-168

Turner B J, D J Gosse 1980 Trophic differentiation in *Ilyodon*, a genus of stream-dwelling goodeid fishes: Speciation versus ecological polymorphism. *Evolution* 34:259-270

Veeh H H, J J Veevers 1970 Sea level at -175m off the Great Barrier Reef, 13,600 to 17,000 years ago. *Nature* 226:536-537

Villaluz D K 1966 The Lake Lanao fisheries and their conservation. Bureau of Printing, Manila

Wahl E 1972 Biostatistische Untersuchungen externer Merkmale an Lanao-Cypriniden. *Mitt Hamburg Zool Mus Inst* 68:177-194

_____ 1976 Morphologisch-Okologische Untersuchungene zur Systematik der Lanao-Cypriniden. D Sc thesis. Univ Hamburg

Washburn L 1977 Our lake for others? The Maranao and the Agus River hydroelectric project. *Res Bull Dansalan Res Ctr* 3:1-16

Whitt G S 1970 Genetic variation in supernatant and mitochondrial malate dehydrogenase isozymes in the teleost *Fundulus heteroclitus. Experientia* 26:734-736

Woltereck R 1941 Dei Seen un Inseln der ''Wallacea'' - Zwischen-region und ihre endemische Tierwelt. II. Teil: Inseln und Seen dur Philippinen. *Int Rev ges Hydrobiol Hydrogr*41:37-176

Wood C E 1966 Two species of Cyprinidae from north central Mindanao. *Philipp J Sci* 95:411-423

Wood C E, J C Wood 1963 A monograph of the fishes of Lake Lanao. Unpubl ms

Zaide G F, M Saber 1975 How the angels built Lake

Lanao. 1-3 M Saber, A T Madale eds. *The Maranao*. Solidaridad, Manila

BIOGEOGRAPHY OF THE ANDEAN KILLIFISH GENUS *ORESTIAS* WITH COMMENTS ON THE SPECIES FLOCK CONCEPT

LYNNE R PARENTI

INTRODUCTION

Orestias (Cyprinodontiformes, Cyprinodontidae) is a speciose and diverse killifish genus distributed throughout the central Andean highlands from Ancash Province, northern Peru to Antofagasta Province, northern Chile (Figure 1). Forty three species of *Orestias* are recognized in the most recent revision (Parenti 1984). Over half of the species are endemic to Lago Titicaca, the large high altitude freshwater lake of the Titicaca basin on the Peruvian and Bolivian Altiplano.

In Tchernavin's (1944) revision of *Orestias*, 20 species and 5 subspecies were recognized. Since then, the *Orestias* species in Lago Titicaca have been referred to as a species flock (Kosswig, Villwock 1964; Richardson et al. 1977; Villwock 1962, 1972). The concept of a species flock as applied to the phylogeny and biogeography of the genus has not been evaluated critically. However, the term *species flock* has been applied casually to relatively speciose taxa exhibiting a variety of diet and habitat preferences that are restricted geographically, usually in a lacustrine environment, and one that may or may not form a monophyletic group. Further, the last of these criteria, monophyly, has not been applied critically in discussions of the utility of the term *species flock*. This criterion is examined in detail with respect to the distribution of *Orestias* throughout its range to determine in which cases, if any, the concept of a species flock can be used to aid in the understanding of the evolution of a particular group.

Orestias is a well-defined monophyletic genus; seven synapomorphic characters, including the absence of a vomer, absence of pelvic fins and fin girdles, a unique squamation and head pore pattern, and specializations of the lower jaw were elaborated by

Parenti (1984). In that revision, definitions of four monophyletic groups of species and their interrelationships were hypothesized. The present investigation follows directly from that revision, and it constitutes a historical biogeographic analysis of the genus *Orestias*.

THE GENUS *ORESTIAS*

The 43 species of *Orestias* were divided into four monophyletic species complexes, referred to as the *cuvieri, mulleri, gilsoni,* and *agassii* complexes. The *cuvieri* complex comprises four species, defined as monophyletic by a series of gill arch characters, and an increase in the modal number of anal fin rays. Included are the predator *O. cuvieri* Valenciennes, and the planktivore *O. pentlandii* Valenciennes, the two large midwater species of Lago Titicaca. *Orestias cuvieri* has a relatively large mouth with large recurved teeth, whereas *O. pentlandii* has few or no teeth in the outer jaws.

The *mulleri* complex comprises the five species defined as monophyletic by a derived dorsal fin structure. Included are *O. crawfordi* Tchernavin and *O. incae* Garman, the relatively deepwater species of Lago Titicaca, taken at depths to 38 meters. These two species have molariform pharyngeal teeth used for crushing prey.

The *gilsoni* complex comprises ten species defined as monophyletic by a derived caudal fin structure. They comprise a group of small inshore species of Lago Titicaca, some of which are known to be sexually mature at under 30 mm SL.

The *agassii* complex comprises 24 species defined as monophyletic by a series of derived squamation characters and a relatively deep caudal peduncle. Species of this complex exhibit a diversity of body shapes, and include

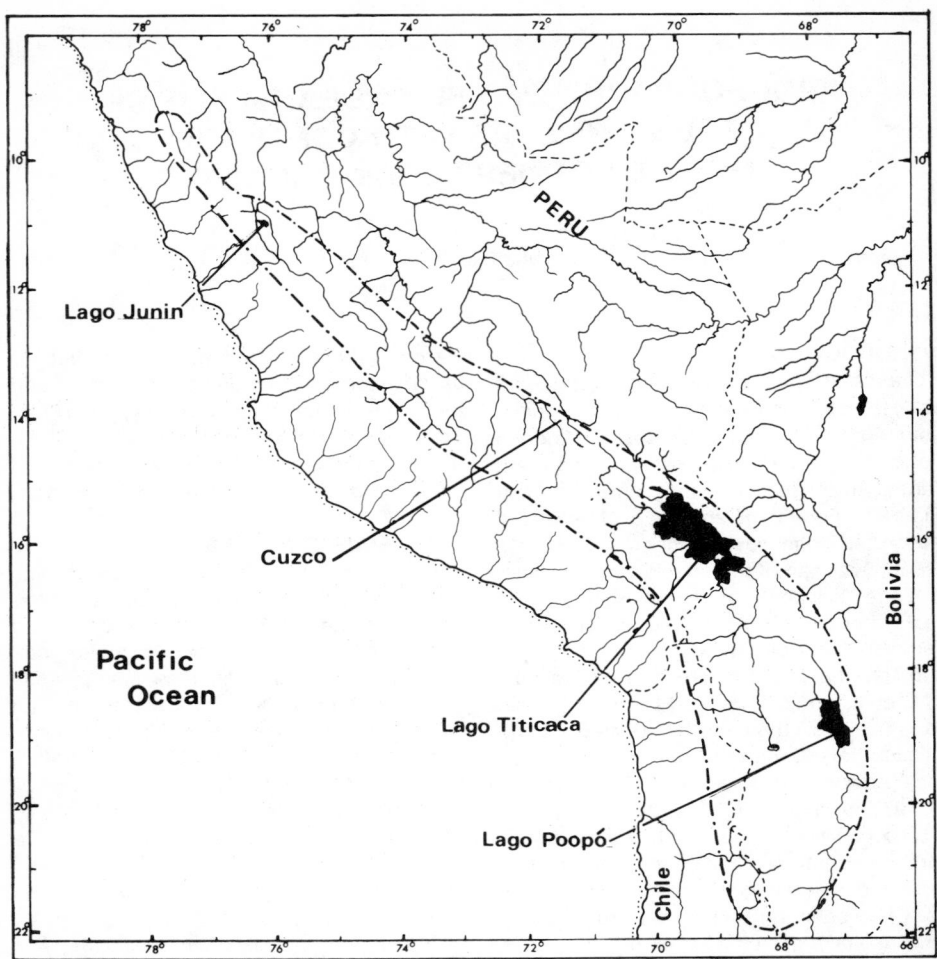

Figure 1. Distributional limits of the genus *Orestias* outlined by the dashed line.

the wide-bodied and wide-headed *O. luteus* Valenciennes, in which some individuals have heads wider than long.

The four species complexes are hypothesized to be related as in the cladogram of Figure 2a. The *mulleri* and *gilsoni* complexes are sister groups. They in turn are most closely related to the *cuvieri* complex. These three complexes together form the sister group of the *agassii* complex. Characters in support of this hypothesis of relationship are given by Parenti (1984).

DISTRIBUTION PATTERNS WITHIN *ORESTIAS*

A study of the relationships and distribution of *Orestias* species should include a historical biogeographic analysis to determine the underlying patterns of distribution. Such an analysis is an evaluation of the species flock concept only insofar as it tests whether or not monophyletic groups of species occur in restricted areas such as a lake or lake basin.

To determine what the cladogram of Figure 2a tells about the history of *Orestias* species complex distribution, it is converted into an

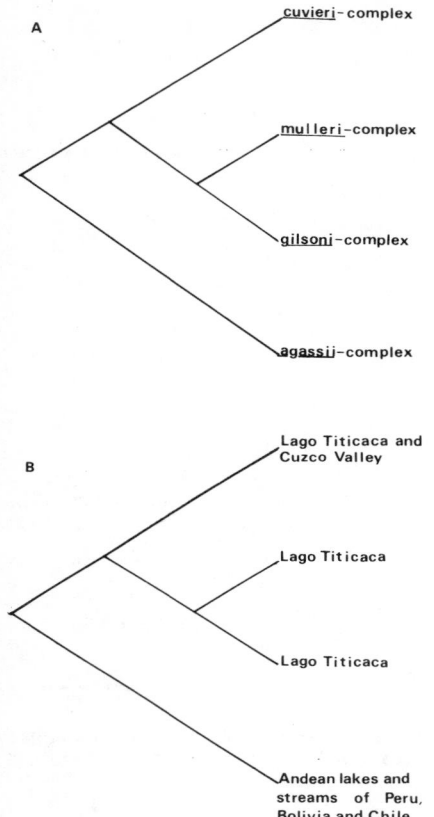

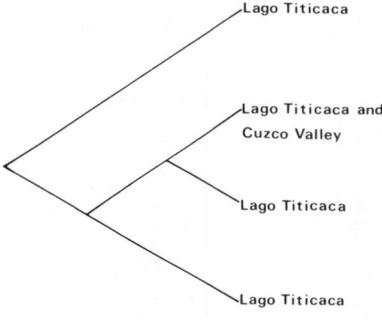

Figure 3. Area cladogram for the species of the cuvieri complex

Figure 2. A: Cladogram of relationships of the four monophyletic species complexes of *Orestias*. B: Area cladogram of *Orestias* species complexes. Derived from 2A by replacing name of complex with range of complex.

area cladogram (Figure 2b) following methodology of cladistic biogeography (Platnick, Nelson 1978; Rosen 1978). Names of species complexes were replaced by their distribution areas. Both *mulleri* and *gilsoni* complexes are known only from Lago Titicaca, and the Cuzco Valley in the Urubamba basin.

The distribution of *Orestias* species within Lago Titicaca is not known with great precision. Many areas of the lake, particularly the relatively deep and steep-sided Lago Grande, have not been collected, or have been collected only in part. Because such little information is available in the Titicaca Basin and the

Poopo Basin, both will be treated as one biogeographic region, the Titicaca Basin. No area cladograms are given for the *mulleri* and *gilsoni* complexes because their species are not found outside that region.

The area cladogram for the *cuvieri* complex (Figure 3) is derived from a cladogram of the four species (Parenti 1984). Just one species, *O. pentlandii* is found outside the Titicaca Basin, in the Cuzco Valley of the Urubamba Basin.

The *agassii* complex is very widespread throughout the central Andes, with species found throughout the Altiplano in the Urubamba and Titicaca basins, southern Bolivia and northern Chile, and outside the Altiplano in northern Peru. Because the distribution of the *agassii* complex includes Lago Titicaca and the Cuzco Valley, areas occupied by the *cuvieri, mulleri* and *gilsoni* complexes, it is only necessary to examine the distribution pattern of the complex to discover the general

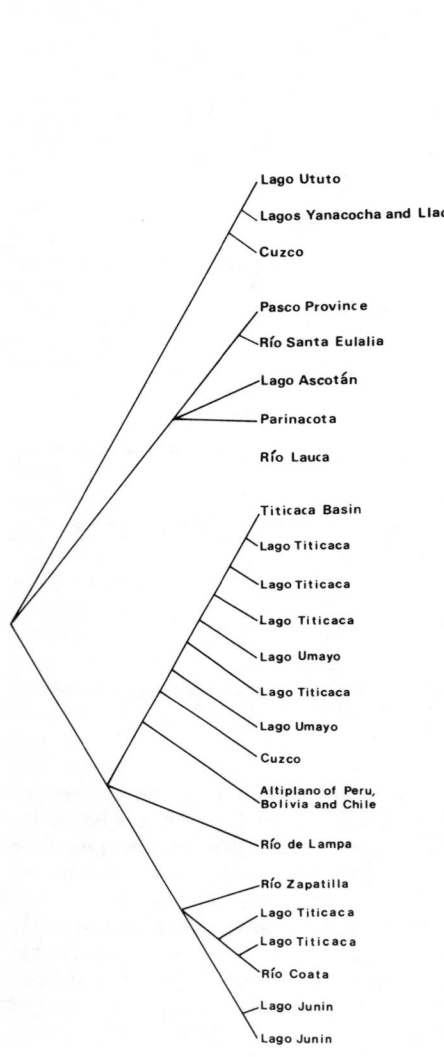

Figure 4. Area cladogram for the 24 species of the *agassii* complex.

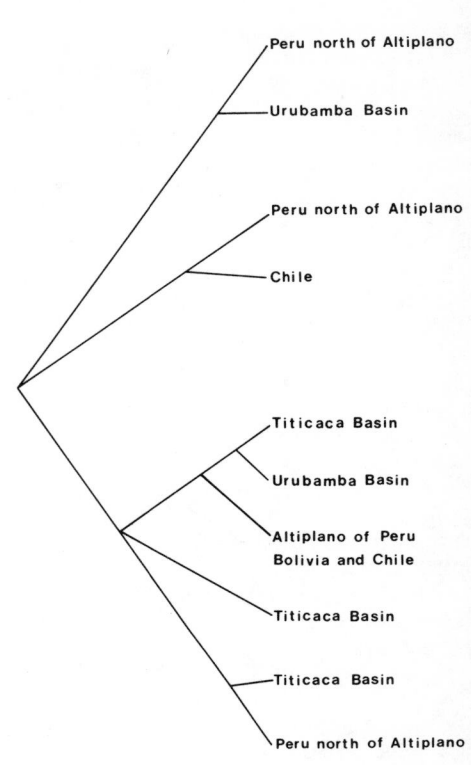

Figure 5. Simplified area cladogram for *agassii* complex. Derived from Figure 4.

patterns of distribution within the genus.

Figure 4 is the area cladogram for the 24 species of the *agassii* complex. Localities appear in Figure 1. We may simplify this figure to that of Figure 5 by using a single designation for all groups of species in one area. For example, the three areas of Lago Ascotan,

Parinacota and Rio Lauca are all in northern Chile. And the three *agassii* complex species in these areas form a monophyletic group. Therefore, in Figure 5, the three areas are combined into one area, Chile. Similarly, Lagos Ututo, Yanacocha and Llacsha contain two sister species found north of the Peruvian Altiplano. The three lakes are combined into one area, Peru north of the Altiplano. Lago Junin and Pasco Province and the Rio Santa Eulalia are also in Peru north of the Altiplano, although they do not contain species closely related to those in the lakes previously named. For simplicity, one area, called Peru north of the Altiplano will be recognized although it is obvious from Figure 5 that the area does not contain one monophyletic group of *Orestias* species.

In Figure 5, four general areas are recognized: the Titicaca basin, the Urubamba basin, Chile, and Peru north of the Altiplano. One variable species, *O. agassii* Valenciennes is found throughout the central Andes in three of these areas, the Urubamba and Titicaca basins and Chile. None of the four areas contains a single monophyletic group of *Orestias* species. Information about the distribution of the species in the four areas included in Figures 2b, 3, and 5 can be summarized as follows:

1) Species in the Titicaca basin are most closely related to other Titicaca basin species, or to species in the Urubamba basin, or Peru north of the Altiplano, but not to Chilean species.

2) Species in Peru north of the Altiplano are most closely related to other species in that area, or to species in the Titicaca and Urubamba basins or Chile.

3) Species in the Urubamba basin are most closely related to other species in that basin, or to species in the Titicaca basin or Peru north of the Altiplano, but not to Chilean species.

4) Chilean species are related to other Chilean species or to species in Peru north of the Altiplano, but not to Titicaca or Urubamba basin species.

DISCUSSION

To determine what groups of *Orestias* species may be referred to as species flocks, we must assess which lakes support monophyletic groups of species. Of the major lakes within the distributional limits of *Orestias*, just one, Lago Junin in northern Peru (Figure 1) supports a monophyletic group of species, but there are only two nominal forms for which the term *species flock* may be appropriate.

The *Orestias* species in Lago Titicaca do not constitute a monophyletic group. There are at least 11 distinct monophyletic lineages in Lago Titicaca (Figures 2 through 4, the *mulleri* and *gilsoni* complexes considered together). Many of these lineages are closely related to a group of species outside the lake, and in several cases, outside the Titicaca basin. The *mulleri* and *gilsoni* complexes together may be considered a flock of 15 species in Lago Titicaca. However, calling them a flock is little more informative than calling them a monophyletic group.

The treatment of *Orestias* species as a

species flock in previous studies has depended on the assumption that one variable and relatively primitive species, *O. agassii*, was the species from which a so-called "adaptive radiation" of *Orestias* emerged in Lago Titicaca (Villwock 1962). Although the taxonomic limit of *O. agassii* was restricted in the latest revision, it remains the most widespread and variable species recognized (Parenti 1984). Nevertheless, the position of *O. agassii* in a phylogeny of *Orestias* species precludes it from consideration as the most primitive or ancestral species. It does not occupy the most plesiomorphic position in a phylogeny of the genus or in the *agassii* complex. Even if it did occupy such a plesiomorphic position, the idea that the present day species is most like the ancestral species that entered or evolved in Lago Titicaca and then gave rise to all other *Orestias* species in the lake is precluded by the fact that all *Orestias* in the lake do not form a monophyletic group.

To provide a hypothesis of the historical biogeography of the genus *Orestias* I seek to answer the question: "Is the distribution pattern of Figure 5 unique to *Orestias* species, or is it found within other taxa of the central Andean biota?" There are several repeated patterns among Figures 2b, 3, and 5, but these involve only two-taxon statements, such as, part of the Urubamba basin fauna has its closest relatives in the Titicaca basin. Only one other fish genus, the catfish *Trichomycterus*, has endemic representatives in Lago Titicaca. Other endemic Altiplano fish include the catfish *Astroblepus*, and possibly the silverside *Basilichthys* (Parenti 1984). Unfortunately, little or no data are available on the definition and relationships of these three genera. It is even doubtful that each of the three is monophyletic. Therefore, at this stage, a statement on the origin and distribution of the central Andean fish fauna must concern solely *Orestias*.

Although we are hampered by lack of knowledge of the geological history of the Andes, we may speculate on how several distantly related groups of *Orestias* species came to be widely distributed throughout the Titicaca and Urubamba basins. The Altiplano is hypothesized to have formed during the Miocene, as the already existing eastern range of the Andes was displaced farther eastward by the formation of the western range by subduction of the Pacific plate (James 1971, 1973). By the end of the Pliocene, beginning of the Pleistocene, a large lake, Ballivian, had formed, and it extended over the area that includes the present Lagos Titicaca and Poopo. These two smaller lakes formed as the water fell in Ballivian. Lakes of the Altiplano have been isolated from ocean and other continental drainages since the Pliocene (Newell 1949). However, since monophyletic groups of species exist with members in and outside of the Titicaca basin, the formation of Lake Ballivian apparently had little to do with the origin of the diversity of the genus *Orestias* as a whole.

The traditional explanation for a high degree of endemism in the Andes is that the biota was separated from that of the rest of South America at least by the time postulated for the last rise of the Andes, generally given as mid-Oligocene. However, this explanation for endemism depends on at least three assumptions: 1) that all elements of the Andean biota can be defined together as unique; 2) that members of the Andean biota have their closest relatives in tropical lowland South America; and 3) that the Andes form a homogeneous, continuous mountain range.

These assumptions hold only in part. The first concerns a definition of the Andes in terms of its biotic elements, which is beyond the scope of this paper. The second assumption concerns the identification of the closest relatives of members of the Andean biota, which can only be discussed for *Orestias*. In most classifications of killifish *Orestias* has been placed in its own tribe or subfamily without a statement about a possible closest relative. Eigenmann (1920, 1927) considered *Orestias* to be most closely related to the North American killifish genus *Empetrichthys*, but most workers did not follow his suggestion. Foster (1967) proposed that *Orestias* is the sister group of the tropical lowland South American killifish *Cynolebias*. In a recent cyprinodontiform revision Parenti (1981) proposed that *Orestias* has no close South American relatives, but is instead a member of a monophyletic group comprising Anatolian and New World cyprinodontines, and most closely related to the Anatolian genus *Kosswigichthys*. The killifishes were classified in two suborders with *Orestias* and *Cynolebias* judged to be members of the cyprinodontoid and aplocheiloid suborders, respectively.

The hypotheses of Foster and Parenti for the relationship of *Orestias* to other killifish genera support divergent explanations for the history of the Altiplano ichthyofauna. The proposal of *Cynolebias* as a close relative of *Orestias* is compatible with the idea that the common ancestor of *Orestias* and *Cynolebias* was a tropical lowland killifish in existence prior to formation of the Altiplano. When the Altiplano was formed, part of this ancestral group would have been separated to give rise to the genus *Orestias*. The alternative hypothesis, that *Orestias* is most closely related to the Eurasian genus *Kosswigichthys* does not necessarily conflict with the traditional interpretation of the formation of the Altiplano given here, but neither is it enhanced by it.

Zeil (1979) proposed an alternative view of the development of the central Andes. He suggests that the Andes do not form one continuous orogen, and therefore that the development of South America does not necessarily conform to the current plate tectonic theories. The consequences of such an alternative proposal for the biogeographic history of South America are discussed elsewhere (Humphries, Parenti In press). Zeil's argument is that the Andes of Colombia, Ecuador and Venezuela fit the plate tectonic model and correlate with hypotheses of Andean mountain building requiring the subduction of a Pacific plate. The Andes of Peru, Chile, Bolivia and Argentina, Zeil contends, do not correspond to the plate tectonic model. He characterized the Andes in these regions as relatively old Jurassic rock lying on top of continental South America. He postulates an extra-continental origin for this rock.

The two hypotheses for the development of the Altiplano give different estimates for the minimum age of the biota. The plate tectonic model requires that if *Orestias* was in place just prior to the lifting of the Altiplano, it was in existence by the Miocene. Zeil's model requires that *Orestias* be older if it entered South America with the anomalous rock. One way of deciding between the two hypotheses based on geology would be to determine if the distribution pattern of *Orestias* (Figure 5) is part of a general pattern for the central Andean biota. *Orestias* has no fossil representatives. Therefore, its minimum age based on biological evidence must be estimated as the age of its sister taxon, also without fossil representatives, or as the age of the biogeographic pattern of which it is a part (Humphries, Parenti In press). Phylogenetic analyses of other elements of the central Andean biota must be carried out before it can be stated whether *Orestias* species conform to a general biogeographic pattern.

The evolution of certain groups of species referred to as flocks, such as the African Rift Lake cichlids, has been termed "explosive" because the ages of the lake basins are estimated to be relatively young. The high number of species and low age of the basins together are interpreted as evidence of a rapid rate of evolution (e.g., Greenwood 1981). The evolution of the genus *Orestias* is not termed rapid because the genus is not considered to have evolved from a single ancestral species or group of species in the Titicaca or any other single Altiplano basin. The evolution of Lake Ballivian in the Plio-Pleistocene had little to do with the origin of diversity of *Orestias* because several species groups in Lago Titicaca have close relatives outside the lake basin. This pattern is repeated several times within the genus, indicating that the geological history of the central Andes, not simply that of individual lake basins, has been critical to the evolution of this very diverse killifish genus.

SUMMARY

Since Tchernavin's revision of the genus *Orestias*, the species have been referred to as a species flock. They may be considered a flock only in that they are diverse, comprising 43 species, and occupy a circumscribed area. The *Orestias* species of Lago Titicaca, the large high altitude lake of the Altiplano of Peru and Bolivia, have been the focus of many comments on the application of the species flock concept. The assumption has been that one widespread, relatively primitive species, *O. agassii*, is a present day representative of the ancestral *Orestias* species that entered or evolved in Lago Titicaca and gave rise to the other *Orestias* species in the lake. However, the *Orestias* of Lago Titicaca do not comprise a monophyletic group of species. If monophyly

is one criterion for application of the term *species flock*, then *Orestias* in Titicaca cannot properly be called a flock.

Monophyletic groups of *Orestias* do occur in Lago Titicaca, such as the 15 species of the *mulleri* and *gilsoni* complexes. Thus, *Orestias* in the lake may be considered an assemblage that includes, in part, several species flocks.

Orestias species may be considered to occupy four general biogeographic regions: the Titicaca basin, including Lagos Titicaca and Poopo; the Urubamba basin; Chile, and ·Peru north of the Altiplano. None of these regions contains a single, monophyletic group of *Orestias* species. The relationships of the species in these regions suggest a complicated history of the central Andes. Relatively recent, Plio-Pleistocene events concerning the evolution of lake basins on the Altiplano most likely occurred after *Orestias* was distributed throughout the Altiplano and beyond, because several species groups have representatives inside and outside the major lake basins. The evolution of *Orestias* is not considered "explosive." Therefore, the uniqueness of the distributional history of *Orestias* can only be assessed when information on the phylogeny of more elements of the Andean biota becomes available.

ACKNOWLEDGMENTS

I thank P H Greenwood for reviewing this paper, and Bernice E Brewster for technical assistance with the manuscript. The work on *Orestias* was supported by a Smithsonian Postdoctoral Fellowship at the National Museum of Natural History and a NATO Postdoctoral Fellowship at the British Museum (Natural History).

REFERENCES

Eigenmann C H 1920 On the genera *Orestias* and *Empetrichthys*. *Copeia* 1920:103-106
_____ 1927 The freshwater fishes of Chile. *Mem Natl Acad Sci* 22:1-80
Foster N R 1967 Trends in the evolution of reproductive behavior in killifishes. *Stud Trop Oceanog* 5:546-566
Greenwood P H 1981 Species flocks and explosive speciation. 61-74 P L Forey ed. *The Evolving Biosphere*. British Museum (Nat Hist). London
Humphries C J, L R Parenti In press. *Cladistic Biogeography*. Oxford University Press Monographs on Biogeography

James D E 1971 Plate tectonic model for the evolution of the central Andes. *Geol Soc Amer Bull* 82: 3325-3346

———— 1973 The evolution of the Andes. *Sci Amer* 229:60-69

Kosswig C, W Villwock 1964 Das Problem der intralakustrischen Speziation im Titicaca und Lanaosee. *Verh Dtsch Zool Ges Keil* 1964 *Zool Anz Suppl* 28:95-102

Newell N D 1949 Geology of Lake Titicaca region, Peru and Bolivia. *Mem Geol Soc Amer* 36:1-111

Parenti L R 1981 A phylogenetic and biogeographic analysis of cyprinodontiform fishes. (Teleostei, Atherinomorpha). *Bull Amer Mus Nat Hist* 168: 335-557

———— 1984 A taxonomic revision of the Andean killifish genus *Orestias* (Cyprinodontiformes, Cyprinodontidae). *Bull Amer Mus Nat Hist* Vol 178:2:107-214

Platnick N I, G Nelson 1978 A method of analysis for historical biogeography. *Syst Zool* 27:1-16

Richardson P J, C Widmer, T Kittel 1977 The limnology of Lake Titicaca (Peru-Bolivia), a large high altitude tropical lake. Institute of Ecology Publ 14 Univ California, Davis.

Rosen D E 1978 Vicariant patterns and historical explanation in biogeography. *Syst Zool* 27:159-188

Tchernavin V V 1944 A revision of the subfamily Orestiinae. *Proc Zool Soc London* 114:140-233

Villwock W 1962 Die Gattung *Orestias* (Pisces, Microcyprini) und die Frage der intralakustrischen Speziation im Titicaca-Seegebeit. *Verh Dtsch Zool Ges Wien* 54:610-624

———— 1972 Gefahren fur endemische Fisch faune durch Einburgerungsversuche und Akklimatization von Freunfischen am Beispiel des Titicaca- Sees (Peru/Bolivia) und des Lanao-Sees (Mindanao/Philippinen). *Verh Internat Ver Limnol* 18:1227-1234

Zeil W 1979 *The Andes, a Geological Review.* Gebruder Borntraeger. Berlin-Stuttgart

EVOLUTIONARY GENETICS OF A "SPECIES FLOCK:" ATHERINID FISHES ON THE MESA CENTRAL OF MEXICO

ANTHONY A ECHELLE and ALICE F ECHELLE

INTRODUCTION

The fish fauna of the Mesa Central, the southernmost portion of the Mexican Plateau, is dominated by three groups: the cyprinodontoid family Goodeidae (35 species), the atherinid genus *Chirostoma* (18 species), and the cyprinid genus *Algansea* (7 species). Nearly all species of these groups are endemic to the Mesa Central. Because of the unusually high endemism and otherwise depauperate nature of the ichthyofauna, Regan (1906-1908) recognized the relatively tiny Mesa Central area as one of three biogeographic subregions of North America. Such an interesting biogeographic pattern begs explanation. In this paper we contribute toward such an explanation for the atherinid component of the fauna, *Chirostoma* and *Poblana*, the latter a nominal genus of two or three species restricted to the Mesa Central.

As a group, atherinids are small silvery-sided predators on invertebrates. *Chirostoma* is included in the subfamily Menidiinae with a variety of genera whose distributions indicate a tendency for coastal forms to invade and speciate in freshwater environments. Single species typically represent the menidiine component of local faunas, and when two occur together, they generally are not congeneric, (e.g., *Menidia* and *Labidesthes* in North America, and *Xenatherina* and *Archomenidia* in Mexico). Outside the Mesa Central, the only North American exceptions to this rule are coastal areas of overlap between species with largely non-overlapping ranges, such as the inland form, *Menidia beryllina*, and the more marine species, *M. peninsulae*.

On the Mesa Central, however, local faunas including two to eight *Chirostoma* species are commonly encountered (Barbour 1973a). The richer assemblages of *Chirostoma* occur in the large lakes, Patzcuaro and Chapala (respec-

tively, 4 and 8 species now or once common), where the various species range in maximum standard length from about 60 mm to menidiine giants of 250-300 mm. Barbour (1973a,b) emphasized that the most apparent morphological differences among species are in body size and trophic characters such as jaw lengths and tooth patterns.

Barbour (1973a) referred to *Chirostoma* as a species flock, and the group certainly has most of the attributes ascribed to species flocks (Greenwood 1984; Ribbink 1984). Compared with other menidiines, they show exceptionally high levels of local sympatry, extreme examples of morphological evolution (large body size), and unusually high speciosity in a relatively small geographic area. Thus, most questions relevant to species flocks apply to *Chirostoma*: 1) Is the group monophyletic? 2) What are the closest relatives outside the Mesa Central? 3) Is there evidence of rapid evolution? 4) How much of the observed morphological radiation is a result of speciation, and how much results from intraspecific polymorphism? With these questions in mind we performed an electrophoretic survey of *Chirostoma*, *Poblana* and representatives of six other menidiine genera that occur in the freshwaters of North America. The results add a genetic perspective to questions of evolutionary dynamics and relationships of atherinids on the Mexican Plateau.

TAXONOMY AND DISTRIBUTION

The taxonomy of *Chirostoma* has had a turbulent history (Barbour 1973b). At various times the genus has been split into two or three genera (Jordan, Evermann 1896; Alvarez 1970). Meek (1904) recognized only one genus and placed the species in three subgenera, whereas other workers (e.g., Jordan, Hubbs 1919) found no basis for recognition of subgenera. De Buen (1945), however, recognized three genera and six subgenera. More recently, Barbour (1973b, 1974) recognized a single genus with 18 species in two groups: the *jordani* species group and the *arge* species group. Finally, Barbour, in an unpublished paper, considers the atherinid genus *Poblana* to be a junior synonym of *Chirostoma*, a decision that would bring the number of *Chirostoma* species to 20, all of which are endemic to the Mesa Central except for *C. humboldtianum* and *C. jordani*. The distribution of *C. humboldtianum* extends to just beyond the western boundary of the Mesa Central. *Chirostoma jordani* occurs in two drainages outside the Mesa Central, the Rio Mezquital in Durango (northern Mexico), and the Rio Panuco at the eastern edge of the Mesa Central.

GEOLOGICAL SETTING

The geological history of the Mesa Central was reviewed by Barbour (1973b) and Smith (1980). We present here a synopsis of their reviews. The Mesa Central is at the southern extreme of the Mexican Plateau (Figure 1). The plateau lies between the great eastern and western mountain ranges of Mexico, the Sierra Madre Oriental and the Sierra Madre Occidental, and extends from the Rio Grande (Rio Bravo del Norte) of Texas and Mexico south to an east-west volcanic axis which extends across Mexico between the 19th and 20th parallels.

The eastern boundary of the Mesa Central developed during the Eocene as the Hidalgoan orogeny produced the Sierra Madre Oriental. The western boundary, the Sierra Madre Occidental, was formed primarily by volcanic activity in the Oligocene and Miocene. Periodic uplifting of the mountains and the plateau has continued into the Quaternary. Beginning as early as the Eocene and continuing to the present, the plateau area has been a region of endorheic drainages (Smith 1980). This, with the erosion of the mountains, has contributed thick sediments to the plateau, and during wetter periods would have produced large numbers of lakes (see Miller 1981).

During mid-Cretaceous times, the area presently occupied by the Mesa Central was flooded by a marine transgression, the Balsas Portal. More southerly connections between the east and west coasts of Mexico, the Tehuantepec and Central American canals, were open as late as the Miocene. The southern boundary of the central plateau and present landforms of the Mesa Central itself are largely due to vulcanism which began in the Miocene, reached a peak in the Pleistocene, and continues intermittently to the present. Regional uplift occurred during early to mid-Pleistocene. Because of vulcanism, uplifting, and some degree of block-faulting, lakes, including some very large ones, have formed and then filled with alluvium, or they have been drained by eroding streams.

At present, the Lerma-Santiago river system, a Pacific drainage, is the major basin of the Mesa Central. The Rio Grande de Santiago, which empties into the Pacific Ocean, originates at the eastern end of Lake Chapala, only 16 km from the mouth of the Rio Lerma, which drains into Lake Chapala from the east. The Mesa Central also possesses a number of interior basins. These include the Valley of Mexico and the Llanos de Puebla in the east, Lakes Santa Maria and Juanacatlan in the west, the Valley of Tocumbo near the basin of Lake Chapala, and the closely associated lakes Zirahuen, Patzcuaro, and Cuitzeo. As indicated by fish distributions (Meek 1904), all of these apparently have had past connections with the Lerma-Santiago system, probably as late as the Pleistocene (Barbour 1973a; Smith 1980).

METHODS AND MATERIALS

Samples from 21 populations of *Chirostoma*, including 15 of the 18 extant species, were processed for electrophoresis. Populations examined and collection localities were as follows, where capital letters and numbers in parentheses refer respectively to codes used to identify populations in Appendix 1-2 and locality numbers given in Figure 1: *C. humboldtianum* (HUM 10), *C. lucius* (LUC 3,5,6), *C. sphyraena* (SPH 2,4,5), *C. promelas* (PRO 2,4), *C. chapalae* (CHA 4,6), *C. consocium consocium* (CONC 3,6), *C. c. reseratum* (CONR 9), *C. grandocule* (GRA 11), *C. estor copandaro* (EST 12), *C. jordani* (JORD 1; JORC 8; JORS 13; JORY 14; JORP 17), *C. labarcae* (LAB 4), *C. aculeatum* (ACU 14),

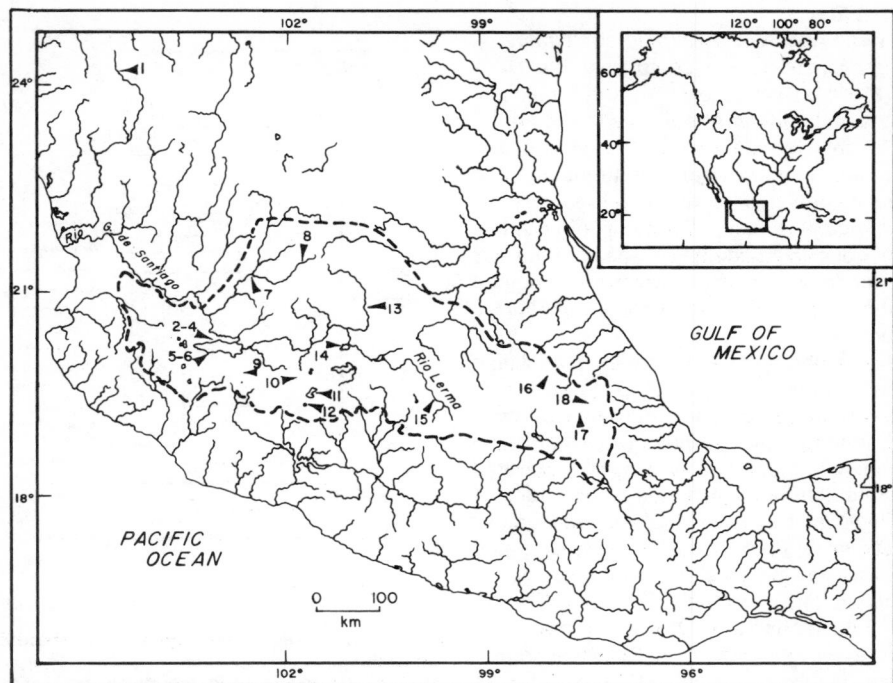

Figure 1. Collection localities in Mexico. Dashed line encloses Mesa Central; numbers show collection localities of Appendix 1.

C. attenuatum (ATT 12), *C. arge* (ARGC 4; ARGY 7), *C. melanoccus* (MEL 9), *C. riojai* (RIO 15). Additional menidiines examined were two species of *Poblana* from the Mesa Central, *P. alchichica* (ALC 18) and *P. ferde-bueni* (FER 16), two species of *Menidia* from Texas, *M. peninsulae* (PEN), and *M. beryllina* (BER), one population each of *Labidesthes sicculus* from Oklahoma and *Membras martinica* from Texas, and samples from the following Mexican forms: *Melaniris crystallina, M. alvarezi, Melaniris* sp., *Xenatherina schultzi, Xenatherina* sp., *Archomenidia sallei* (two populations). For added outgroup perspective, we also examined a sample of *Atherinomorus stipes*, a marine atherinid of the subfamily Taeniomembrasinae.

All specimens were stored on dry ice, transported to the laboratory, and maintained at -60 C. Protein extracts were obtained from liver, epaxial muscle, and combinations of eye and brain. Standard methods of horizontal starch-gel electrophoresis

(Selander et al. 1971; Siciliano, Shaw 1976) were used to examine 22 presumptive gene loci. Abbreviations for presumptive loci are as given by Echelle et al. (1983), as are the locus/buffer systems with the following exceptions: *Aat-1, Gpi-1* and *-2, Pgm-2,* and *Gp-5* were resolved better with a lithium hydroxide system (Selander et al. 1971) than with the systems used earlier, and *G3pdh-1,* and *Ldh-1* and *-2* were resolved best on the tris versene borate system (Siciliano, Shaw 1976). Loci for multilocus systems are assigned numerals 1, 2, 3, etc., in the order of decreasing anodal mobility; on the same basis, alleles within loci are designated a, b, c, etc. Heterozygosities, polymorphism, genetic similarities, and UPGMA clustering were computed with BIOSYS-1 (Swofford, Selander 1981). Voucher specimens for each sample are catalogued in the fish collection at the Oklahoma State University Museum of Cultural and Natural History.

RESULTS

Phenetic Analysis. Electromorph frequencies (Appendix 2) were used to compute Nei's (1978) index (I) of genetic similarity for all pairwise combinations of samples. Here we include only the details for *Chirostoma* and *Menidia*; electromorph frequencies and genetic identities for other genera will be presented elsewhere. The sample-by-sample matrix of genetic identities was summarized in a dendrogram by the unweighted pair-group method with arithmetic averaging (UPGMA, Figure 2).

The members of the subfamily Menidiinae cluster together, and separately from *Atherinomorus* of the marine subfamily Taeniomenbrasinae (Schultz 1948). Within the menidiines there are two well separated clusters, one which includes the two species of *Menidia*, all species of *Chirostoma* (unless otherwise indicated, our use of the name *Chirostoma* includes the nominal genus *Poblana*), and *Labidesthes sicculus*. The second cluster includes the other menidiine genera we examined. Within the former cluster, the two species of *Menidia* fall in an assemblage that encompassed *Chirostoma* and is strongly divergent from *Labidesthes* (mean I = .42; range = .23-.51).

Within the *Menidia-Chirostoma* assemblage, *C. riojai*, *Poblana ferdebueni*, and *P. alchichica* form a cluster associated with *C. riojai* and more loosely with *C. attenuatum*. A second cluster consists of the five populations of *C. jordani*: the three populations from the Rio Lerma were highly similar (locations 8, 13, 14, Figure 1; I = .999-1.0), while samples from two disjunct populations (locations 1, 17) outside the Rio Lerma are somewhat more divergent from each other (I = .93) and had intermediate levels of similarity to the other three (.96-.97). The third cluster consists of the highly similar (I = .999) *C. labarcae* and *C. aculeatum*, linked with *C. arge* (I = .87), and at a lower level (.79), with the two species of *Menidia*. Finally, a fourth cluster includes a group of eight highly similar taxa: *C. lucius*, *C. sphyraena*, *C. chapalae*, *C. promelas*, *C. grandocule*, *C. humboldtianum*, and both subspecies of *C. consocium*, with genetic identities ranging from .960-1.00 (mean, .985); this group clustered at a slightly lower level (I = .895-.941; mean, .916) with *C. estor*

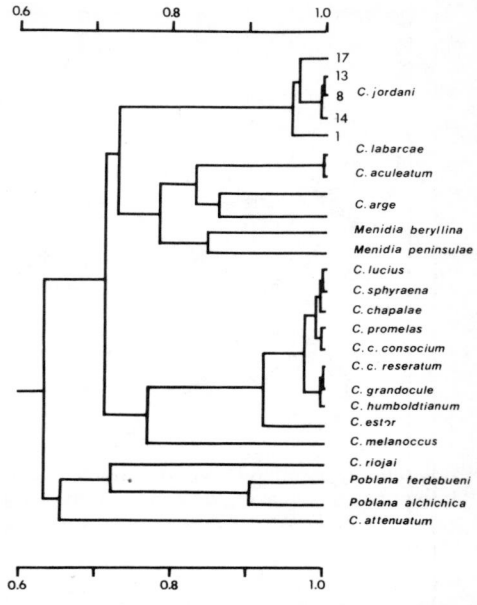

Figure 2. UPGMA dendrogram of phenetic similarities. Based on Nei's index of genetic identity. Numbers for *C. jordani* identify populations as numbered in Figure 1.

and more distantly (mean, .766) with *C. melanoccus*.

Cladistic Analysis. The UPGMA dendrogram (Figure 2) represents a phenetic analysis which is meant only to summarize genetic identities. Each character (locus) is weighted equally in the assessment of genetic similarity, with no attention to whether the various character states (alleles) were present ancestrally (plesiomorphic) or whether they are derived (apomorphic) states. We next present a cladistic analysis (Hennig 1966) of the presence-absence of electromorphs in which only the pattern of shared derived electromorphs (synapomorphies) is used to infer the branching sequence of the phylogeny.

Methods of inferring branching sequences (cladograms) from electrophoretic data are controversial. We chose the method used by Avise et al. (1980) and Buth et al. (1980) in which presence-absence of alleles is used to infer cladistic relationships. This method usually does not give fully resolved trees with electrophoretic data. However, it was chosen over the various methods using genetic distances because of the controversy over that approach (e.g., Farris 1981, Swofford 1981). Transformation Series Analysis (Michevich, Mitter 1981) was not available to us; also, it may have its own theoretical problems (D. Buth pers. comm.).

Atherinomorus was used as the initial outgroup. Of the six known species of *Menidia* (Echelle et al. 1983) only *M. beryllina* and *M. peninsulae* were compared on the same gels with the other taxa we examined. However, previous electrophoretic studies have examined large numbers of samples of all known species of *Menidia* over large portions of their geographic ranges (Johnson 1975; Duggins 1980; Echelle et al. 1983). This allows inferences regarding relationships of species not directly examined in this study. Such inferences are weakened because 1) they are based only on the taxonomic distribution of electromorphs we detected in *M. beryllina* and/or *M. peninsulae* and 2) different laboratories will differ in electromorph resolution, thus producing different patterns of character state distribution. In general, our data were consistent with those presented by Johnson (1975) and Duggins (1980), especially for the more common electromorphs in *M. beryllina* and *M. peninsulae*.

Interestingly, no electrophoretic synapomorphy identifies *M. beryllina* and *M. peninsulae* as parts of a clade distinct from *Chirostoma*. Rather, the most parsimonious hypothesis that we were able to resolve (Figure 3) suggests that *M. peninsulae* may be closer to *Chirostoma* than to any species of *Menidia*. For each of four loci (*Pgm-2, Gpi-1, Es-1, 6Pgd*) *Menidia peninsulae* shares with a variety of *Chirostoma* species an apparently derived electromorph that is not shared with *M. beryllina*. Gpi-1g is absent in all other bisexual (gonochoristic) species of *Menidia* (Johnson 1975; Echelle et al. 1983). However, it does occur in the recently discovered form, *M. clarkhubbsi*, which apparently arose as an interspecific hybrid with an *M. peninsulae*-like parent (Echelle et al. 1983). Pgm-2h occurs at low frequencies in Texas populations of *M. beryllina*, a pattern which may reflect ongoing hybridization with *M. peninsulae* and/or the presence of a second unisexual form (Echelle et al. 1983; unpubl); this allele appears

absent in all other species of *Menidia*. The two remaining synapomorphies between *M. peninsulae* and *Chirostoma*, Es-1f and 6Pgd-g, have not been detected in *M. beryllina* or in *M. colei* (Echelle et al. 1983), but we cannot assess their occurrence in *M. menidia* and *M. extensa*.

Four additional electromorphs, Aat-1a, Aat-2e, Gpi-1e, and Idh-1b, further indicate close cladistic relationship between *Menidia* and *Chirostoma*. Idh-1b, which occurs at high frequencies in most *Menidia* species and all atherinids on the Mesa Central except *P. ferdebueni* and *P. alchichica*, is especially supportive of this hypothesis. Aat-2e provides similar support; it occurs at high frequencies in all species of *Menidia* and in all Mesa Central atherinids. Neither Idh-1b nor Aat-2e was detected in the other menidiine genera we examined

Poblana ferdebueni and *P. alchichica* are fixed for two apparent synapomorphies, Idh-1f and Ldh-1f, which define a clade restricted to the eastern portion of the Mesa Central. Relatively close cladistic relationship between these two species and *C. riojai*, a species from the eastern extreme of the Rio Lerma drainage, is suggested by the occurrence, to the exclusion of other electromorphs, of Mdh-2e in all three species. *Mdh-2* is highly conservative in the taxa we examined. All other species of the *Labidesthes-Menidia-Chirostoma* assemblage are essentially fixed for the Mdh-2f electromorph. Therefore, Mdh-2e provides convincing evidence for close cladistic association between *C. riojai* and the *P. ferdebueni-P. alchichica* clade. A similar argument based on Idh-2f indicates, however, that *C. riojai* may share an ancestor with *C. melanoccus* that is not shared with *P. ferdebueni* or *P. alchichica*. Taking Mdh-2e and Idh-2f together, there are two equally parsimonious hypotheses: 1) *C. melanoccus* is the sister species of a *riojai-ferdebueni-alchichica* clade; 2), *C. melanoccus* and *C. riojai* represent a separate clade which diverged from a common ancestor with *P. ferdebueni* and *P. alchichica*. The first hypothesis requires loss of Idh-2f in an ancestor of the *ferdebueni-alchichica* lineage; the second hypothesis requires loss of Mdh-2e in *C. melanoccus*. One electromorph, Gpi-1h, weakly suggests a cladistic relationship

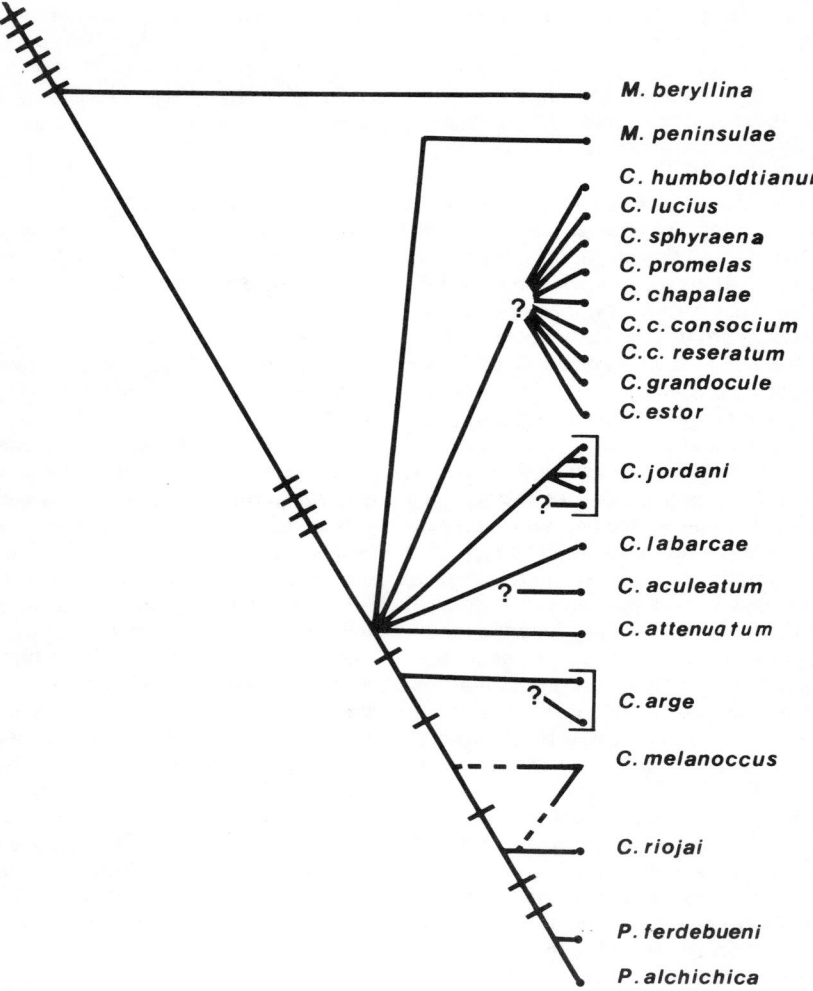

Figure 3. Cladogram showing the most parsimonious hypothesis of the evolution of synapomorphic electromorphs. Tick marks identify positions of synapomorphic alleles. Question marks identify presumed monophyletic clusters of taxa/populations which show no synapomorphies but have unusually high interspecific genetic identities (*C. labarcae-C. aculeatum* and the 9 species connected by single nodes) or are considered members of the same species (*C. arge, C. jordani*). Dotted lines for *melanoccus* show equally plausible plcaements.

between *C. arge* and the four species just discussed. This requires loss of the allele in the ancestor of *P. ferdebueni* and *P. alchichica.*

The approach we have used allows no further inferences regarding interspecific cladistic relationships in the *Menidia-Chirostoma* assemblage. All other derived electromorphs in this assemblage are either autapomorphies of a single species or they occur with no revealing patterns of distribution in one or more species of the clade just described, in one or both of the *Menidia* species we examined, and in a variety of the remaining species of *Chirostoma*. The lack of cladistic information extends even to populations considered conspecific under the current taxonomy (Barbour 1973b). For example, in both *C. consocium* and *C. arge*, the two different populations share no synapomorphy that indicates monophyly. In *C. jordani* there is no synapomorphy that unites the disjunct population from Durango with the populations from the Mesa Central. One synapomorphy, Gpi-2f, occurred in the four *C. jordani* samples from the Mesa Central but was absent in the sample from Durango. This indicates that the Durango form, which some refer to as *C. mezquital* (Miller 1981), may be cladistically distinct from the Mesa Central populations, but this hypothesis is only slightly more parsimonious than the possibility that Gpi-2f occurred in the common ancestor and subsequently was lost in the Durango form. A second derived electromorph, Pgm-2j, occurred in the two samples from the eastern reaches of the Rio Lerma (locations 13,14), but only at very low frequencies (.05), providing very weak support for cladistic relationship.

Species-Level Status of Sympatric Forms. Between most forms that we examined from the same lake, there were fixed allelic differences at two or more loci, leaving no doubt that the forms involved are good biological species. However, five species from Lake Chapala, *C. promelas, C. lucius, C. sphyraena, C. chapalae,* and *C. consocium consocium,* showed unusually high levels of interspecific genetic identity (Figure 2) and, correspondingly, no fixed allelic differences (Appendix 2). However, G-tests of independence (Sokal, Rohlf 1969:599) of allelic distributions between paired species revealed significant (alpha = .05) allele frequency differences at 1-3 loci in all cases (see Figure 4). The loci and taxa showing significant differences were *Es* (*C. c. consocium* versus each of three species, *C. lucius, C. sphyraena,* and *C. promelas*; each of the last three vs *C. chapalae*); Gpi-1 (*C. c. consocium* vs *C. chapalae, C. lucius,* and *C. sphyraena; C. chapalae* vs *C. promelas*); and G3pdh-1 (*C. sphyraena* vs each of the other four).

Because of small sample sizes the computations for most of these species were based on pooled samples from 2-3 locations. All samples were from the western end of Lake Chapala where they were separated by less than 15 km. No significant intraspecific differences in allelic frequencies occurred in samples from different locations within the

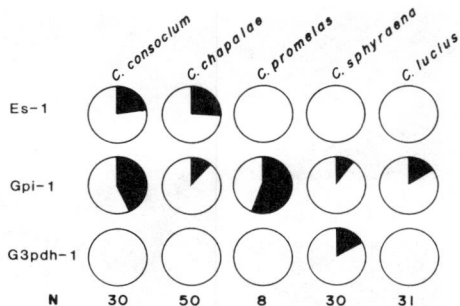

Figure 4. Pie diagrams show frequencies of alternative alleles at 3 loci in 5 species of the *C. humboldtianim* group from Lake Chapala.

lake. However, in all cases the samples were small; larger samples may have revealed minor differences. Figure 5 shows the range in body size and form of these five species.

Because they are believed to hybridize (Barbour 1973b; Barbour, Chernoff 1984), documentation of genetic discontinuity between the two large species, *C. sphyraena* and *C. lucius* is especially critical. To avoid including hybrids in our analysis, we used only those specimens showing the extremes of jaw morphology for the two species (Barbour 1973b); the carcasses of the specimens are catalogued as voucher specimens in the Oklahoma State University Museum of Natural and Cultural History. Only two loci were polymorphic (.05 criterion) among these two fishes, and allelic frequencies for one, *Gpi-1*, were essentially identical. At *G3pdh-1*, however, one third of the *C. sphyraena* specimens were heterozygous for a fast electromorph that was absent in *C. lucius* and the three other Lake Chapala species examined for this enzyme. Furthermore, this electromorph was present in the single specimen of *C. sphyraena* examined from the south shore of the lake where most of the *C. lucius* were taken. These data provide genetic support for the conclusion based on morphological data (Barbour, Chernoff 1984) that *C. sphyraena* and *C. lucius* are reproductively isolated.

Genetic Variability. Three measures of genetic variability were computed in this analysis: average heterozygosity per locus (*H*, from direct counts); percent of loci polymorphic (*P*, .99 and .95 criteria); and number

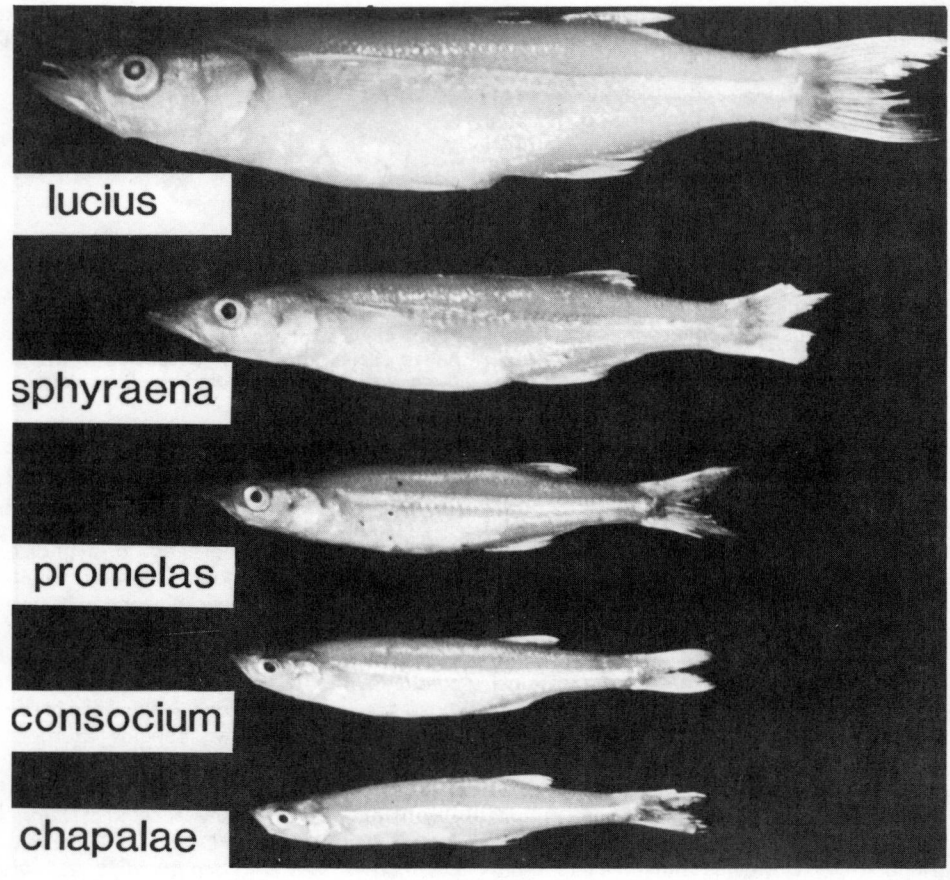

Figure 5. Five species of *C. humboldtianum* group from Lake Chapala, photographed on a single frame. Largest specimen is 180 mm standard length.

of unique alleles (*A*, .01 criterion for presence). The first two are commonly used measures. The third is the number of autapomorphies, and can be considered a measure of added allelic variety since origin of the species. These three measures were computed from those 22 genetic loci examined in all populations of the study. Mann-Whitney U-tests were used in all tests of significance.

Plots of *P*, *H*, and *A* among species of *Chirostoma* and *Poblana* are shown in Figure 6. Average polymorphism (.95 criterion = $P_{.95}$ = 13.6), and heterozygosity (.042) were slightly below the averages in Nevo's (1978) review of 51 species of fishes (15.2 and .051 respectively). There were no significant

differences for *H*, $P_{.95}$ or $P_{.99}$ between *Chirostoma* and the other atherinid genera we examined. There was none between the *jordani* and *arge* species groups for any of the genetic variability measures.

As a group, the eight genetically most similar taxa of the *C. jordani* group showed low averages for genetic variability compared with those for the 15 remaining *Chirostoma* populations ($P_{.95}$ = 10.2 vs 15.4; $P_{.99}$ = 14.2 vs 19.7; *H* = .024 vs .052; *A* = .50 vs 1.4). The differences in $P_{.99}$, *H* and *A* were statistically significant (n = 15,8; U = 89; p ◄ .05; U = 96.5; p ◄ .01; and U = 88.5; p ◄ .05, respectively). That for $P_{.95}$ was near the level required for significance (U = 82; .10 ► p ► .05).

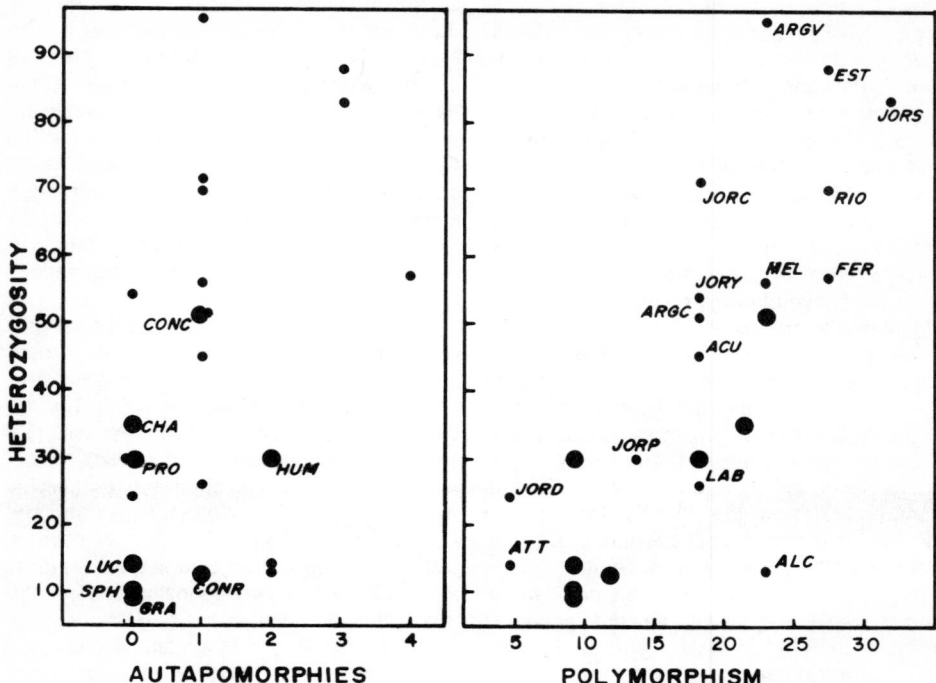

Figure 6. Genetic variability in *Chirostoma*. Points representing the *C. humboldtianum* group are labeled in the left panel. Those for other species are labeled in the right panel. Values for heterozygosity x1000; polymorphism x100.

The differences remain significant, and in general were heightened when these measures were expressed as the average for the species in cases where more than one population was examined (*C. concosium, C. arge,* and *C. jordani*).

DISCUSSION

Barbour (1973b) originally hypothesized that *Chirostoma* is diphyletic, with the *jordani* and *arge* species groups arising from *Menidia*-like and *Melaniris*-like ancestors, respectively. Our data support the recent conclusion from osteology that all species of *Chirostoma* share with *Menidia* a common ancestor not shared with *Melaniris* (Chernoff 1983; Barbour pers. comm.).

Our results are consistent with the suggestion (Miller, Chernoff 1979) that, based on a lack of diagnostic features, perhaps *Chirostoma* and the nominal genus *Poblana* should be subsumed under the name *Menidia*.

Average genetic identity (I) between *Chirostoma* and *Menidia beryllina* and *M. peninsulae* is .71 (.53-.85; SD = .08), a value which, based on calculations from Johnson's (1975) data, is higher than the maxima observed between these species of *Menidia* and two congeners, *M. menidia* (range = .61-.70) and *M. extensa* (.54-.70). The average value between *Chirostoma* and *M. beryllina* and *M. peninsulae* is well above that usually found in congeneric comparisons and markedly higher than usually found in intergeneric comparisons (Avise, Aquadro 1982). Finally, the cladistic analysis indicates that *Menidia* is paraphyletic. Taxonomic revision may therefore be desirable, but the final decision awaits further information and completion of the review by Barbour and Chernoff of the systematics of New World atherinids.

The distributions of extant and fossil fishes suggest three "biogeographic tracks" (Croizat 1974) that help explain the evolution of the

Mesa Central ichthyofauna (Miller, Smith in press): a Pacific coastal track, a western mountain track along the Sierra Madre Occidental, and a Plateau track. The latter suggests past connections between the Mesa Central and the Rio Grande drainage of Texas and northern Mexico. Miller and Smith (in press) noted that the present distributions of Menidia and Chirostoma may represent a part of the Plateau track. Further evidence of this track consists of the presence on the Mesa Central of several forms having evolutionary affinities with groups now restricted primarily to eastern North America: two catfishes of the subgenus Ictalurus (I. dugesi and an extinct species), two suckers (Moxostoma mascotae and an extinct congener), and an extinct black bass (Micropterus relictus). The oldest fossils indicative of this track on the Mesa Central apparently date to the Plio-Pleistocene and include an indeterminate species of Chirostoma from the Chapala area (Smith 1980).

The present distributions of M. peninsulae and M. beryllina extend southward in coastal areas of the Gulf of Mexico and associated streams to latitudes occupied by the Mesa Central. However, development of the Sierra Madre Oriental in early Eocene apparently precludes the possibility that ancestral atherinids invaded the Mesa Central by a southern route. Using, to our knowledge, the slowest molecular clock thus far calibrated for electrophoretic data (Sarich 1977; Nei's D of $1.0 = 30$ million years), the computed mean of 10.8 million years (SD = 2.0 my) and the upper bound of 20.1 million years of divergence time between species of Menidia and Chirostoma fall far short of the Eocene. As emphasized by Avise and Aquadro (1982), the accuracy of such estimates of absolute divergence times is highly questionable. Published calibrations range from 0.7 million to 30 million years per unit of Nei's genetic distance, a 43-fold difference. We chose the slowest clock to indicate that even the longest available estimate of divergence time sheds doubt on the possibility that the ancestral Menidia-like form(s) invaded the Mesa Central via Gulf Coast rivers south of the Rio Grande. Also unlikely is the suggestion (Barbour 1973a) that in early Tertiary times and by way of a marine transgression (Balsas Portal), the ancestral form(s) invaded the Mesa

Central from the Pacific coast. The genetic data seem more compatible with the hypothesis that the distribution of Menidia-Chirostoma is part of a relatively recent (Plio-Pleistocene?) biogeographic track connecting the Lerma-Santiago river system with the Rio Grande. Interestingly, the original Menidia-like occupants of this track may have been cladistically closer to the coastal form, M. peninsulae than to the inland form, M. beryllina, which now occupies inland reaches of the Rio Grande.

Our data provide little insight into the integrity of Barbour's (1973b) arge species group, which, adding Poblana, encompasses 10-11 species, of which we examined nine. Though the branching pattern is not well resolved, four species, P. ferdebueni, P. alchichica, C. riojai and C. melanoccus show synapomorphic evidence of relatively close relationship, and C. arge shows weak evidence of association with this group. Some osteological features also indicate a close relationship between the first three of those species (Barbour pers. comm.). Two morphologically similar and genetically almost identical (I = .999) species of the arge group, C. labarcae and C. aculeatum, even in the absence of synapomorphies, are obviously each other's closest extant relatives among the species we examined. Although some contact apparently occurs in Lake Chapala at the mouth of the Rio Lerma (Barbour 1973b), these two forms are effectively allopatric and species-level status is not well established. However, genetic and morphological differentiation (max. SL = 85 mm, C. labarcae, 130-140 mm, C. aculeatum; differences in length and shape of head and jaws) seem no less than that between certain pairs of apparently reproductively isolated forms in Lake Chapala, such as C. lucius and C. sphyraena.

Barbour (1973b) proposed that 10 Chirostoma species form a monophyletic assemblage, the jordani group. These include nine of the species we examined: C. jordani, C. humboldtianum, C. chapalae, C. consocium, C. lucius, C. sphyraena, C. promelas, C. grandocule, and C. estor. Excepting C. jordani, a small (80 mm max. SL), wide-ranging species quite Menidia-like in appearance, the members of this group are moderate (85-100 mm) to large-sized (125-300 mm) species primarily occurring in lacustrine environments. Although our data provide no synapomorphies for this group, the unusually high interspecific genetic identities among the eight species exclusive of *C. jordani* (I = .89-1.00) indicate recent descent from a common ancestor. Genetic identities between the most divergent of these, *C. estor,* and the other seven species (.89-.94; mean = .92) was higher than that between the two populations of *C. arge* (.87), and in some cases was as high as that between two geographically isolated populations of *C. jordani.* Osteological evidence for monophyly of these eight species is presented by Barbour (1973b) and Barbour and Chernoff (1984). Barbour (1973b) suggested that several of these lake-dwellers are descendents of a *C. humboldtianum*-like ancestor. The high levels of genetic identity are certainly compatible with this hypothesis. For convenience, we therefore refer to the *jordani* group, exclusive of *C. jordani* itself, as the "*humboldtianum*" group.

The levels of genetic identity between the Lake Chapala members of the *humboldtianum* group indicate a virtual lack of differentiation in enzyme-coding loci (e.g., in 7 of the paired combinations, I ► .99). We cannot exclude the possibility that these forms are morphs of a single species. However, unlike the two examples of Mexican fishes —goodeids of the genus *Ilyodon* (Turner, Grosse 1980) and Cuatro Cienegas cichlids (Sage, Selander 1975; Kornfield et al. 1982)— in which differentiated morphs comprise a single species, the Lake Chapala forms show weak evidence of intermorph genetic discontinuities. Given the lack of evidence to the contrary, we assume that the genetic/morphologic discontinuities reflect reproductive isolation. Presumably, this is a striking example of the increasing evidence that speciation and considerable morphological evolution can occur with little or no change in structural genes.

A consistent pattern emerging from biochemical-genetic studies of fish species flocks is that such assemblages typically include clusters of species having inordinately high levels of interspecific genetic similarity. Examples, in addition to the *humboldtianum* group, include African cichlids from Lake Malawi (Kornfield 1978) and Lake Victoria (Sage et al. 1984), pupfishes from Lake Chichancanab, Mexico (Humphries 1984), and cyprinids from Lake Lanao, Phillipines (Kornfield, Carpenter 1984). The genetic similarities and levels of morphological divergence among members of such groups suggest repeated speciation events and striking morphological evolution in relatively recent times. Though no precise time frame can be given, genetic analyses have provided a new modality of observation which is providing data consistent with the view derived from zoogeography, that these species flocks or major components of them are products of relatively recent evolution.

The members of the *C. humboldtianum* group exclusive of *C. estor* may be an extreme case of recent evolution. Genetic identity between these species (.960-1.00) is as high or higher than that obtained in comparisons of Rio Lerma populations of *C. jordani* with two isolated conspecific populations, JORD and JORP (.960-.972), which may have been isolated from the Rio Lerma only since the late Pleistocene (Barbour 1973a). However, consideration of other aspects of genetic structure confounds this interpretation. Reduced variability in these seven species of the *C. humboldtianum* group seems best explained by the hypothesis that low variability occurred originally in a common ancestor and was passed on to the descendent species. Reduction of variability as a result of stochastic processes in different lineages leading to the seven species seems inconsistent with the unusually high levels of interspecific genetic identity; some random fixation of alternate alleles would be expected. Natural selection seems unlikely as a causative agent, as it would have had to cause reduction across a large number of loci in this group of species, but not in other lake dwellers such as *C. estor.* (If reduced variability is passed on from a common ancestor, then we cannot employ Templeton's [1980] models relating heterozygosities and genetic distances to mechanisms of speciation.) If this group evolved from a genetically depauperate ancestor, or if reduced variability occurred at the time of speciation, then both the reduced heterozygosity and the virtual lack of apomorphic alleles (Figure 6) can be explained by the hypothesis that the group is relatively young (Nei et al. 1975; Soule 1976). Time of divergence from such an ancestor has not been sufficient for accrual of the genetic variability seen in other species of *Chirostoma*. Under a model of neutral genetic variability, the return of heterozygosity to "normal" levels is extremely slow, while the gain of new alleles is faster (Nei et al. 1975). Thus, the paucity of apomorphies in this group potentially indicates youthful age.

However, the possibility exists that low numbers of unique alleles have been maintained by recurrent bottlenecking. Newly arisen alleles will be rare and, in the absence of strong selection, easily lost during severe bottlenecks. Enzyme-coding loci presumably are not active in genetic transiliences associated with speciation (Templeton 1979), and such mechanisms would not cause new alleles to spread rapidly through a population. Thus, the group of seven genetically depauperate species could be old, and yet, because of repeated bottlenecking and maintenance of near genetic monomorphism, show low numbers of unique alleles and high levels of genetic identity.

Nonetheless, the *humboldtianum* group still seems relatively youthful compared with other *Chirostoma*. One member of the group, *C. estor*, has high genetic variability and presumably has undergone no severe bottleneck. Yet, as previously noted, genetic identities between it and the seven genetically depauperate species were near the level seen in conspecific *Chirostoma* comparisons.

Based on extant forms, the rate of evolution among *Menidia*-like forms of the Mesa Central appears to have been much faster than that in other areas. *Menidia* occurs in coastal areas and associated rivers from Nova Scotia to the Yucatan Peninsula. Over this vast range there are only 6-7 recognized species, and these show little morphological differentiation. Two primary factors may account for the heightened evolutionary pace on the Mesa Central. 1) The intensely active volcanic/tectonic history of the area provides ample evidence that an unusually large number of drainages and lakes have evolved and then become fragmented and/or integrated with other drainages (Barbour 1973a; Smith 1980; Smith et al. 1975). As Barbour emphasized, opportunities for allopatric speciation and secondary contact have been extensive. 2) The ichthyofauna of the Mesa Central is

depauperate, both in numbers of species and in diversity of higher taxa. For example, Lake Chapala is 30 times larger than Lake Waccamaw, North Carolina (1080 vs 36 km^2), but has only 38 percent as many native species —14 vs 36 (Barbour, Brown 1974). Thus, regardless of mode of speciation, the likelihood is greater that newly arisen species will find niche space and become successfully established.

On the Mesa Central and elsewhere, lakes seem especially prone to support clusters of species with unusually high levels of genetic similarity. Examples, in addition to African cichlids, Lake Chichancanab pupfishes, Lake Lanao cyprinids, and the *humboldtianum* group of *Chirostoma*, include reproductively isolated pairs of brown trout in Sweden (I = .975; Ryman et al. 1979), and whitefish in the Allegash Basin, Maine (I = .988-.996; Kirkpatrick, Selander 1979). Most lake faunas occur in strongly isolated lake basins, and, because of the ephemeral nature of lakes, have geologically short life expectancies. For example, a distinct form of *Chirostoma grandocule* apparently was driven to extinction when a temporary drying of Lake Cuitzeo occurred in 1941 (Barbour 1974). Opportunities for interbasin colonizations by lake-adapted forms are limited by time and space. Consequently, local speciation may play a greater role in the development of lake faunas than, for example, in that of stream faunas where, through time, colonization by diverse lineages may be more important. Lakes, then, would have a higher incidence of genetically similar forms.

At the same time, lake faunas are also more susceptible to bottlenecking due to short- and/or long-term dewatering (see McCune et al., 1984, for a striking example of supposed drying of lakes on a 21,000 year cycle). This would retard the evolution of electrophoretically detectable apomorphies. Perhaps this, taken with the depauperate ichthyofauna and opportunities for rapid diversification, helps explain the relative lack of synapomorphies in our data for the *Menidia-Chirostoma* assemblage. *Chirostoma* may well be monophyletic, but if the original *Menidia*-like ancestor diversified into several lineages prior to establishment of new alleles at relatively high frequencies, attempts to demonstrate monophyly will be frustrated by a lack of corroborating synapomorphies. This would also explain the general lack of well corroborated branching patterns at other levels in our cladistic analysis. McCune et al. (1984) argue convincingly that so-called "unresolved polytomies" may be especially common in cladistic analyses of species flocks. They argue, as did Barbour and Miller (1978), that some polytomies may be true representations of phyletic pattern.

SUMMARY

The Mesa Central, the southernmost area of the great Mexican Plateau, supports a species flock of atherinid fishes which includes 18 species of *Chirostoma* and two species of the nominal genus *Poblana*. Eighteen of the species are endemic to the Mesa Central where, compared with atherinids elsewhere, they show high levels of local sympatry and extreme examples of morphological evolution.

Phenetic and cladistic analyses of electrophoretic data are consistent with the following suggestions of other workers: 1) *Chirostoma* and *Poblana* share with *Menidia* an ancestor not shared with other atherinid genera; 2) *Chirostoma* and *Poblana* should be subsumed under the name *Menidia*; and 3) *Menidia* and the Mesa Central atherinids comprise part of the "Plateau" biogeographic track which indicates a Plio-Pleistocene connection with northern faunas such as that of the Rio Grande drainage of Texas and Mexico.

The occurrence of synapomorphic electromorphs indicates that the atherinids on the Mesa Central comprise, with *Menidia peninsulae*, a monophyletic group that excludes all other species of *Menidia*. The data provided very little additional information on cladistic relationships within the *Menidia-Chirostoma-Poblana* assemblage. This may be reflective of rapid diversification when the *Menidia*-like ancestor(s) invaded the ichthyologically depauperate, geologically active Mesa Central. Speciation may have occurred faster than unique electromorphs could become established in ancestral species.

One group of eight species, the "*humboldtianum* group," has such high interspecific genetic identities (Nei's I = 0.89-1.0) that,

even without corroborating synapomorphies, it apparently represents a monophyletic group. Seven extremely similar (I = 0.96-1.0) members of this group had low populational heterozygosity, low genic polymorphism, and low numbers of autapomorphic electromorphs. The genetic structure suggests that these seven forms have descended from a genetically depauperate ancestor. Genetic discontinuities among sympatric forms in Lake Chapala suggest that these highly similar forms are biological species and not intraspecific morphs. High genetic identities (.89-.94) between these seven species and *C. estor*, a genetically variable member of the group, suggest a relatively recent evolutionary history for the *humboldtianum* group.

Lakes seem especially prone to support flocks of genetically similar forms. Because lake basins are relatively short-lived and geographically isolated, opportunities for interbasin colonization by lake-adapted forms are limited. Local speciation may play a greater role in the development of lake faunas than in that of stream faunas where colonization by diverse lineages may be more important.

ACKNOWLEDGMENTS

We thank M E Douglas for computer expertise, O T Lind for introducing us to the Mesa Central, C D Barbour for many forms of aid and encouragement, B Chernoff and R R Miller for reviewing the manuscript, D Buth and D Morizot for advice with electrophoresis, S Contreras-Balderas and D Hillis for helpful comments, D L Swofford for help with BIOSYS-1, D Mosier and B J Turner for specimens, and T S Echelle, and L M Echelle for tolerance. Supported by the National Science Foundation (DEB-7912227, DEB-8023838) and a Dean's Incentive Grant, Oklahoma State Univ. We also thank the Departamento de Pesca, Mexico, for permission to collect specimens.

REFERENCES

Alvarez J 1970 Pesces Mexicano (claves). Inst nac invest biol Pesqueras Mexico, D F, Ser invest Pesquera No 1:1-166

Avise J C, C F Aquadro 1982 A comparative summary of genetic distances in the vertebrates. *Evol Biol* 15:151-185

Avise J C, J C Patton, C F Aquadro 1980 Evolutionary genetics of birds I. Relationships among North American thrushes and allies. *Auk* 97:135-147

Barbour C D 1973a A biogeographical history of *Chirostoma* (Pisces: Atherinidae): A species flock from the Mexican Plateau. *Copeia* 1973:533-556

_____ 1973b The systematics and evolution of the genus *Chirostoma* Swainson (Pisces, Atherinidae). *Tulane Stud Zool Bot* 18:97-141

_____ 1974 Redescription and taxonomic status of *Chirostoma compressum*, a Mexican atherinid fish. *Copeia* 1974:277-279

Barbour C D, J H Brown 1974 Fish species diversity in lakes. *Am Nat* 108:473-489

Barbour C D, B Chernoff 1984 Comparative morphology and morphometrics of the pescados blancos (genus *Chirostoma*) from Lake Chapala, Mexico. 111-128 A A Echelle, I Kornfield eds. *Evolution of Fish Species Flocks*. Univ Maine Press at Orono

Barbour C D, R R Miller 1978 A revision of the Mexican cyprinid fish genus *Algansea*. *Misc Publ Mus Zool Univ Michigan* 155:1-72

Buth D G, B M Burr, J M Schenck 1980 Electrophoretic evidence for relationships and differentiation among members of the percid subgenus *Microperca*. *Biochem Syst Ecol* 8:297-304

Chernoff B 1983 Revision of American atherinid fishes, subfamily Menidiinae, and systematics of the nominal subgenus *Atherinella*. PhD thesis, Univ Michigan, Ann Arbor

Croizat L, G Nelson, D E Rosen 1974 Centers of origin and related concepts. *Syst Zool* 23:265-287

De Buen F 1945 Investigaciones sobre ictiologia Mexicana. I. Atherinidae de aguas continentales de Mexico. *An Inst Biol Mex* 16:475-532

Duggins C F 1980 Systematics and zoogeography of *Lucania parva*, *Floridichthys* and *Menidia* (Osteichthyes: Atheriniformes) in Florida, the Gulf of Mexico and Yucatan. PhD thesis, Florida State Univ, Tallahassee

Echelle, A A, A F Echelle, C D Crozier 1983 Evolution of an all-female fish, *Menidia clarkhubbsi* (Atherinidae). *Evolution* 37:772-784

Farris J S 1981 Distance data in phylogenetic analysis. V A Funk, D R Brooks eds. *Advances in Cladistics*. New York Botanical Garden. Bronx

Greenwood P H 1984 What *is* a species flock? 13-20 A A Echelle, I Kornfield eds. *Evolution of Fish Species Flocks*. Univ Maine Press at Orono

Hennig W 1966 *Phylogenetic Systematics* Univ Illinois Press

Humphries J M 1984 Genetics of speciation in pupfish from Laguna Chichancanab, Mexico. 129-140 A A Echelle, I Kornfield eds. *Evolution of Fish Species Flocks*. Univ Maine Press at Orono

Johnson M S 1975 Biochemical systematics of the atherinid genus *Menidia*. *Copeia* 1975:662-691

Jordan D S, B W Evermann 1896-1900 *The Fishes of North and Middle America*. *Bull U S Nat Mus* 47:1-3136

Jordan D S, C L Hubbs 1919 A monographic review of the family Atherinidae or silversides. Stanford Univ Publ

Kirkpatrick M, R K Selander 1979 Genetics of speciation in lake whitefishes in the Allegash Basin. *Evolution* 33:478-485

Kornfield I L 1978 Evidence for rapid speciation in African cichlid fishes. *Experientia* 34:335-336

Kornfield I, K Carpenter 1984 Cyprinids of Lake Lanao, Phillipines: Taxonomic validity, evolutionary rates and speciation scenarios. 69-84 A A Echelle, I Kornfield eds. *Evolution of Fish Species Flocks*. Univ Maine Press at Orono

Kornfield I, D C Smith, P S Gagnon 1982 The cichlid fish of Cuatro Cienegas, Mexico: direct evidence of conspecificity among distinct trophic morphs. *Evolution* 36:658-664

McCune A R, K S Thomson, P E Olsen 1984 Semionotid fishes of the Mesozoic Great Lakes of North America. 27-46 A A Echelle, I Kornfield eds. *Evolution of Fish Species Flocks*. Univ Maine Press at Orono

Meek S E 1904 The freshwater fishes of Mexico north of the Isthmus of Tehauantepec. *Field Col Mus Pub 93 (Zool)* 5:1-252

Mickevich M E, C Mitter 1981 Treating polymorphic characters in systematics: a phylogenetic treatment of electrophoretic data. 45-58 V A Funk, D R Brooks eds. *Advances in Cladistics*. New York Botanical Garden. Bronx

Miller R R 1981 Coevolution of deserts and pupfishes (genus *Cyprinodon*) in the American southwest. 39-94 R J Naiman, D L Soltz eds. *Fishes in North American Deserts*. Wiley. New York

Miller R R, B Chernoff 1979 What is *Menidia*? Abstracts, Ann Mtg American Soc of Ichthyologists and Herpetologists, Orono Maine

Miller R R, M L Smith (in press) Origin and geography of the fish fauna of central Mexico. C H Hocutt, E O Wiley eds. *Zoogeography of North American Fishes*. Wiley-Interscience. New York

Nei M 1978 Estimation of average heterozygosity and genetic distance from a small number of individuals. *Genetics* 89:583-590

Nei M, T Maruyama, R Chakraborty 1975 The bottleneck effect and genetic variability in populations. *Evolution* 29:1-10

Nevo E 1978 Genetic variation in natural populations: patterns and theory. *Theoret Pop Biol* 13:121-177

Regan C T 1906-08 Pisces. *Biologia Centrali-Americana* 8:1-203

Ribbink A J 1984 Is the species flock concept tenable? 21-26 A A Echelle, I Kornfield eds. *Evolution of Fish Species Flocks*. Univ Maine Press at Orono

Ryman N, F W Allendorf, G Stahl 1979 Reproductive isolation with little genetic divergence in sympatric populations of brown trout (*Salmo trutta*). *Genetics* 92:247-262

Sage R D, R K Selander 1975 Trophic radiation through polymorphism in cichlid fishes. *Proc Nat Acad Sci* 72:4669-4673

Sage R D, P V Loiselle, P Basasibwaki, A C Wilson 1984 Molecular versus morphological change among cichlid fishes (Pisces: Cichlidae) of Lake Victoria. 185-202 A A Echelle, I Kornfield eds.

Evolution of Fish Species Flocks. Univ Maine Press at Orono

Sarich V M 1977 Rates, sample sizes and the neutrality hypothesis for electrophoresis in evolutionary studies. *Nature*(London) 265:24-28

Schultz L P 1948 A revision of six subfamilies of atherine fishes with descriptions of new genera and species. *Proc U S Nat Mus* 98:1-48

Selander R, M Smith, S Yang, W Johnson, J Gentry 1971 Biochemical polymorphism and systematics in the genus *Peromyscus*. Variation in the old-field mouse (*Peromyscus polionotus*). *Studies in Genetics VI* Univ Texas Publ 7103:49-90

Siciliano M J, C R Shaw 1976 Separation and visualization of enzymes on gels. *Chromatographic and Electrophoretic Techniques*. 2:185-209

Smith M L 1980 The evolutionary and ecological history of the fish fauna of the Rio Lerma basin, Mexico. PhD thesis, Univ Michigan, Ann Arbor

Smith M L, T Cavender, R R Miller 1975 The climatic and biogeographic significance of a fossil fish fauna from the late Pliocene-Early PLeistocene of the Laka Chapala basin (Jalisco, Mexico). *Univ Michigan Mus Paleont Pap Paleont* 12:29-38

Sokal R R, F J Rohlf 1969 *Biometry*. Freeman. San Francisco

Soule M 1976 Allozyme variation: its determinants in space and time. F Ayala ed. *Molecular Evolution*. Sinauer. Sunderland Mass

Swofford D L 1981 On the utility of the distance Wagner procedure. V A Funk, D R Brooks eds. *Advances in Cladistics*. New York Botanical Garden. Bronx

Swofford D L, R B Selander 1981 BIOSYS-1: a FORTRAN program for comprehensive analysis of electrophoretic data in population genetics and systematics. *J Hered* 72:281-282

Templeton A R 1979 The unit of selection in *Drosophila mercatorum*. II. Genetic revolutions and the origin of coadapted genomes in parthenogenetic strains. *Genetics* 92:1265-1282

———— 1980 Modes of speciation and inferences based on genetic distances. *Evolution* 34:719-720

Turner B J, D J Grosse 1980 Trophic differentiation in *Ilyodon*, a genus of stream dwelling goodeid fishes: speciation versus ecological polymorphism. *Evolution* 34:259-270

Appendix 1. Collection Localities in Mexico and as otherwise noted.

Sites for *Chirostoma* and *Poblana*:

1. Presa Pena del Aguila 20 km N Durango, Durango
2. Lake Chapala near San Juan Cosala, Jalisco
3. Lake Chapala near San Pablo, Jalisco
4. Lake Chapala 5 km E Chapala, Jalisco
5. Lake Chapala near Puerto Corona, 10 km W Tuxcueca, Jalisco
6. Lake Chapala near Tuxcueca, Jalisco
7. Rio Verde at San Nicolas de las Flores 32 km NW San Juan de los Lagos, Jalisco
8. Tributary, Rio Verde near Cuarenta 24 km NE Lagos de Moreno, Jalisco
9. Outlet from San Juanico Dam on Rio Cotija 8 km SE Cotija, Michoacan
10. Outlet from Laguna de Zacapu, Zacapu, Michoacan
11. Lake Patzcuaro at Ihuatzio, Michoacan
12. N shore, Lake Zirahuen, Michoacan
13. Presa Ignacio Allende 11 km SW San Miguel de Allende, Guanajuato
14. W shore Lake Yuriria, Guanajuato
15. Presa Ignacio Ramirez 19 km N, 4 km W Toluca, Mexico
16. Lake Chignahuapan, Chignahuapan, Puebla
17. Potholes 2 km E El Carmen, Tlaxcala
18. Lake Alchichica, Zalayeta, Puebla

Sites for other genera:

19. *Labidesthes sicculus*; Carter Lake, Madill, Marshall Co., Oklahoma, USA
20. *Menidia beryllina*; brackish pool, Live Oak Point, Rockport, Aransas Co., Texas, USA
21. *M. peninsulae*; Aransas Bay, Rockport, Aransas Co., Texas, USA
22. *Membras martinica*; SE end, Falcon Reservoir, Zapata Co., Texas USA
23. *Melaniris crystallina*; Rio Grande de Santiago, Highway 15 bridge, 50 km NNW Tepic, Nayarit
24. *M. alvarezi*; Rio Jaltepec, Highway 185 bridge, Veracruz-Oaxaca
25. *Melaniris* sp.?; Rio Hueyapan, Hueyapan, Veracruz
26. *Xenatherina schultzi*; trib. Rio Jaltepec, Highway 185, 2.7 km N bridge over Rio Jaltepec, Veracruz
27. *Xenatherina* sp.?; trib. Rio Mamantel, Highway 186 ca. 4.8 km SW Ejido Pital, 32 km SW Escarcega, Campeche
28. *Archomenidia sallei*; two samples, localities as described for *X. schultzi* and *X.* sp.
29. *Atherinomorus stipes*; canal to Barnes Sound, NW end Key Largo, Marion Co., Florida USA

APPENDIX 2: ALLELIC FREQUENCIES: *

		Chirostoma																				Poblana		Menidia	
Locus	allele	JORD	JORC	JORY	JORS	JORP	LUC	PRO	SPH	CONR	CONC	GRA	HUM	CHA	EST	LAB	ACU	ATT	ARGV	ARGC	MEL	RIO	FER	ALC	PEN	BER
Sample n:		*10*	*10*	*15*	*10*	*28*	*31*	*8*	*30*	*28*	*30*	*10*	*10*	*25*	*10*	*13*	*13*	*12*	*10*	*10*	*10*	*23*	*20*	*25*	*10*	*10*
Ak-2a	a	100	100	100	100	100	100	100	100	100	100	100	100	100	100	100	100		100	100	100	100	100	100		100
Ck-1	b	100	100	100		100	100	100	100	100		100	100	100	30			100	100	100	90		61	94	5	5
	c				95						93				60	73	77				10	45	34	6	75	5
	d				5						7				10	27	23					17	5		20	35
	e																									55
Ck-2	b	100	100	100	95	100	100	100	100	100	100	100	100	100	100	100	100	100	100	100	100	100	100	100	100	100
	d				5																					
Es-1	a	100	100	100	95	100	100	100	98	89	77	95	100	74	100	100	15	100	15	83	100	100	23	100	94	
	b				5				2	11	23	5		26			85		75	55			78		6	
	d																		10							
Gadph-1	b	100	100	100	100	100	100	100	100	100	100	100	100	100	100	100	100	100	100	100	100	100	100	100	100	100
Gadph-2	a	100	100	100	100	100	100	100	100	100	95	100	100	100	100	100	100	100	100	100	100	100	100	100	100	100
	b										5															
Aat-1	a	100	100	100	100	100	100	100	98	95	95	100	100	98	100	100	100	100	100	100	100	100	3	98	95	
	b								2	5	5			2									98	2	5	
Aat-2	d	100	100	100	100	100	100	100	100	100	100	100	15	88	100	96	100	82	20	100	40	14	100	100	95	20
	e												85	13		4		18	20		50	84				40
	f																				10	2				15
Aat-3	c	100	100	100	5	84	56	90	100	57					100					5			97			
Gp-5	a				95	16	44	10		43	100									95			3			
Gpi-1	a	100	100	100	100	100	100	100	100	100	100	100	100	100	100			100	20					100	25	
	b																		20						20	10
	c																		60						40	90
	d																								15	
	e																									

*

Locus	Allele	JORD	JORC	JORY	JORS	JORP	LUC	PRO	SPH	CONR	CONC	GRA	HUM	CHA	EST	LAB	ACU	ATT	ARGV	ARGC	MEL	RIO	FER	ALC	PEN	BER
Gpi-2	b	55	50	67	45	100													40	100	5	20		2	10	
	d		15																60		95	78	100	98	90	100
	e			33	55																	2				
	f	45	35																							
	g																									
Idh-1	a						100	100	100	98	100	100	100	98	100	96	100	100	65	75	95		100			
	b	100	100	100	100					2				2		4			35	25	5					
	c																									
	f																									
Idh-2	h	100	100	100	100	100	100	100	100	100	100	100	100	100	100	100	100	100	100	100	100	100	100	100	100	100
	e																							98		
	f																							2		
Ipo	b	100	100	100	100	100	100	100	100	100	100	100	94	100	83	100	100	100	100	100	13	98	100		100	100
	d																				88	2				
	e					98							6		17											
	f					2																				
Ldh-1	b	100	100	100	100	100	100	100	100	100	100	100	90	100	100	100	100	100	100	5	100	100	100	100	100	100
	c											2	10							95						
	d																									
	f																									
Ldh-2	h		5	5	5																					
	a	100	95	95	95	100	98	94	100	98	100	100	100	100	95	100	100	100	100	100	100	100	100	100	100	100
	b														5											
	c						2	6		2																
	e																									
Ldh-3	b	100	100	100	100	100	100	100	100	100	100	100	100	100	100	100	100	100	100	100	100	100	100	100	100	100
Mdh-1	b	100	100	100	100	100	100	100	100	100	100	95	100	100	100	100	100	100	100	100	95	100	100	100	100	100
	d											5	95								5					
	e												5													
Mdh-2	f	100	100	100	100	100	100	100	100	100	100	100	100	100	100	100	100	100	100	100	100	100	100	100	100	100
6Pgd	b	100	100	100	100	100	100	100	100	100	100	100	100	100	100	100	96	100	100	100	95	100	100	100	100	100
	c																					2				5
	d																4					7	3			
	e																12					67				
	f																	100				4				
	g														35		84		100				98		85	85
	h														65											15
	j																									
	k																									
Pgm-2	b	10	5	5	17									2	25				5		20	2	15		5	5
	e	75	75	75	65	70								98	75	100	100	100	95	100	98		85	98	85	95
	f					30																		2		
	h	15	10	10	10																					
	j		5	5	5																					
	k				3																					
αGdph-1	a	100				100	100	100	17																	
	b								83	100	100		100													

* Frequencies multiplied by 100. Boldface numbers signify apomorphies.

APPENDIX 3. Nei's (1978) Genetic Identities (x100) Among Populations
of *Chirostoma* (1-21), *Poblana* (22, 23) and *Menidia* (24, 25)

(Population abbreviations are as given in *Methods and Materials*. Values of 100 result from rounding error. No two samples were genetically identical.)

Population	1	2	3	4	5	6	7	8	9	10	11	12	13	14	15	16	17	18	19	20	21	22	23	24
1 JORD	-																							
2 JORC	97	-																						
3 JORY	96	100	-																					
4 JORS	97	100	100	-																				
5 JORP	93	97	97	96	-																			
6 LUC	69	73	74	74	69	-																		
7 PRO	70	74	74	74	69	100	-																	
8 SPH	69	73	73	73	69	100	100	-																
9 CONR	69	73	73	73	69	97	99	96	-															
10 CONC	71	75	75	75	70	99	100	99	99	-														
11 GRA	69	73	73	73	68	97	99	96	100	98	-													
12 HUM	69	73	73	73	68	98	99	97	100	99	100	-												
13 CHA	70	74	74	74	70	100	99	100	96	100	96	97	-											
14 EST	68	66	66	67	62	90	92	90	93	91	94	93	89	-										
15 LAB	66	70	71	70	66	81	80	81	77	82	77	78	83	71	-									
16 ACU	66	70	70	70	66	82	81	82	78	83	78	79	84	72	100	-								
17 ATT	56	52	52	53	49	66	68	66	70	68	70	69	67	79	64	65	-							
18 ARGV	72	71	71	71	75	68	68	68	68	68	67	68	68	67	80	81	60	-						
19 ARGC	74	78	79	78	76	72	72	71	71	72	71	72	72	65	85	85	58	87	-					
20 MEL	60	56	55	56	52	75	76	75	77	76	77	77	75	81	73	75	73	73	69	-				
21 RIO	53	56	57	56	52	71	72	71	71	72	72	71	72	69	65	66	65	55	65	73	-			
22 FER	58	54	54	55	51	71	71	71	71	71	71	71	71	69	64	64	63	63	59	67	74	-		
23 ALC	55	51	51	51	48	65	67	65	69	67	69	69	66	69	63	64	69	61	59	68	70	90	-	
24 PEN	78	74	74	75	71	62	62	62	61	63	62	62	62	66	77	77	73	76	73	68	53	58	61	-
25 BER	76	80	80	80	76	72	73	72	72	73	72	73	72	66	84	85	59	77	82	70	59	58	57	85

COMPARATIVE MORPHOLOGY AND MORPHOMETRICS OF THE PESCADOS BLANCOS (GENUS *CHIROSTOMA*) FROM LAKE CHAPALA, MEXICO

CLYDE D BARBOUR AND BARRY CHERNOFF

INTRODUCTION

In this paper we consider the importance of differential growth and development to the differentiation of a small species flock on the Central Plateau of Mexico. Before addressing this issue, we first examine the current criteria for recognizing species flocks and then present an alternative perspective.

The observation that closely related species may be restricted to the same geographic area has been as interesting to evolutionary biologists as its explanation has been elusive. Early attempts to explain the origin of what have come to be regarded as classic examples of species flocks, the cichlids of lakes Tanganyika and Victoria and the gammarid crustaceans of Lake Baikal, included the suggestion that speciation had occurred sympatrically (see Greenwood 1984). This idea was forcefully countered by the argument (e.g. Mayr 1942, 1963), later somewhat softened (Mayr 1970, 1978), that the genetic differentiation of populations that leads to speciation can take place only under conditions of geographic isolation. Genetic evidence suggesting non-allopatric modes of speciation was sparse at first, and later was explained away as unnecessary, incomplete, improbable, or involving too many unrealistic assumptions. Past considerations of these evolutionary processes and their relationship to the origin of species flocks have suffered not from an emphasis on process (Greenwood 1984), but from the dogmatic fashion in which almost all presumed events of speciation, mostly intralacustrine, were explained by evoking some sort of physical fragmentation of the ecosystem (White 1978:12). Other workers sought to explain the multiple differentiation of lineages by focusing on intrinsic changes within populations and individuals

and the relationship of these to the environment (Gosline 1968; Rosenzweig 1978; White 1978; Templeton 1981); however, until recently, these explanations have not received the attention they deserve. We agree with White (1978) that it is no longer possible to exclude differentiation within a continuous and unimpeded environment as an important, if not major source of the forms that comprise species flocks.

Mayr (1963) stabilized the meaning of the term *species flock* by writing: "Special situations occur in various parts of the world where a considerable number of closely related species, so-called species swarms [= species flocks], are confined to a narrowly circumscribed area, no close relatives occurring elsewhere." This statement, though useful to some in the past, appears inadequate to us because it does not convey a sense of the biological significance of species flocks. Nor does it clearly state those intrinsic qualities which separate species flocks from species assemblages. Mayr clearly states that species flocks are special, and it is this aspect of his definition that we emphasize.

We agree with Greenwood (1984) that the central criterion for defining a species flock must be monophyly *sensu* Hennig (1966). The suspicion raised historically, that these faunas might result from a nonclassical mode of speciation cannot be properly addressed except within the context of monophyletic groups. Although the last clause of Mayr's statement can be interpreted to mean that the entire flock shares a most recent common ancestor with respect to all other relatives, his subsequent (Mayr 1974) discussion of the terms "relationship" and "monophyly" argues for the necessity of accepting paraphyletic taxa. Because the species flock concept deals with the diversification of lineages, it would be biologically misleading to exclude species from consideration simply because they had diverged dramatically from those forms with which they share a most recent common ancestor.

We eschew Mayr's additional criterion of a "considerable number" of species because it interferes unnecessarily with our understanding of the unique character of species flocks. The species flock concept should concern processes of diversification under particular environmental conditions. We emphasize diversification within or between species because the two are qualitatively the same (Charlesworth et al. 1982). Our concept focuses on evolutionary processes and is independent of taxonomists' assessments of species boundaries and hence species numbers. For this reason, we hold, as do Liem and Kaufman (1984), that the smallest species flock is a single, polymorphic species, such as the Cuatro Cienegas cichlid, *Cichlasoma minckleyi* (Sage, Selander 1975; Kornfield et al. 1982; Kornfield, Taylor 1983). We consider the term *species flock* still useful because most of the examples do involve a number of different species. We have substituted the more general terms *diversification* and *differentiation* for *speciation* in order to accommodate the inclusion of polymorphisms except where, for historical or other reasons, it does not seem appropriate.

We also believe that basin size must be subordinated to considerations of the historical aspects of biogeography and geomorphology. Without the occurrence of sister species, living or extinct, in other basins or behind some other physical barrier, the question of modes of differentiation must remain open. The common lack of such evidence suggests to us that allopatric (geographic) modes of differentiation are inadequate to account for the existence and extent of many presumed flocks. To us it appears fruitful to inquire whether these species might have originated by means of microallopatric (Dominey 1984) or nonallopatric mechanisms (Endler 1977; White 1978; Templeton 1981).

In fishes, almost all known or suspected species flocks are lacustrine. Only a few lotic flocks are known (*Caecomastacembelus*; Travers, pers comm) or suspected (*Ilyodon*, Turner, Grosse 1980; *Saccodon* Roberts 1974). The apparent predominance of physically unimpeded, *in situ* differentiation in lakes rather than rivers and streams may be a function of complex interactions between environmental stability, habitat heterogeneity, and species diversity.

Although portions of the fluvial environment may be heterogeneous in features such as thermal regime, depth profile, bottom type, and current patterns, and seemingly favorable for within-lineage differentiation, they may be too unpredictable within some time frame for differentiation to take place. Lakes, on the other hand, are heterogeneous, and the larger systems may be stable or predictable enough for diversification to occur by means of niche partitioning or competition (Smith, Todd 1984). We also note that lakes are initially colonized largely from rivers and streams, and not all species present in these habitats are capable of surviving in a lacustrine environment. This filtering effect should result in unoccupied peaks on the lake's "Wrightian surface", thus creating an opportunity for competitive speciation (Rosenzweig 1978).

Definitions and descriptions of phenomena may influence the approaches taken to explain them. Here, we have tried to eliminate criteria which we consider tangential to the importance of species flocks that "... lies in the empirical evidence which can be derived from them and applied in a broader evolutionary context" (Greenwood 1984). We consider species flocks to be monophyletic groups for which an origin in a physically unimpeded environment cannot be ruled out. When species flocks are viewed in this light, the patterns which will be uncovered by future work should reflect not only the efficacy of the various mechanisms of *in situ* differentiation, but also clarify the biological and physiographic conditions under which differentiation is likely to shift on the continuum from allopatric to sympatric processes and vice versa. We agree with Rosenzweig (1979) that the relationship between these two processes is complex and the boundary is fuzzy. As he points out (p 284), it is not possible neatly to delineate sympatric and allopatric differentiation; past semantic arguments on what constitutes geographic isolation have only served to cloud more basic issues.

Species flocks as we define them may be lacustrine or fluviatile. However, our definition excludes lineages whose members inhabit and are endemic to portions of larger drainages that have been fragmented by past climatic and tectonic events, such as *Chirostoma* on the Mesa Central of Mexico, and *Orestias* on the Andean Altiplano. Although these groups may include species flocks, the remaining forms show varying degrees of distributional overlap, suggesting that differentiation has followed the classic allopatric model. We do not want to diminish the importance of the contributions that studies of such groups can make to our knowledge of evolutionary patterns and processes. Nor do we deny that differentiation by means of extrinsic geographic isolation is probably the most common source of phenotypic diversity. Our purpose in reorienting the term *species flock* is to highlight those special cases where intrinsic mechanisms play the major role in the diversification of populations.

What we find most compelling about species flocks, large or small, is that they serve as flags attracting the attention of biologists to unique opportunities to study the relationship between organism and environment. Rates and mechanisms of differentiation and speciation and the historical aspects of ecological interactions are but a few topics about which species flocks may prove informative. Within this framework we view studies of species flocks along with studies of geographic variation to be fundamental to the understanding of biological diversification.

PESCADOS BLANCOS OF LAKE CHAPALA

Lake Chapala, the largest body of fresh water in Mexico, lies on the Mesa Central between two east-west trending mountain ranges about 50 km south of Guadalajara, Jalisco (Barbour 1973b Fig 6). The geological history of this lake and its former drainage connections have been reviewed (Barbour 1973b; Barbour, Miller 1978; Smith 1980). Among the fishes in Lake Chapala, atherinids (silversides) of the genus *Chirostoma* are prominent with eight species present (Barbour 1973a).

Although many atherinids are typically small (◀100 mm SL), the *jordani* group of

Chirostoma contains six species that have evolved into relatively large piscivorous forms. Four of these large *Chirostoma* are presently sympatric in Lake Chapala. The largest three, *C. sphyraena, C. lucius,* and *C. promelas* attain standard lengths (SL) greater than 150 mm, and *C. lucius* is known to exceed 300 mm SL (Barbour 1973b). Because of their large size and quality of flesh, they form the basis of a substantial commercial fishery in the lake. In Mexico, the term *pescados blancos* refers to the large, commercially important silversides of the Mesa Central. It is used here specifically to refer to *C. sphyraena, C. lucius,* and *C. promelas,* which we recognize as a species flock.

Reports that pescados blancos were formerly present in Lake Santa Magdalena (Tamayo, West 1964; Smith, Miller 1980) may be references to *C. consocium, C. humboldtianum,* or *C. estor,* large species whose distributions could have included this part of western Jalisco. We assume that *C. sphyraena, C. lucius,* and *C. promelas* are restricted to Lake Chapala until specimens or fossils demonstrate otherwise.

Considerable confusion has existed regarding the characters used to define these species. For this reason, estimates of the number of large atherinids present in the lake have always been regarded with some suspicion. Each of the three recognized forms tends to have a qualitatively distinctive appearance, such as the blackened snout of *C. promelas* or the protruding lower jaw of *C. lucius,* but the quantitative characters are more variable than recognized by earlier workers. Barbour (1973a) noted that no single meristic or morphometric character will unequivocally separate the species. Not surprisingly, the group as recognized at present contains a number of names placed in synonymy.

Here we provide preliminary evidence of the monophyly of *C. sphyraena, C. lucius,* and *C. promelas* and their relationship to other species of *Chirostoma.* This phylogenetic analysis will provide the basis for interpreting the morphometrics in an evolutionary context. We address three questions on morphological variation: 1) What are the morphological boundaries between these three pescados blancos? 2) To what extent do differences in allometric growth within species contribute

Table 1. Morphometric data on *Chirostoma lucius, C. sphyraena* and *C. promelas*. (Values in thousandths of the standard length.)

Variables	C. lucius (n=78)					C. sphyraena (n=44)					C. promelas (n=24)			
	holotype[2]	small (n=23) range	mean	large (n=54) range	mean	holotype[3]	small (n=21) range	mean	large (n=22) range	mean	holotype[4]	small (n=16) range	mean	large (n=7) range
Standard length mm	172.8	37.8-99.7	72.9	100.4-302.6	155.5	202.9	29.3-93.0	71.5	106.2-206.3	143.7	154.0	50.0-94.7	68.2	114.0-164.8
Predorsal 1 length	563	504-552	529	516-570	538	556	507-539	523	520-592	542	582	519-579	543	539-588
Predorsal 2 length	694	621-676	653	639-717	663	705	649-679	663	661-720	681	698	641-683	665	660-700
Prepelvic length	448	400-448	422	415-490	453	482	408-445	427	420-503	454	484	412-463	432	447-467
Preanal length	616	561-619	582	577-658	608	632	553-611	582	591-679	617	644	569-641	598	611-649
Body depth	241	172-206	188	178-247	208	186	158-186	172	197-222	209	209	166-209	190	184-239
Caudal peduncle length	196	196-233	213	179-221	201	173	188-225	206	177-207	197	178	184-227	209	173-206
least depth	96	83-101	90	80-95	88	84	74-90	82	75-92	84	86	84-102	93	85-95
Head length	308	280-316	295	294-350	312	320	282-302	293	288-323	305	308	282-311	296	291-313
Orbit diameter	53	65-98	78	49-76	64	50	55-99	66	44-61	53	51	58-77	70	49-60
Postorbital head length	144	116-136	128	129-166	143	142	118-136	128	129-143	137	127	118-143	126	125-147
Interorbital width	73	64-87	74	61-76	67	67	68-83	74	66-80	72	68	68-79	73	63-72
Snout length	112	82-102	95	99-124	109	127	95-118	106	110-129	120	123	93-117	104	109-124
Mandible length	139	112-134	124	124-179	147	155	116-136	128	132-159	142	128	109-130	117	112-134
Second dorsal fin length of base	120	106-133	123	108-165	126	110	94-123	111	100-128	113	110	101-122	114	104-134
height	141	160-187	171	138-186	162	134	144-171	153	121-148	138	147	151-180	166	135-151
Anal fin length of base	206	209-237	225	201-251	219	204	202-235	219	197-222	209	200	184-230	209	188-224
height	146	172-207	189	147-194	171	130	156-181	166	128-161	149	158	160-196	182	144-166
Pectoral fin length	174	181-220	196	168-228	198	188	162-186	173	153-194	176	184	174-200	189	162-188
Pelvic fin length	116	120-145	130	114-140	125	95	108-124	115	95-122	109	127	117-136	125	107-126

[1] small ▼ 100 mm SL, large ▲ 100 mm SL; [2] BM 92.2.8.75. Hybrid with *C. sphyraena* excluded from species summary statistics;
[3] BM 92.2.8.77   [4] CAS 6156.

to shape differences between species? 3) Is there any correspondence between differentiation in these fishes and other lacustrine species flocks?

METHODS

As listed in the Appendix, 161 specimens representing *C. sphyraena, C. lucius,* and *C. promelas* were examined for seven meristic, and, excepting one analysis, 19 morphometric variables. Standard length was added to our suite of measurements for the multivariate analysis of allometry (Table 3). Counts and measurements were made according to Barbour (1973a) although our terminology differs in that "predorsal 1 length" and "predorsal 2 length" refer to the distances between the tip of the snout and the origins of the first and second dorsal fins, respectively. All measurements were made with dial calipers accurate to 0.1 mm.

Phylogenetic analyses were carried out according to the method of Hennig (1966). Only character-states hypothesized to be evolutionarily derived were used to infer genealogical relationships. The polarity of each transition series was determined from morphological comparisons with outgroups. Because Chernoff (1983) has shown that *Menidia, Labidesthes,* and *Poblana* are most closely related to *Chirostoma,* these genera were used as primary outgroups. In addition, other representative menidiines, such as *Melaniris, Xenomelaniris,* and *Archomenidia,* more distantly related to *Chirostoma* were also included as outgroups.

Principal components analysis (PCA) was used to identify suites of characters contributing to morphological trends present in the data as shown in Table 2. Because PCA does not require specimens to be identified *a priori,* we can examine scatterplots of principal component scores for specimens to evaluate the initial premise that all individuals belong to a single population. Meristic and mensural data were analyzed separately. Principal components were calculated from the covariance matrix of log10 transformed morphometric variables and from the correlation matrix of meristic variables. Individuals lacking complete data were not included in particular analyses. The method of Humphries et al. (1981) was used to better estimate size-free

Table 2. Principal components analysis of *Chirostoma lucius* (n = 78), *C. sphyraena* (n = 44) and their presumed hybrids. Variable loadings for components, PC I and PC II, and the sheared second component, H. ($\alpha = .013$; $\beta_1 = .999$; $\beta_2 = .013$)

Variables	PC I	PC II	H
Percent variance	98.8	.6	
Predorsal 1 length	.23	.15	.14
Predorsal 2 length	.23	.20	.20
Prepelvic length	.24	.16	.16
Preanal length	.24	.17	.17
Body depth	.26	-.18	-.19
Caudal peduncle			
length	.21	.02	.02
least depth	.22	-.04	-.04
Head length	.24	.09	.09
Orbit diameter	.17	-.46	-.46
Postorbital head length	.25	.09	.08
Interorbital width	.21	.28	.27
Snout length	.26	.43	.42
Mandible length	.27	.12	.12
Second dorsal fin			
length of base	.23	-.18	-.19
height	.21	-.30	-.31
Anal fin			
length of base	.21	.00	.00
height	.19	-.30	-.31
Pectoral-fin length	.23	-.22	-.22
Pelvic-fin length	.22	-.27	-.27

shape differences between species.

The rosette diagram of character-vectors (Figures 5, 7, 8) are Cartesian coordinate plots of variable coefficients for the components illustrated that have been aligned with the axes of the principal components. These vectors display magnitude and direction of character change and, hence, the relative correlations among the variables and of each variable with the axes.

The multivariate allometry of each species was estimated from the largest within-group principal component. Variable coefficients of each component were normalized to a mean square of one. That is, if all coefficients were identical they would have a value of one. Traditional allometric studies compare the scaling of various body dimensions to some

Table 3. Principal components analysis of *Chirostoma lucius* (n = 78), *C. sphyraena* (n = 44), and *C. promelas* (n = 23). Between species analysis includes presumed hybrids, with loadings for PC I, PC II, PC III. Within species analyses with normalized loadings (allometric coefficients) for first principal component. Subscripts indicate significance level of difference among allometric coefficients (decimals omitted).

Variables	Between Species			Within Species		
	PC I	PC II	PC III	*lucius*	*sphyraena*	*promelas*
SL				.99	1.00	1.01
Predorsal 1 length	.23	.16	.09	1.01	1.04	1.05_{001}
Predorsal 2 length	.23	.20	.05	1.01	1.04	1.04_{01}
Prepelvic length	.24	.17	.02	1.07	1.09	1.09
Preanal length	.24	.18	.08	1.04	1.08	1.09_{001}
Body depth	.26	-.17	.27	1.13	1.05	1.13_{001}
Caudal peduncle						
length	.21	.00	-.06	.89	.93	.88
least depth	.22	-.03	.27	.95	1.01	$.97_{01}$
Head length	.24	.09	-.01	1.05	1.05	1.05
Orbit diameter	.17	-.47	-.18	.71	.69	.65
Postorbital head length	.25	.08	-.02	1.12	1.08	1.12_{05}
Interorbital width	.21	.26	-.04	.89	.95	$.95_{01}$
Snout length	.26	.44	.15	1.14	1.17	1.19_{05}
Mandible length	.27	.09	-.41	1.19	1.15	1.10_{001}
Second dorsal fin						
length of base	.23	-.18	-.62	1.01	1.00	1.06
height	.21	-.30	.17	.91	.86	$.84_{001}$
Anal fin						
length of base	.22	-.01	-.28	.95	.93	.97
height	.20	-.30	.17	.85	.84	.80
Pectoral fin length	.23	-.22	.17	1.00	1.02	$.94_{05}$
Pelvic fin length	.22	-.26	.21	.95	.90	$.92_{05}$

measure of body length. Here, however, we are estimating the scaling of each measured length, including standard length, with an estimate of general body size, that is, the first within-group principal component (Jolicoeur 1963; Chernoff, Miller 1982). In this study, the relationship between standard length and general size is essentially isometric. Therefore, character allometries based on standard length as a proxy for size would approximate the coefficients of general size (Table 3) up to the error inherent in measures of standard length. However, there is no *a priori* reason to expect standard length to scale isometrically with general size in every instance (Chernoff, Miller 1982). Because each allometry coefficient is the slope of the regression of a particular variable upon the general size factor, significant differences among the species for the allometry coefficients were discovered from analyses of covariances. Calculations were performed at the computer centers of the University of Michigan and Wright State University using the Michigan Interactive Data Analysis System

(MIDAS) and the Statistical Analysis System (SAS).

RESULTS AND DISCUSSION

Evidence of Monophyly. We tentatively accept Barbour's (1973a) division of *Chirostoma* into the *arge* and *jordani* species groups. Although his analysis was based on external characters and not subjected to a rigorous phylogenetic analysis, preliminary osteological studies suggest that the dichotomy is real. In all members of the *jordani* group, the ventral margin of the anterior ceratohyal grades smoothly from the narrow anterior arm to the posterior spatulate portion. We consider this condition to be derived and evidence of the monophyly of the *jordani* species group. In all our outgroups, including here the *arge* group, the transition from the narrow anterior arm occurs abruptly, in a step-like fashion. No derived character states have been found which unite the species of the *arge* group, although some species clusters are becoming apparent. For example, *C. arge*, *C. labarcae* and *C. aculeatum* all have a rectangular coronoid process (derived).

Barbour (1973a) concluded that *C. sphyraena*, *C. lucius*, and *C. promelas* are the most derived members of the *jordani* group, and that they share an immediate common ancestor. Although his phylogeny left many unanswered questions regarding the cladistic relationships within the groups, we offer preliminary osteological evidence in support of this hypothesis, as follows:

1) Width of the shaft of the parasphenoid in ventral view (Figure 1). In *C. sphyraena* and *C. lucius* the anterior strut-like portion of the parasphenoid is much thickened posterior to the vomer. From anterior to posterior, the whole strut tapers smoothly without any, or with only a very slight constriction about 2/3 to 3/4 of the distance along its length. Because the parasphenoid of our largest skeletal specimen of *C. promelas* (55.3 mm SL) has the same relative conformation of a larger *C. lucius* (ca. 98 mm SL), we anticipate the adult parasphenoid of *C. promelas* to be at least as robust as that found in *C. sphyraena*, and *C. lucius*. We consider this character state to be derived and an indicator of the monophyly of the pescado blancos, because it is not found in any of the outgroups. A superficially similar condition is found in *C. aculeatum* of the *arge* group. Here, however, the shape of the shaft is different: the sides are parallel until the posterior margin of the vomer is reached. It is only at this point that it begins to narrow. Within the *jordani* group (Figure 1) the narrowest parasphenoid is found in *C. jordani* and *C. chapalae* (not figured). Of the other species not illustrated, that of *C. grandocule* is similar to *C. humboldtianum*, and *C. patzcuaro* and *C. consocium* have parasphenoids intermediate in width to those of *C. jordani* and *C. humboldtianum*.

2) Median curvature of the gnathic ramus. The arc described by the gnathic ramus as it curves medially to meet its opposite member at the symphysis is much reduced in *C. estor*, *C. sphyraena*, *C. lucius*, and *C. promelas*. We consider this condition to be derived, because in almost all other menidiines, the curvature is more acute, giving the anterior circumference of the lower jaw a more abruptly rounded aspect.

3) Shape of the lacrimal (Figure 2). An increase in the length of the lacrimal forms a morphocline paralleling the thickening of the shaft of the parasphenoid. The longest lacrimals, unique among the Menidiinae and

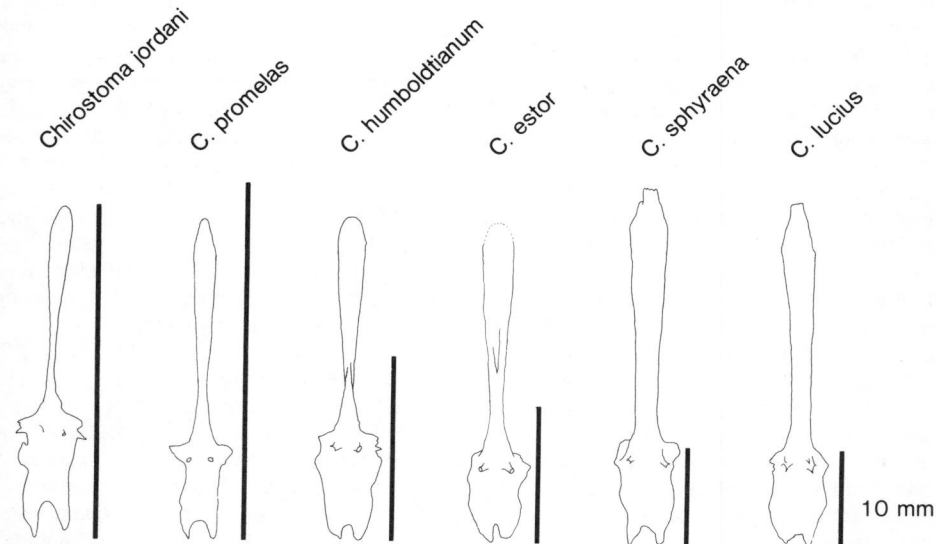

Figure 1. Ventral view of the parasphenoids of 6 species of *Chirostoma*. The anterior portion is directed upward. From left to right: TU 37761, 510, 513, 40863, UMMZ 209279, 192549.

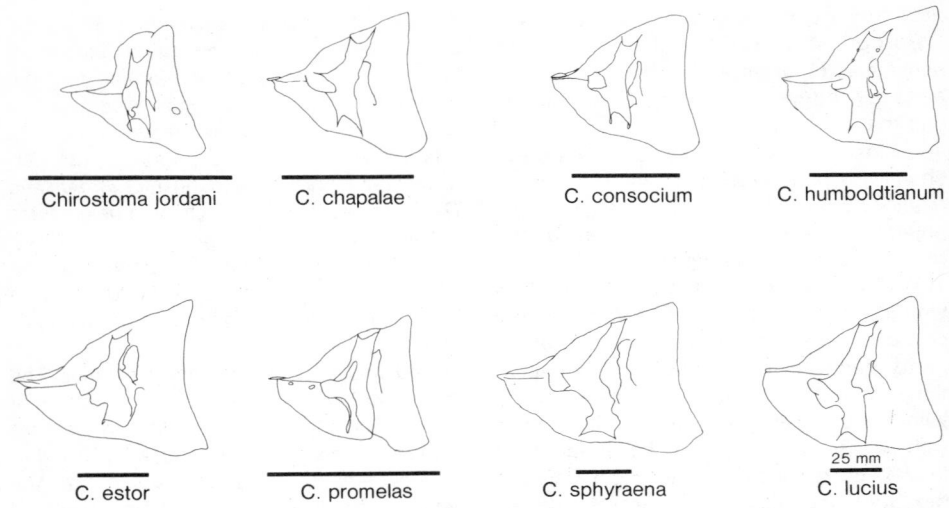

Chirostoma jordani   C. chapalae   C. consocium   C. humboldtianum

C. estor   C. promelas   C. sphyraena   C. lucius

25 mm

Figure 2. Lateral view of the lacrymals of 8 species of *Chirostoma*. Anterior to the left. From left to right, upper row: TU 37761, 40836, 40845, 513; lower row: TU 40863, 510, UMMZ 209279, 192549.

thus derived, are found in *C. estor, C. sphyraena, C. lucius,* and *C. promelas.* Lacrimals of intermediate size are present in *C. humboldtianum, C. grandocule, C. consocium, C. patzcuaro,* and *C. chapalae.* The smallest is found in *C. jordani.* We also note that an enlarged lacrimal is found in *C. aculeatum.*

A summary of our phylogenetic conclusions is presented in Figure 3. Although we noted that the lacrimal of *C. aculeatum* is elongated, we reject any hypothesis of relationship between this species and the pescados blancos. Such a proposal would require three character-state reversals involving the shapes of: the ventral margin of the anterior ceratohyal, the coronoid process (rounded in the pescados blancos, Figure 3), and the shaft of the parashenoid. Clearly, the relationship of *C. aculeatum* lies outside the *jordani* group.

We also reject a possible sister species relationship between *C. estor* and *C. sphyraena* based on the occurrence of very large lacrimals in both forms. The lacrimal morphocline also parallels one of inceasing body and snout length, suggesting that the length of the bone is partly size-related. It is not surprising that *C. estor,* which may reach 273 mm SL, also has a very long lacrimal.

Finally, we note that Barbour (1973a, figure 15) suggested that *C. estor,* the lineage leading to the pescados blancos, and the remaining larger members of the *jordani* group, were all independently derived from different populations of a widespread lacustrine species similar to *C. humboldtianum.* He also pointed out that *C. estor* and the pescados blancos share a derived character-state, small scales, and that alternatively, *C. estor* may be the primitive sister species of the remaining forms. Our study has corroborated this latter view. It is also of interest that *C. estor,* now restricted to Lakes Patzcuaro and Zirahuen, was collected from Lake Chapala by Meek and Lutz in 1901. The divergence of the lineage leading to *C. sphyraena, C. lucius,* and *C. promelas* could have occurred within any of the basins surrounding Lake Chapala, or perhaps, within the lake itself.

Analysis of *C. lucius* and *C. sphyraena*

The first two principal components account for 99.4 percent of the variance and describe morphometric trends consistent with some of Barbour's (1973a) statements concerning species differences. The first component (Table 2) accounts for almost 99 percent and is close to being an estimate of pooled within-group size. The second component presents a contrast between two suites of characters,

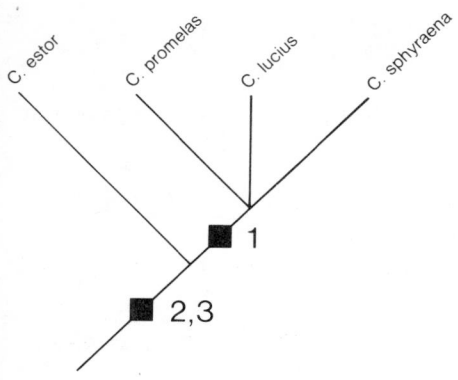

Figure 3. Cladistic relationships of 4 species of *Chirostoma*. Black squares indicate derived character states as follows: 1. shaft of parasphenoid in ventral view much thickened posterior to vomer, tapers more or less smoothly to base; 2. arc described by curvature of gnathic ramus much reduced; 3. lacrimal greatly elongated.

and is very close to being a shape difference factor. The scatter of scores in the first two principal components resulted in two dense, elongate clusters separated by a relatively sparse band of individuals. The clusters and band have an oblique orientation indicating that the first and second components were confounded by shape and size differences respectively.

When the effects of size were regressed from the second component and scores were scattered on the first component and shape-difference factor (H, Figure 4), the clusters and band reside horizontally on the plot. The loadings on the shape-difference factor are almost identical to those of the second principal component, indicating that the confounding effect of size was relatively minor. Note also that α and β_2 are close to zero in Table 2. Nonetheless, the useful discrimination is now contained entirely on H.

The large clusters in Figure 4 correspond to *C. sphyraena* and *C. lucius* as recognized by Barbour (1973a). The two species are quite distinct from each other morphometrically, despite their arbitrary separation from the band of intervening individuals. *C. lucius* has a larger orbit and larger second dorsal, anal, pectoral, and pelvic fins. *C. sphyraena*, on the other hand, has a relatively longer snout,

longer predorsal 2 length, and greater interorbital width.

We interpret the specimens between the species clusters as hybrids. This conclusion is based on the fact that we were unable to find any derived character-states which would distinguish them from *C. sphyraena* and *C. lucius* and unite them as a group. The hybrids are mosaics. No two have the same appearance. Their intermediate positions result from combinations of characters which differ in every specimen. The uniqueness of each individual was confirmed when additional PCA's were performed on the hybrids with each of the species separately. The hybrids did not cluster as a group. Each specimen maintained the same relative position with respect to each of the species boundaries, and the major axes emphasized the same variables. Other hybrids may also be present among our specimens which are not intermediate and inadvertently assigned by us to one or the other parental clusters (see Neff, Smith 1979). Specimens near the arbitrary species boundaries do not appear to be mosaics, and perhaps they are the result of introgression. If gene flow is occurring, then it is apparently insufficient to obliterate the two clusters even though hybridization between the species is not a recent phenomenon. For example, the type of *C. lucius* was collected in 1892. It is for this reason, that we regard *C. sphyraena* and *C. lucius* as distinct at the species level.

Figure 4 also illustrates the difference between the typological and biological species concepts (Mayr 1963). *C. lucius* and *C. sphyraena* were described six times in 1900, and the type specimens (circled numbers in Figure 4) span almost the entire range of variation for the two species and their hybrids.

In his brief discussion of the nominal forms, Barbour (1973a) correctly assessed the nature of the problem but was unable to recognize hybrids consistently. Thus, he synonymized *C. crystallina*, *C. lermae*, and *C. diazi* with *C. sphyraena* on the one hand, and *C. ocotlanae* with *C. lucius* on the other. Even though the holotype of *C. lucius* is almost exactly intermediate between *C. lucius* and *C. sphyraena*, we retain Barbour's choice of species-group name under the authority of Article 17, paragraph 2 of the International

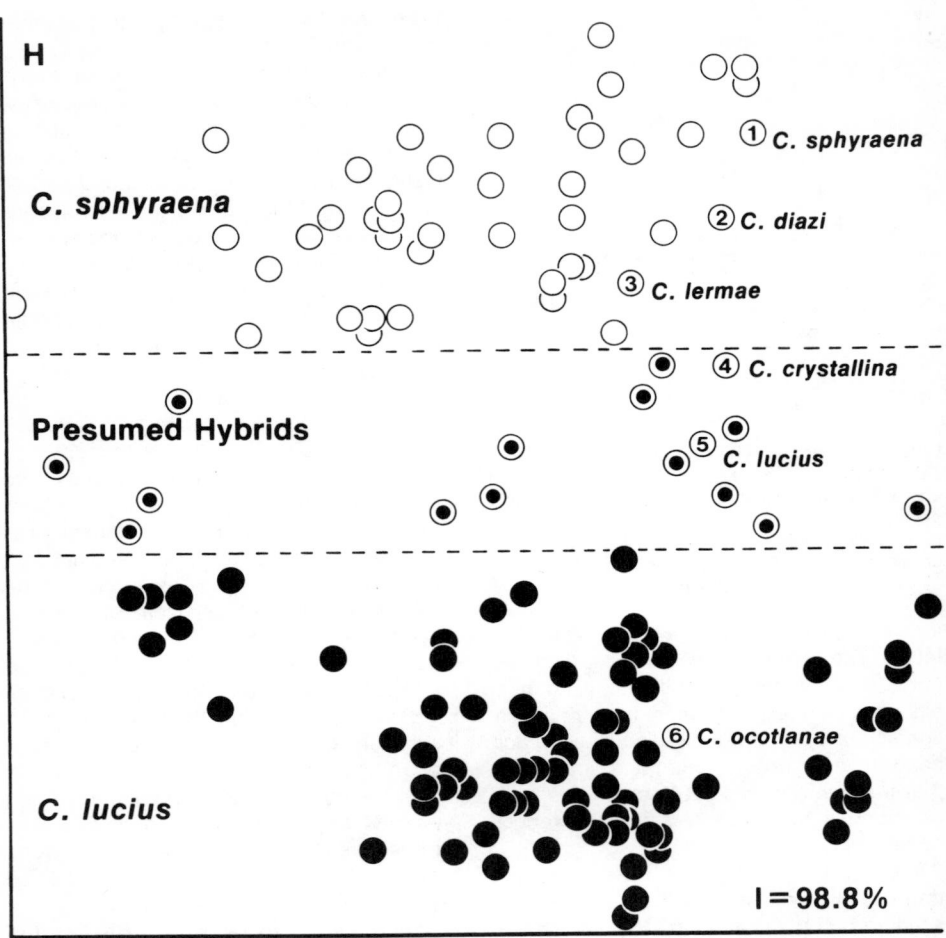

Figure 4. Scatter diagram of scores on size-independent shape factor (H) and first morphometric principal component (I) for *Chirostoma sphyraena* (n = 44) and *C. lucius* (n = 78). Specimens falling between the arbitrarily drawn dashed lines are presumed hybrids. Circled numbers represent holotypes of species placed in synonymy with either *C. lucius* or *C. sphyraena*. Each symbol represents at least one specimen.

Code of Zoological Nomenclature.

Principal components analysis of seven meristic characters (Figure 5) indicates much less divergence between the two species than was apparent in our morphometric analysis. Nor do the hybrids cluster discretely either within the zone of overlap or elsewhere in the space of the first two principal components. As noted by Barbour (1973a, figures 5, 7), the number of median lateral (MLS) and interdorsal scales (IDS) are best for distinguishing *C. lucius* and *C. sphyraena*, even though there

is considerable overlap. *C. lucius* also has a slightly greater number of gill rakers and second dorsal fin rays.

Of the holotypes for which we have complete meristic data, *C. lucius*, *C. ocotlanae* and *C. sphyraena* fall well within the ranges of their respective species. The type of *C. crystallina*, which is intermediate morphometrically, although closer to *C. sphyraena* (Figure 4), is meristically more distant from *C. sphyraena* than are specimens of *C. lucius*. This may be an example of extralimital variation of some hybrids (Neff, Smith 1979). The remaining hybrids show a range of variation on component I equal to that of both *C. lucius* and *C. sphyraena*. When the individuals plotted in Figure 5 are projected on the axis of component I, 8 of the hybrids have scores in the zone of overlap between the two species; 6 of the remaining 8 have scores within the range exhibited by *C. lucius*.

Analyses of *C. lucius*, *C. sphyraena* and *C. promelas*.

The best resolution of the morphometric differences separating the three species of pescados blancos, including the types and hybrids, is obtained when component III is plotted against II. Eight characters show relatively high loadings (▶ .2) on the second axis and separate the forms into three overlapping clusters (Figure 6, Table 3): interorbital width (IO), orbit diameter (ORB), snout length (SN), predorsal 2 length (PD2), pectoral fin length (P1L), pelvic fin length (P2L), second dorsal fin height (D2H) and anal fin height (AH). Both *C. promelas* and the hybrids fall

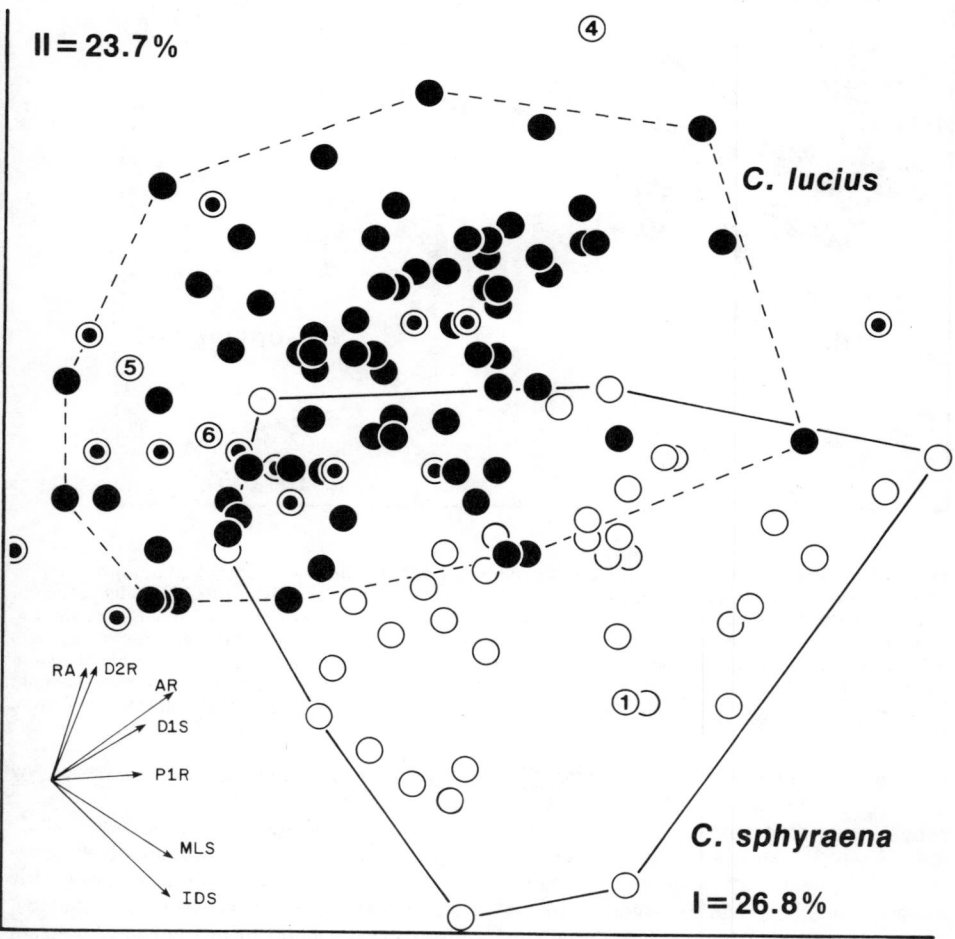

Figure 5. Scatter diagram of scores for *Chirostoma lucius* (n = 37) and *C. sphyraena* (n = 37) on first two meristic principal components (I, II). The circled dots represent presumed hybrids. The circled numbers represent holotypes placed in synonymy with either *C. lucius* or *C. sphyraena* (See Figure 4). Each symbol represents at least one specimen. Abbreviations: AR, anal fin rays; D1S, first dorsal fin spines; D2R, second dorsal fin rays; IDS, interdorsal scales; MLS, median lateral scales; P1R, pectoral fin rays; RA, gill rakers in the lateral row of the first gill arch.

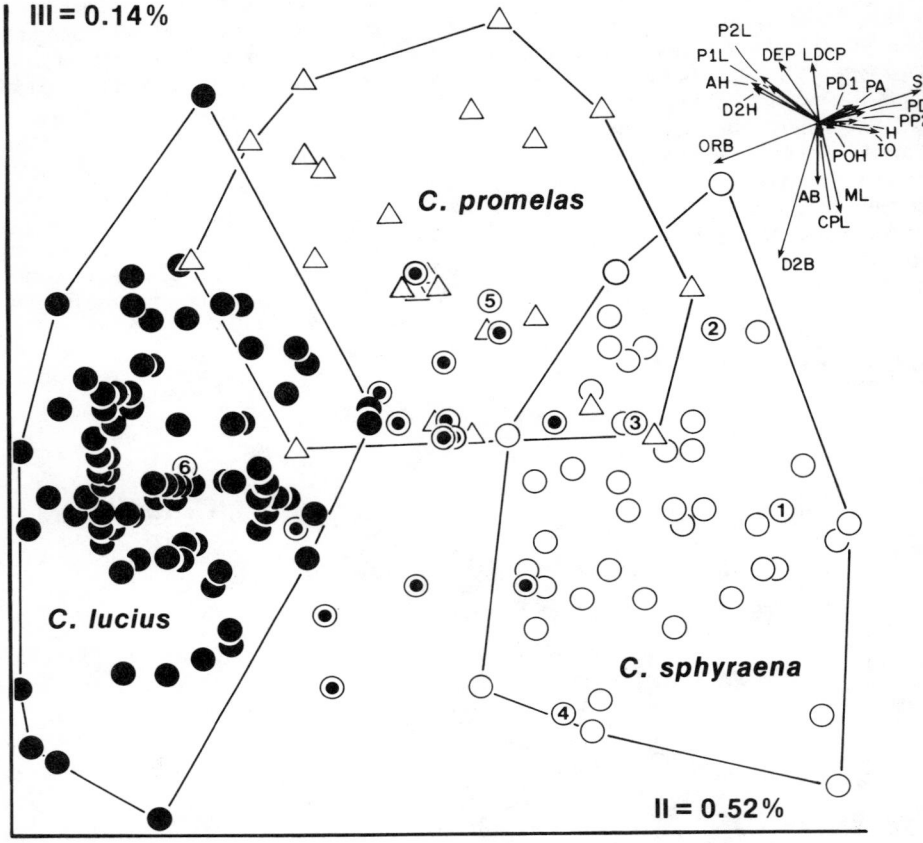

III = 0.14%

II = 0.52%

C. promelas

C. lucius

C. sphyraena

P2L
P1L — DEP LDCP
AH — PD1 PA SN
D2H — PD2
PP2
ORB — H
POH IO
AB — ML
CPL
D2B

Figure 6. Scatter diagram of scores on second and third morphometric principal components (II, III) for *Chirostoma lucius* (n = 78), *C sphyraena* (n = 44), and *C. promelas* (n = 23). Other symbols as in Figure 4. Each symbol represents at least one specimen. Abbreviations: AB, anal fin base; AH, anal fin height; CPL, caudal peduncle length; D2B, second dorsal fin base; DEP, body depth; D2H, second dorsal fin height; H, head length; IO, interorbital width; ML, mandible length; LDCP, caudal peduncle least depth; ORB, obit diameter; PA, preanal length; PD1, predorsal 1 length; PD2, predorsal 2 length; P1L, pectoral fin length; P2L, pelvic fin length; POH, postorbital head length; PP2, prepelvic fin length; SN, snout length.

between *C. lucius* and *C. sphyraena*. The pescados blancos are not distinguished entirely by the scatter of their scores on component III although *C. promelas* falls at one end of the range of variation. *C. promelas* and certain other individuals exhibit relatively short anal and second dorsal fin bases (AB, D2B), short mandible lengths (ML), large body and caudal peduncle depths (DEP, LDCP) and pelvic fin lengths (P2L). The extension of the mandible beyond the snout varies among the three species from very protrusive in some

C. lucius to being included by the upper jaw in *C. promelas*.

As in the analysis of *C. lucius* and *C. sphyraena*, meristic variables are not as effective as the morphometric ones for distinguishing all three pescados blancos. Here also the same seven meristic characters load heavily on the first two principal components. When the scatter plots on components I and II for the two analyses are compared (Figures 5, 7), an interesting pattern of variation emerges. The positions of *C. lucius*,

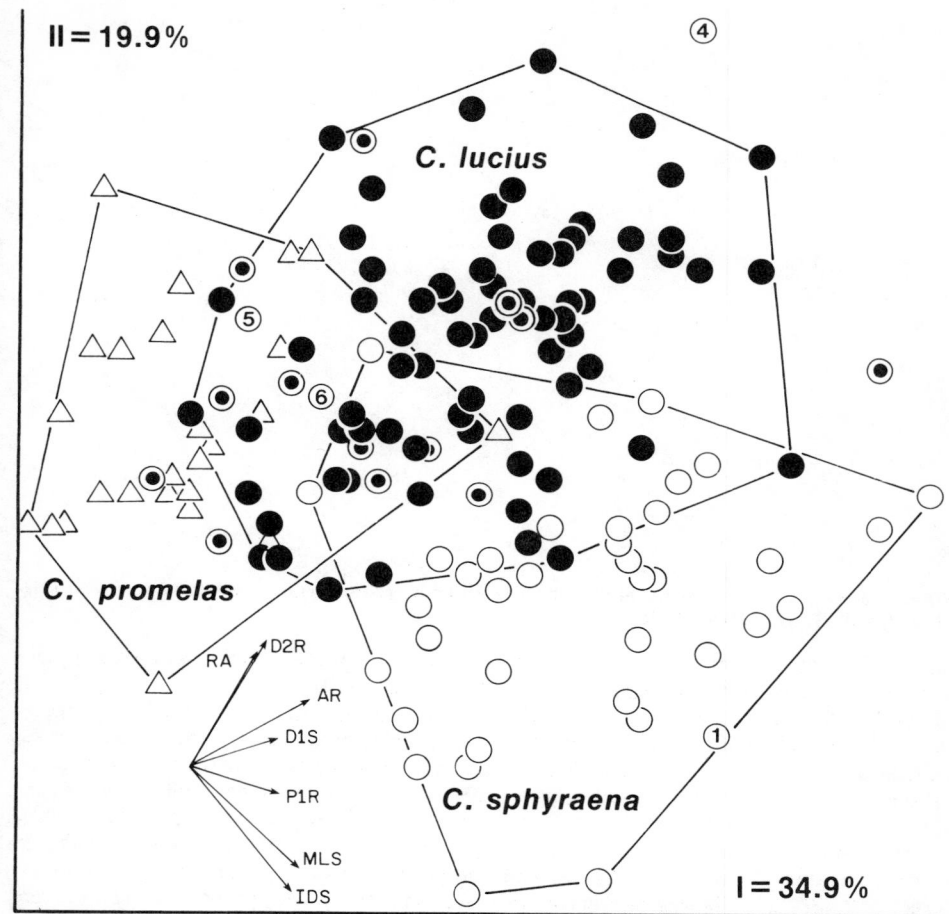

Figure 7. Scatter diagram of scores on first two meristic principal components (I, II) for *Chirostoma lucius* (n = 70), *C. sphyraena* (n = 37), and *C. promelas* (n = 25). Other symbols and abbreviations as in Figure 5. Each symbol represents at elast one specimen.

C. sphyraena and the types and hybrids remain unaltered. *C. promelas* differs from both species on component I because of its fewer anal fin rays (order of modal values for each character: *C. promelas* 19; *C. lucius* 20; *C. sphyraena*, 20 — multiple modes are possible), pectoral fin rays (15, 16, 16), second dorsal fin rays (11, 12, 12) and first dorsal fin spines (4, 5, 5). On the second component *C. promelas* tends to resemble *C. sphyraena* by having a lower number of gill rakers (24, 26, 25) and *C. lucius* by having fewer median lateral scales (57, 64, 72) and interdorsal scales (8, 11, 15 and 22). The separation is

not absolute. *C. promelas* overlaps both *C. lucius* and *C. sphyraena*, but more so with *C. lucius* and with the presumed hybrids.

Several osteological features also distinguish *C. lucius* from *C. sphyraena*. The neck of the premaxilla (Figure 8), that portion lying between the ascending process and the alveolar process, is narrow and elongate in *C. sphyraena*. This condition is rare in the outgroups and is considered derived. The relatively primitive state, a widened neck, is similar to that found in *C. lucius*. The upright coronoid process seen in *C. sphyraena* (Figure 8) is most commonly found among the

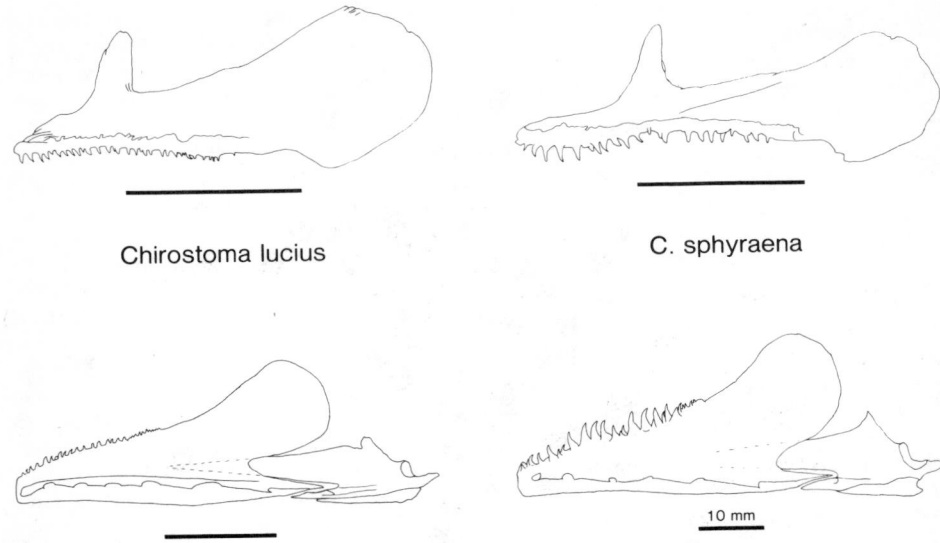

Chirostoma lucius   C. sphyraena

10 mm

Figure 8. Lateral views of premaxillae (upper) and mandibles (lower) of *Chirostoma lucius*, UMMZ 192549, and *C. sphyraena*, UMMZ 290279.

outgroups. In the derived condition exhibited by *C. lucius*, the coronoid process is produced posteriorly much more strongly. In *C. lucius*, the anteroventral quadrant of the lacrimal is broad and rounded (Figure 2). This condition is unique with respect to other silversides and is considered derived. The relatively primitive condition consists of a narrower and more pointed anterior edge (Figure 2; also see Parenti 1981:370). *C. promelas* is primitive for the latter two characters although it has a slightly elongated narrow neck of the premaxilla. This condition is not as extreme as that found in *C. sphyraena*. Uniquely derived states of *C. promelas* include a black muzzle, and in adults an upper jaw projecting well beyond the tip of the mandible. The distribution of these character states leads to an unresolved trichotomy for the pescados blancos (Figure 3). Further study may reveal characters useful for resolving this problem. On the other hand, the morphometric intermediacy of *C. promelas* raises the possibility of a hybrid origin for this species. In addition, there is no *a priori* reason to assume that speciation events cannot occur simultaneously, especially in tectonically active areas, giving rise to real polychotomies, or that they have occurred so close together in time as to

be unresolvable by current methodology (Barbour, Miller 1978).

Our analyses have indicated that morphometric variables are powerful discriminators among the pescados blancos. Examination of Table 1 also reveals that adult body proportions of one species may occur in juvenile specimens of another; e.g., the predorsal 2 length characteristic of large *C. lucius* also occurs in small *C. sphyraena* and *C. promelas*. To what degree, then, are the observed differences between species explainable in terms of differences in rates of relative growth? When growth rates are compared, widespread discordances are noted (Table 3) indicating that at least in part the divergence in shape among species is explainable by differences in allometry. For example, snout length and mandible length are positively allometric in each of the species, but their relative strengths vary significantly.

The predominant trend within *C. lucius* is to develop a relatively short upper jaw and an enlarged, protruding mandible. *C. lucius* has the largest allometric coefficient for mandible length (1.19) and the smallest coefficient for snout length (1.14). Conversely, *C. promelas* has a relatively long snout and shorter mandible, which is usually included in the upper jaw. It has the largest allometric coefficient for snout length (1.19) and the smallest for mandible length (1.10). Thus, some morphological differences among the species result from evolutionary modification of developmental rates. Such is probably the case for the large mandible of *C. lucius*, an expression of accelerated rate, which is clearly derived. The lengths of the mandibles of the other large members of the *jordani* group, *C. grandocule*, *C. humboldtianum* and the sister species to the pescados blancos, *C. estor*, are less than or equal to those of *C. sphyraena* and *C. promelas* (Barbour 1973a, figure 12). On the other hand, not all interspecific morphometric differences discovered by principal components analysis are explainable by differences in the allometric coefficients. These may be the result of changes in the timing of specific developmental events. For example, *C. lucius* tends to have a proportionately larger pectoral fin length (mean = 198, Table 1) than does *C. sphyraena* (mean = 176) but the allometries of the fins are not significantly different.

It is important to realize that a within-species principal components analysis may reveal differences in growth rates for characters which do not load heavily on among-species principal components analysis. If the regression lines for a character on size cross, then principal components analysis will not distinguish the different clouds of points despite the differences in the slopes. Therefore, even though in the three-species analysis (Table 3) predorsal 1, preanal, prepelvic and postorbital head lengths have extremely low coefficients, they should not be discounted as meaningful discriptors of biological divergence because allometries differ significantly (Table 3).

In conclusion we address our original questions. We regard *C. lucius* and *C. sphyraena* as separate species. Osteologically they remain distinct and morphologically they remain distinct despite apparent hybridization. Nonetheless, the possiblility that they might be morphs of a single species cannot be ruled out entirely. Echelle and Echelle (1984) were able to provide only weak evidence of genetic discontinuity between these pescados blancos. Other recent studies (Turner, Grosse 1980; Kornfield, Koehn 1975; Sage, Selander 1975; Kornfield et al. 1982) indicate that

differentiation or diversification might proceed in some instances by means of polymorphism rather than speciation; however, in those cases there is no evidence of genetic discontinuity between morphs.

We also regard *C. promelas* as a separate species because of its derived conditions (e.g., black snout, allometries). Its morphological boundary overlaps those of *C. sphyraena* and *C. lucius* in the planes of the meristic and morphometric principal component analyses. The intermediacy of *C. promelas* with respect to the other pescados blancos in some morphospaces suggests the possibility of hybrid origin. If so, then *C. promelas* has certainly transcended the mosaicism of hybrids —no doubt a function of its genetic coalescence.

Gould (1977) and Alberch et al. (1979) have discussed the importance of heterochrony to the evolution of form. Minor alterations in the rate and timing of developmental processes may have profound effects on the resultant morphologies. In the light of our phylogenetic hypothesis, we have interpreted the divergence in allometries among *C. lucius*, *C. sphyraena*, and *C. promelas* as they relate to the observable shape differences. Our data demonstrate that changes in relative rates of growth as well as timing of development result in differences in the morphology of adult organisms. Greenwood (1979, 1981) has previously attributed large scale morphological change within the lineages of African cichlids to small scale allometric and allochronic changes (also, see Strauss 1984). Greenwood (1979:218) stated:

Essentially these changes [trophic radiation of Lake Victoria cichlids] are reducible to differential growth in parts of the skull and other skeletal elements, their associated musculature, and in the dentition (both oral and pharyngeal). Seen in isolation, at any level of their differentiation, the changes would not be considered major ones. Yet cumulatively within a lineage, and comparatively between lineages, the effect has been to produce an ecologically, morphologically, and functionally diverse assemblage of species that is truely outstanding.

With this statement, Greenwood characterizes not only the evolution of the cichlids of Lake Victoria, but also the diversification of *Chirostoma*, including the pescados blancos of Lake Chapala, and probably the ciscoes of the American Great Lakes and the Cuatro Cienegas cichlid. Greenwood's assertions and our data lead us to suspect that alterations of the genetic parameters of growth have played a major role in the formation of species flocks.

SUMMARY

The current species flock concept is re-evaluated and placed in an alternate framework. We restrict the term to those monophyletic groups for which an origin in a physically unimpeded environment cannot be ruled out. Three species of large atherinid fishes of the genus *Chirostoma*, known as pescados blancos, which inhabit Lake Chapala, Mexico, are recognized as a small species flock. Evidence for monophyly is presented and the integrity of the three forms is confirmed, despite putative hybridization, by osteological comparisons and principal components analyses of meristic and morphometric data. Analysis of the multivariate allometric coefficients suggest that diversification of the pescados blancos is in large measure described by differences in rates of relative growth and timing of development. Such changes in differential growth and development have been implicated in the evolution of other species flocks as well.

ACKNOWLEDGMENTS

We thank J M Humphries, S G Poss, M L Smith for comments during our work, T Iwamoto for measurements on the type of *Chirostoma diazi*, P H Greenwood for sharing knowledge of species flocks, R K Johnson and W F Smith-Vaniz for reviewing our introduction, R J Jensen for aid in collecting specimens and in preliminary data analysis, and J Taylor for typing the manuscript. The following curators allowed one of us (CDB) to examine specimens of *Chirostoma*: L P Woods, R K Johnson, Field Museum of Natural History (FMNH); R D Suttkus, Museum of Natural History, Tulane University (TU); R R Miller, Museum of Zoology, University of Michigan (UMMZ); G S Myers, W Freihoffer, W N Eschmeyer, California Academy of Sciences (CASSU); A Solorzano, formerly of Laboratorio Biologico, Oficina de Pesca, Mexico (LB); P H Greenwood, British Museum of Natural History (BM). Research was aided by a Research Development Grant to CDB from the College of Science and Engineering, Wright State University and an Edwin C Hinsdale Scholarship, Museum of Zoology, University of Michigan to BC.

REFERENCES

Alberch P S, S J Gould, G F Oster, D B Wake 1979 Size and shape in ontogeny and phylogeny. *Paleobiology* 5:296-317

Barbour C D 1973a The systematics and evolution of the genus *Chirostoma* Swainson (Pisces: Atherinidae) *Tulane Stud Zool Bot* 18:97-114

———— 1973b A biogeographical history of *Chirostoma* (Pisces: Atherinidae): a species flock from the Mexican Plateau. *Copeia* 1973:533-556

Barbour C L, R R Miller 1978 A revision of the Mexican cyprinid fish genus *Algansea*. Misc Publ Mus Zool, Univ of Michigan No 155, 72p

Charlesworth B, R Lande, M Slatkin 1982 A neo-darwinian commentary on macroevolution. *Evolution* 36:474-498

Chernoff B 1983 Revision of American atherinid fishes, subfamily Menidiinae, and systematics of the nominal subgenus *Atherinella*. PhD thesis. Univ Michigan. Ann Arbor

Chernoff B, R R Miller 1982 Mexican freshwater silversides (Pisces: Atherinidae) of the genus *Archomenidia*, with the description of a new species. *Proc Biol Soc Wash* 95:428-439

Dominey W J 1984 Effects of sexual selection and life history on speciation: Species flocks in African cichlids and Hawaiian *Drosophila*. 231-250 A A Echelle, I Kornfield eds. *Evolution of Fish Species Flocks*. Univ Maine Press at Orono

Echelle, A A, A F Echelle 1983 Evolutionary genetics of a "species flock." atherinid fishes from the Mesa Central of Mexico. 231-250 A A Echelle, I Kornfield eds. *Evolution of Fish Species Flocks*. Univ Maine Press at Orono

Endler J A 1977 *Geographic Variation, Speciation, and Clines*. Princeton Univ Press. New Jersey

Gosline W A 1968 Considerations regarding the evolution of Hawaiian animals. *Pacific Sci* 22:267-273

Gould S J 1977 *Ontogeny and Phylogeny*. Belknap Press. Cambridge, Mass

Greenwood P H 1979 Macroevolution — myth or reality? *Biol J Linnean Soc* 12:293-304

———— 1981 Species flocks and explosive speciation. 819-834 *The Haplochromine Fishes of the East African Lakes: Collected Papers on their Taxonomy, Biology, and Evolution*. Cornell Univ Press. Ithaca New York

———— 1984 What *is* a species flock? 13-20 Echelle A A, I Kornfield eds. *Evolution of Fish Species Flocks*. Univ Maine Press at Orono

Hennig W 1966 *Phylogenetic Systematics*. Univ Illinois Press. Urbana

Humphries J M, F L Bookstein, B Chernoff, G R Smith, R L Elder, S G Poss 1981 Multivariate discrimination by shape in relation to size. *Syst Zool* 30:291-308

Jolicoeur P 1963 The multivariate generalization of the allometry equation. Biometrics 19:497-499

Kornfield I L, R K Koehn 1975 Genetic variation and speciation in New World cichlids. *Evolution* 29:427-437

Kornfield I, D C Smith, P S Gagnon, J N Taylor 1982 The cichlid fish of Cuatro Cienegas, Mexico: direct evidence of conspecificity among distinct trophic morphs. *Evolution* 36:658-664

Kornfield I, J Taylor 1983 A new species of polymorphic fish, *Cichlasoma minckleyi* from Cuatro Cienegas Mexico (Teleostei: Cichlidae). *Proc Biol Soc Wash* 96:253-269

Liem K F, L Kaufman 1984 Intraspecific macroevolution: functional biology of the polymorphic cichlid species *Cichlasoma minckleyi*. 203-216 A A Echelle, I Kornfield eds. *Evolution of Fish Species Flocks*. Univ Maine Press at Orono.

Mayr E 1942 *Systematics and the Origin of Species*. Columbia Univ Press. New York

_____ 1963 *Animal Species and Evolution*. Belknap Press. Cambridge Mass

_____ 1970 *Populations, Species, and Evolution*. Belknap Press. Cambridge Mass

_____ 1974 Cladistic analysis or cladistic classification? *Zool Syst Evol-forsch* 12:94-128

_____ 1978 Review of: *Modes of Speciation* by M J D White. *Syst Zool* 27:478-482

Neff N A, G R Smith 1979 Multivariate analysis of hybrid fishes. *Syst Zool* 28:176-196

Parenti L R 1981 A phylogenetic and biogeographic analysis of cyprinodontiform fishes (Teleostei, Atherinomorpha). *Bull Am Mus Nat Hist* 168:335-557

Roberts T R 1974 Dental polymorphism and systematics in *Saccodon*, a neotropical genus of freshwater fishes (Parodontidae, Characoidei). *J Zool London* 173:303-321

Rosenzweig M L 1978 Competitive speciation. *Biol J Linn Soc* 10:275-289

Sage R D, R K Selander 1975 Trophic radiation through polymorphism in cichlid fishes. *Proc Nat Acad Sci USA* 72:4669-4673

Smith G R, T N Todd 1984 Evolution of species flocks of fishes in north-temperate lakes. 47-68 A A Echelle, I Kornfield eds. *Evolution of Fish Species Flocks*. Univ Maine Press at Orono

Smith M L 1980 The evolutionary and ecological history of the fish fauna of the Rio Lerma basin, Mexico. PhD thesis. Univ Michigan. Ann Arbor

Smith M L, R R Miller 1980 *Allotoca maculata*, a new species of goodeid fish from western Mexico, with comments on *Allotoca dugesi*. *Copeia* 1980:408-417

Strauss R E 1984 Allometry and functional feeding morphology in haplochromine cichlids. 217-230 A A Echelle, I Kornfield eds. *Evolution of Fish Species Flocks*. Univ Maine Press at Orono.

Tamayo J L, R C West 1964 The hydrography of Middle America. 84-121 R Wauchope, R C West eds. *Handbook of Middle American Indians*. Vol 1. Univ Texas Press. Austin

Templeton A R 1981 Mechanisms of speciation — a population genetic approach. *Ann Rev Ecol Syst* 12:23-48

Turner B J, D J Grosse 1980 Trophic differentiation in *Ilyodon*, a genus of stream-dwelling goodeid fishes: speciation versus ecological polymorphism. *Evolution* 34:259-270

White M J D 1978 *Modes of Speciation*. Freeman. San Francisco

APPENDIX

The following specimens were used in our morphometric and meristic analyses: *Chirostoma lucius*: CDB uncat, (2); FMNH 3639(10), 3660(3), 3671(1); TU 31970(1), 31971(24), 31980(3), 31986(12), 40829(2), 40838(19), 40840(6); UMMZ 179741(2), 193464(1). *C. sphyraena*: TU 31985(7), 40837(32), one specimen, No 8, 58.0 mm SL, in TU 31986, erroneously identified as *C. lucius*; UMMZ 167724(1), 192542(1), 192549(1). *C. promelas*: FMNH 3656(3), 3664(9); LB 4116(4); TU 31987(1), 40823(1), 40838(1), 40843(4); UMMZ 167723(1), 179740(1), 193465(1), 197649(1), 201569(1). Holotypes: *C. lucius*, BM 1892.2.8.75; *C. sphyraena*, BM 1892.2.8.77; *C. ocotlanae*, CAS(SU) 6160; *C. crystallina*, CAS(SU) 6158; *C. lermae*, CAS(SU) 6159; *C. diazi*, CAS(SU) 6157; *C. promelas*, CAS(SU) 6156.

GENETICS OF SPECIATION IN PUPFISHES FROM LAGUNA CHICHANCANAB, MEXICO

JULIAN M HUMPHRIES

INTRODUCTION

What genomic modifications occur during speciation? Can any generalizations be made about the relationship between morphological and genetic changes accompanying this process? The predicted correlation between morphological and genetic changes in the genome has been questioned recently (Avise et al. 1975; Carson 1976; Kirkpatrick, Selander 1979; Kornfield 1978; Turner, Grosse 1980). These studies suggest that the traditional view associating speciation with major genomic changes (Mayr 1963; Dobzhansky 1970) may not be a general model, and that the amount of genetic variation accompanying speciation can vary widely (White 1978; Templeton 1980). Because most closely related species of vertebrates are allopatric, the relationships among speciation, phyletic change, geographic variation and environmental induction are confounded. Species flocks, which among vertebrates are most common in fishes, provide exceptions to this generalization. Species flocks allow more direct study of the changes that occur during and after speciation because the effects of geographic and environmental variation are partitioned out. Large species flocks, such as *Haplochromis* in Lake Victoria (Greenwood 1974) or the cottids of the Baikal basin (Berg 1949) are more difficult to study than small flocks. With large flocks, monophyly is difficult to verify, because they cover large geographic areas, and allopatric schemes of speciation are frequently unfalsifiable.

Occasionally a small number of closely related forms occur in relatively limited geographic areas. The best known examples are the polymorphic cichlid, *Cichlasoma minckleyi*, from the Cuatro Cienegas basin of northern Mexico (Sage, Selander 1975; Kornfield et al. 1982; Kornfield, Taylor 1983), whitefishes (*Coregonus*) of the Allegash Basin,

Maine (Kirkpatrick and Selander 1979), *Coregonus* in the Great Lakes of North America (Todd 1981; Bailey, Smith 1981; Smith, Todd 1984), and the goodeid genus *Ilyodon* from central Mexico (Turner, Grosse 1980). These studies demonstrate that despite varying amounts of morphological evolution during the formation of a "species," genetic differentiation as measured by electrophoresis has been minimal or in some cases nonexistent. In contrast, most allopatric species of fishes generally exhibit amounts of genetic differentiation typical for other vertebrates (Avise 1976). This "uncoupling" of genetic and morphological indications of differentiation is interpreted variously by evolutionary biologists. At least three, perhaps nonexclusive, explanations have been offered. 1) Possibly the portion of the genome sampled by electrophoresis is not representative of the genotype as a whole. The structural and/or regulatory genes coding for morphological characters could undergo evolution without a concordant change in the portion of the genome examined by electrophoresis (Turner et al. 1979). 2) Most isozymic variation may be due to neutral substitutions of amino acids. The resulting differences will not reflect natural selection but merely the passing of time (Wilson et al. 1977; Selander 1982). Recently speciated sister taxa would be more similar genetically than older such pairs, but concordance between genetic and morphological similarity would not be expected. 3) Sage and Selander (1975) suggest that large amounts of morphological variation associated with small genetic differences may be due to polymorphism in the structural genes encoding the morphology rather than speciation and phyletic evolution.

At least one other explanation is possible. Small amounts of genetic change might be able to produce large morphological effects.

Little is known of the genetics of morphological characters, especially complex factors such as jaw shape and head shape. However, such differences could be induced by changes in timing during development (Gould 1977). If true, then shape differences among species may be due to a few genes, possibly with pleiotropic effects. This explanation would reconcile the presumed discrepancy between genetic and morphological characters in species flocks. A test of this hypothesis is not simple because of difficulties in determining the genetics of morphological characters. For example, even for a small number of loci, the results of hybrid crosses would be ambiguous due to both interaction terms and effects of dominance. The ease with which hybrids are formed, their viability, and their fertility might provide circumstantial evidence of only minor genetic reorganization.

The genus *Cyprinodon* at first would seem to offer little opportunity for the study of speciation by any means other than vicariance. Almost every species is allopatric and they frequently are restricted to an isolated habitat. Many species undoubtedly represent remnants of more widespread Pleistocene populations (Miller 1981). The level of differentiation in both morphological and genetic (Turner 1974; Turner, Liu 1977) characters is minimal. However, a flock of five species from Laguna Chichancanab on the Yucatan peninsula of Mexico (Humphries, Miller 1981; Humphries 1984) offers a unique opportunity to study mechanisms for evolutionary change.

This system has at least three advantages. 1) Because all five forms live in complete syntopy, differences among taxa cannot be attributed to environmental influences such as temperature or habitat size, nor to varying competitors, predators, or availability of resources. 2) The basin has been isolated since at least the early Pleistocene. Thus it is unlikely that the species result from multiple invasions. This assumption is strengthened by the presence of a single ancestral species or sister group, *C. artifrons* (*C. variegatus artifrons* in recent literature), outside the basin. 3) Because the number of taxa is small, one may accurately assess species relationships.

MATERIALS AND METHODS

I collected specimens of endemic pupfishes

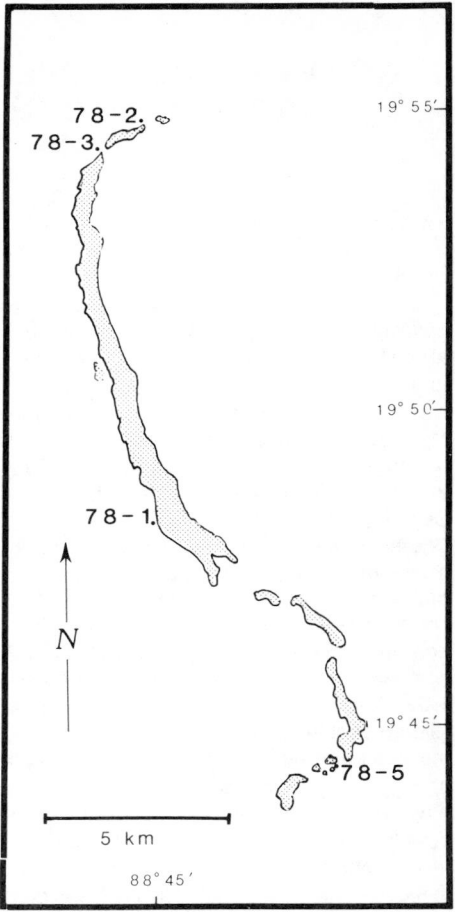

Figure 1. Laguna Chichancanab. Collecting localities for electrophoretic analysis.

at four sites on Laguna Chichancanab (Figure 1) and of *C. artifrons* at one site near Progreso, Yucatan. The first letter of the specific name attached to the associated collecting locality designates each sample; thus *C. beltrani* from the 1978 collection at locality 3 is abbreviated as B783. All fish were frozen on dry ice at the collection site, returned to the laboratory and held at −90° C. Four of the five species endemic to Laguna Chichancanab were examined electrophoretically; *C. verecundus*, a recently discovered, rare species (Humphries 1984) was unavailable. Extracts of individual muscle, liver and eye samples were prepared separately and treated by standard methods

Table 1. Enzymes surveyed: tissue specificities and optimal buffers for resolution.

(xx = easy to score; x = difficult to score)

Enzyme	Abbrev.	Loci	Eye	Liver	Muscle	TCE[1]	EBT[2]	LIOH[3]
Glucosephosphate isomerase	Gpi-2	1	xx	xx	x		+	+
Phosphoglucomutase	Pgm	1		xx	xx			+
Isocitrate dehydrogenase	Idh	1			xx	+		
Mannose 6-phosphate isomerase	M6pi	1			xx	xx	+	+
Malate dehydrogenase	Mdh	2	xx		xx	+		+
Lactate dehydrogenase	Ldh	1	xx	xx	xx	+	+	+
Lactate dehydrogenase	Ldh-eye	2	xx				+	+
Alpha-glycerophosphate dehyrogenase	αGpd	1			xx	+		
Alcohol dehydrogenase	Adh	1		xx	x			+
Superoxide dimutase	Sod	1			xx			+
Aspartate aminotransferase	Aat	2		xx	xx		+	+
Creatine kinase	Ck	2	xx					+
Muscle protein	Mp	4			xx		+	+

[1] Tris-EDTA-citrate, pH 7.0; [2] Tris-EDTA-borate, pH 8.6; [3] LiOH/Tris-citrate, pH 8.0

of horizontal starch-gel electrophoresis and histochemical staining (Shaw, Prasad 1970; Selander et al. 1971; for details of laboratory procedure see Humphries 1981). Nineteen enzyme systems plus general proteins were surveyed. A total of 20 loci representing twelve enzyme systems plus muscle proteins were consistently interpretable and were used in the analyses reported here (Table 1). Interpretation of banding patterns on gels, identification of protein structure and homozygous and heterozygous individuals followed the techniques and assumptions of recent studies on cyprinodontiform fishes (Turner 1974; Darling 1976; Turner, Grosse 1980). Loci and presumptive alleles for each banding system are designated, respectively, by the numerals 1 and 2, and the letters, a, b, c, and d in their order of mobility. Several checks were made on methodology to assure accuracy of genotypic designations for each individual. These included rerunning at least 10 percent of the samples, use of a standard (the goodeid fish, *Xenotoca variata*) on most gels, and checks of some loci by comparing individual patterns from two or three tissues with more than one buffer system. The second determination always confirmed the original designation.

The chi-square goodness-of-fit test (without Yates' correction: Sokal, Rohlf 1969) was used to compare observed genotypic frequencies with their Castle-Hardy-Weinberg (CHW) expectations. Genetic variability was measured as unbiased estimates of heterozygosity, their standard error (Nei 1978), observed heterozygosity, and proportion of loci polymorphic per sample (locality or species.)

Measuring heterogeneity among populations or species is a relatively simple matter when samples differ by complete allelic substitution at one or more loci. This typically is the case in electrophoretic surveys of congeneric species (Avise 1976). When samples possess no diagnostic alleles, the available statistics for measuring differentiation are quite varied. The methods I employ include row-x-column chi-square heterogeneity tests (Workman, Niswander 1970), Wright's F-statistics (Wright 1965) and their multiallelic extension (Nei 1977; Nei, Chakravarti 1977), Nei's (1978) unbiased measure of genetic distance, and Rogers' (1972) measure of genetic distance. The questions to which these analytical tools will be applied concern measurement of both within- and between-species diversity. The heterogeneity chi-square will be used primarily to assess whether gene or genotype frequencies at a collection locality can be determined to have come from a panmictic population. For this test the maximum likelihood ratio, L, was used rather than the standard chi-square statistic. The F-statistics serve a similar purpose: they provide a measure of "population subdivision." In particular, F_{st}, which is also known as the standardized genetic variance, gives the ratio of the actual variance in gene frequencies among groups (s_p^2), to its limiting value, $p_t(1-p_t)$, that would result if there were complete fixation of subpopulations at the same gene frequencies as in the total (t) population (Workman, Niswander 1970). F_{st} has been used conventionally to assess the amount of random genetic drift among populations within a species. In recent years this index has been used to measure differentiation among sympatric morphs (Todd 1981). In either case, values range from zero for exactly equivalent populations to unity for populations fixed at different alleles. G_{st}, a multiallelic formulation of F_{st} (Nei 1977; Nei, Chakravarti 1977) was used in this analysis. The interpretation of this statistic is identical to its F-statistic equivalent.

Two measures of genetic distance or identity are generally used in isozymic studies. The standard genetic distance of Nei, D, is based on a model of accumulated gene substitutions (Nei 1972). An alternative statistic, Rogers' D, is a multiple of the Euclidean distance between populations in multidimensional (multiallelic) space. Although the lack of a biological foundation might at first seem to make Roger's D less desirable than Nei's, the assumptions involved in the latter are rarely tested, or testable. A straight geometric measure may be preferable when it is not known whether the data conform to the assumptions of a more specialized model. For comparative purposes, both measures are used in my analysis. All statistics and genetic distance data were computed by either ALLOZYME, a genetic analysis program written by R. Strauss, or MIDAS, a statistical package at the University of Michigan. I also used a Wagner tree algorithm written by J S Rogers.

RESULTS

Gene frequencies for polymorphic loci appear in Table 2. Muscle proteins, *Mdh-1, Mdh-2, Ldh-2, Aat-1,* and *Ck-2* were fixed in all samples. *Ldh-1, Ldh-3, Aat-2* and *Adh* were polymorphic at frequencies less than .05. Other than their use in estimates of genetic parameters, these systems will not be mentioned further. Only three loci were not in CHW equilibrium (Table 3). No biological significance is placed on the CHW deviation

in the *C. maya* sample. The presence of a single rare homozygote, a/a accounts for almost all of the deviation from equilibrium. The two loci in S785 are only slightly in disagreement with CHW; however, the joint occurrence in a single population may indicate non-panmictic reproduction at that locality.

Within-Species Variation

All four species are relatively homogeneous throughout the basin. Interlocality variation in each species was examined with an R x C test on both genotype and gene frequencies for each of the polymorphic loci. When all localities were considered simultaneously, none of the comparisons in *C. beltrani, maya,* or *labiosus* was significant. When localities were considered pairwise, there were still only two significant comparisons: *C. labiosus,* H78-3 versus H78-5 for *Gpi-2* gene frequencies, Fisher's exact probability = .04; and *C. maya,* H78-1 versus H78-3 for *M6pi* gene frequencies, Fisher's exact probability = .04. Because the heterogeneity test is a relatively insensitive procedure and provides no information on relative variation within and among populations, I turned to Nei's G_{st} statistic; the values ranged from .011 to .020, indicating a relatively small amount of subdivision. In other words, over 98 percent of the genetic diversity in each species occurs within, rather than between samples. This agrees with the contingency analysis, and indicates that intraspecifically, the populations throughout the basin are not far from panmixia.

Among-Species Variation

Although no fixed allelic differences occurred between the four species, one may still test for differentiation among them by comparing gene and genotype frequencies at a location. Of the three locations with more than one species sampled, two, H78-3 and H78-5, were clearly heterogeneous (Table 4). Probabilities of obtaining the observed gene frequencies closely mimic those for observed genotypes. An overall measure of heterogeneity can be obtained by summing the natural logs of the probabilities and multiplying by -2 (Sokal, Rohlf 1969:621). For H78-3 the combined probabiity for genotypes is highly significant when tested as a chi-square (p ◄ .0001, df = 10) as is H78-5 (p ◄ .0002, df = 10).

Table 2. Gene frequencies for polymorphic loci in Chichancanab pupfishes and *C. artifrons* (.95 criterion). *(as percents)*

Locus	Sample size allele	beltrani				M783	maya		labiosus			simus		artifrons
		B782	B783	B781	B785		M781	M785	L783	L781	L785	S783	S785	A789
Gpi-2	n	10	41	40	40	41	23	26	42	23	30	10	33	28
	a	50	51	41	50	34	41	44	61	59	45	70	50	29
	b	50	49	59	49	64	59	56	39	41	55	30	50	66
	c	0	0	0	1	0	0	0	0	0	0	0	0	5
Pgm	n	10	45	40	41	44	23	25	45	23	30	10	33	30
	a	5	0	3	1	2	0	6	0	0	3	15	14	7
	b	95	100	97	99	98	100	94	100	100	97	85	86	93
Idh	n	—	37	40	14	36	23	18	35	23	22	10	24	28
	a	—	4	4	4	6	4	13	0	2	0	25	19	2
	b	—	93	91	96	94	89	81	99	98	100	60	77	98
	c	—	3	5	0	0	7	6	1	0	0	15	4	0
M6pi	n	10	42	40	40	44	23	26	41	23	30	10	32	29
	a	0	0	0	0	4	2	4	0	0	0	0	0	0
	b	0	1	0	0	0	0	0	0	0	2	5	0	2
	c	35	46	44	53	78	63	81	50	61	56	35	41	22
	d	65	52	56	47	18	35	15	50	39	42	60	59	76
αGpd	n	—	37	40	21	44	23	15	35	23	22	10	24	28
	a	—	4	1	0	0	0	0	0	0	0	0	2	29
	b	—	96	99	100	100	100	100	100	100	100	100	98	71

Homogeneity cannot be rejected for the third locality, H78-1. However, when the samples at H78-1 are tested pairwise, each of the three possible pairs is heterogeneous at one or more loci. Failure to reject equal frequencies among all species at this locality may be due to the small sample sizes for *C. maya* and *C. labiosus*.

Table 3. Samples not in CHW equilibrium.

Sample	Locus	a/a	a/b	b/b	b/c	p
M785	*Pgm*	1	1	23		.001
S785	*Gpi-2*	11	11	11		.055
S785	*Idh*	0	9	13	2	.040

The apparent heterogeneity among species suggests that the morphs at a locality are not the result of a polymorphism. A further test of this hypothesis is available. If the gene or genotype frequencies among localities changed in a parallel fashion among morphs, as in the Cuatro Cienegas cichlids (Sage, Selander 1975), the hypothesis of polymorphism with some assortative mating would remain plausible. This is most easily examined by graphing the actual change in gene frequency between pairs of adjacent localities. Figure 2 illustrates the change in gene frequency of the common allele at two loci in each sample for which there are sufficient data. The two loci were chosen because they show the highest level of heterozygosity, and therefore potentially possess the most information. At neither locus is there any obvious pattern of parallel change among localities.

Although the use of F-statistics typically is limited to within-species comparisons, they also have been used to assess heterogeneity between different species at a locality. This is useful only if the species are polymorphic

Table 4. Maximum likelihood estimates, L, of independence of genotypes and gene frequencies.

Gene frequency values calculated from number of gene doses. Rare genotypic classes pooled to reduce cells with expected values less than 5.

Locality	Locus	L for geno-types	Signif. geno-types	Signif. gene freqs.
H78-3	*Gpi-2*	18.5	.0051	.0014
	Pgm	14.9	.0091	.0024
	Idh	32.9	<.0001	.0001
	M6pi	40.1	.0001	<.0001
	αGpd	7.1	.0678	.0705
H78-1[1]	*Gpi-2*	6.8	.1488	.1275
	Pgm	3.1	.2105	.2134
	Idh	5.3	.2582	.2818
	M6pi	9.0	.1719	.0632
	αGpd	1.5	.4620	.4632
H78-5	*Gpi-2*	4.0	.6778	.7850
	Pgm	7.7	.0526	.0126
	Idh	22.0	.0012	.0024
	M6pi	22.3	.0011	.0001
	αGpd	2.5	.4776	.4803

[1] No *C. simus* sampled at this locality.

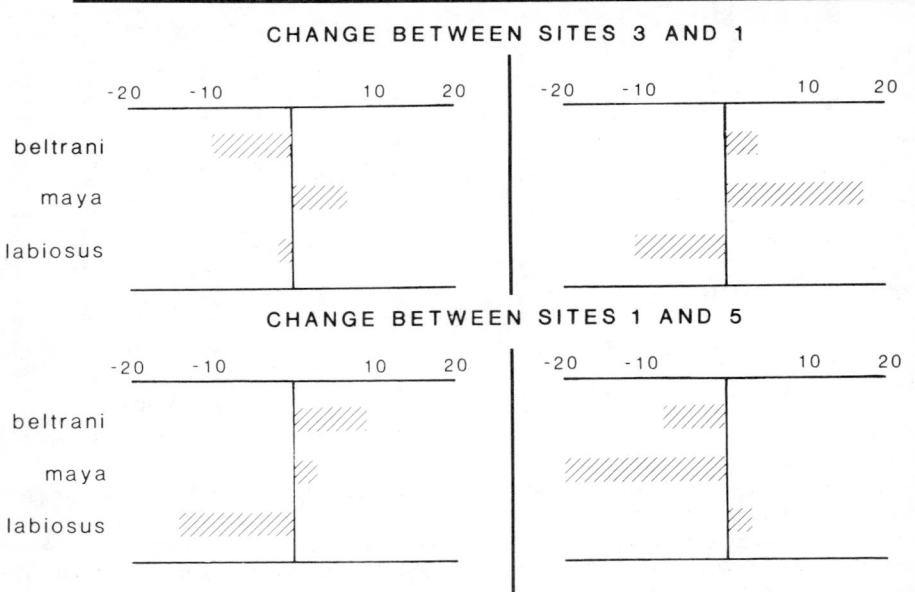

Figure 2. Change in gene frequency from site to adjacent site within species.
Each bar represents the change in frequency for the common allele. The left column is for glucose-phosphate isomerase (GPI), the right column for mannose 6-phosphate dehydrogenase.

Table 5. Estimates of genetic variability.
P = proportion of loci polymorphic;
H_e = Nei's unbiased estimate of heterozygosity;
H_o = observed heterozygosity.

Sample	Loci	P	H_e	± S.E.	H_o
B782	8	.38	.138	.081	.150
B783	20	.20	.062	.003	.059
B781	16	.38	.079	.042	.074
B785	20	.20	.056	.035	.060
Mean	16	.30	.074	.020	.074
M783	20	.30	.053	.028	.054
M781	16	.19	.075	.043	.079
M785	20	.20	.064	.033	.051
Mean	19	.30	.063	.019	.060
L783	20	.25	.054	.034	.057
L781	16	.19	.064	.042	.076
L785	16	.19	.068	.043	.065
Mean	17	.30	.061	.022	.066
S783	20	.20	.092	.044	.115
S785	20	.25	.083	.039	.071
Mean	20	.25	.087	.028	.093
A789	18	.33	.085	.038	.080

for the same alleles. If the species have diagnostic alleles, then statistics such as Nei's or Rogers' D are more appropriate. The relative degree of among-species subdivision at a locality, although higher than the among-population G_{st}, is still exceedingly low. The values for G_{st} range from .03 to .10. To put these numbers in perspective, typical F_{st} values for vertebrate populations *within* a species are .04-.15 (Nei 1977).

Genetic Variability

Nevo (1978) reports an average heterozygosity for fishes of .0513. In this study, estimated heterozygosity ranged from .0529 to .1382 (Table 5). Considering the sample bias towards heterozygous loci, these values are well within the typcial range. Except for B782, which is based on a small sample of loci, the ranges and values are similar both within and among species. *C. simus* has the highest values, but they do not differ markedly from those of the other species.

Table 6. Genetic distances (x 1000)
Nei's identity above the diagonal; Rogers' similarity below diagonal.

Sample	B782	B783	B781	B785	M783	M781	M785	L783	L781	L785	S783	S785	A789
B782	--	999	1000	998	970	989	970	997	992	995	997	1001	995
B783	977	--	999	1000	993	998	993	1000	999	999	993	998	991
B781	975	986	--	999	991	998	990	998	996	999	990	998	990
B785	972	993	983	--	995	999	995	1000	1000	1000	992	998	988
M783	920	968	965	974	--	998	999	992	994	996	979	989	976
M781	946	978	983	982	977	--	998	997	999	1000	986	994	980
M785	932	968	964	973	984	978	--	992	995	995	985	991	972
L783	962	988	977	990	964	974	963	--	1000	999	993	997	986
L781	951	982	971	988	966	982	964	991	--	999	989	994	977
L785	964	981	982	990	972	985	972	982	986	--	986	995	984
S783	956	962	951	960	937	940	951	963	948	943	--	998	979
S785	982	981	975	977	955	962	968	970	960	965	978	--	990
A789	959	953	947	951	943	929	933	944	925	938	929	948	--

The observed heterozygosity, H_o showed no systematic deviation from the unbiased expected value, H_e. The proportion of polymorphic loci (.19-.38) is also typical for surveys of similar loci in other fishes.

Genetic Distance

The gene frequencies at 20 loci were used to calculate coefficients of similarity (Rogers' S, Nei's I) among the 13 samples (Table 6). It should be noted that two of the comparisons with zero distance are with B782, which was sampled at only eight loci. Thus, comparisons made with that sample will have relatively large standard errors. The lowest I's are generally associated with the Chichancanab samples compared against C. artifrons, (I = .976-.995). The overall range of inter-specific values (.972-1.00) is extremely high, and is more typical of conspecific comparisons in other fish genera. The Rogers' S values, although lower than the I's, are also extremely high for vertebrate species.

Evolutionary relationships estimated from a similarity measure such as Nei's or Rogers' index are generally summarized in a dendrogram or network produced by some clustering algorithm. Such a technique allows easy comprehension of a complex matrix of distance values. As mentioned earlier, Nei's I may be flawed for use in most clustering methods. With my data, both Nei's and Rogers' indexes are so high that it is doubtful whether any clustering procedure could produce meaningful results. Figure 3 is a Wagner tree based on Rogers' similarity measure. Although there is some grouping by taxa, the only distinctive feature is the clear

separation of C. artifrons from the other samples.

DISCUSSION

All of the statistical methods portray a consistent pattern of variation in the Chichancanab pupfishes. Only minor intraspecific variation occurs throughout the basin. This result is simiar to that Darling (1976) obtained for closely adjacent populations of another pupfish, C. variegatus.

The interspecific results, however, are atypical for reproductively isolated species. Most studies of interspecific differentiation between congeneric vertebrates show genetic distances to be an order of magnitude higher than those for the Chichancanab pupfishes. In fact, the mean between-species similarity for these pupfishes is very close to the mean within-species value (.962 vs .982, Rogers'; .992 vs .999, Nei's). The genetic similarity of these fishes contrasts markedly with the sharp morphological differences among the species (Figure 4), particularly in shape-related features. This type of morphological divergence is comparable to that found among the morphs of Cichlasoma minckleyi. Here too, characters associated with feeding show the clearest differences. The parallel does not extend, however, to the genetic data. At Cuatro Cienegas there is evidence of within-locality intermorph homogeneity coupled with concordant changes among morphs across localities (Sage, Selander 1975).

The existence of discrete morphotypes in Cichlasoma minckleyi indicates that the genetics of some morphological characters may be simpler than previously suspected.

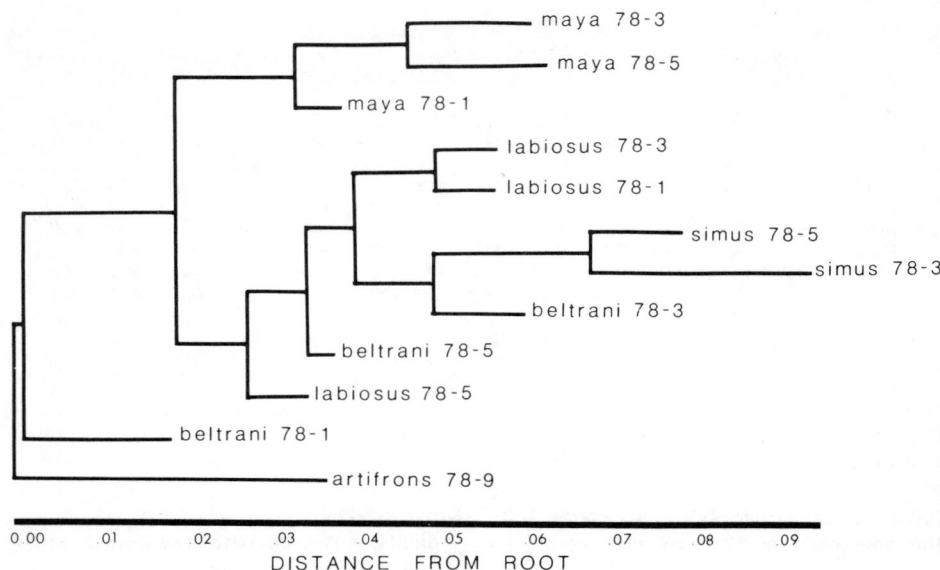

Figure 3. Relationships among the Chichancanab pupfishes as indicated by a Wagner tree based on Rogers' distances. Rooted at *C. artifrons*.

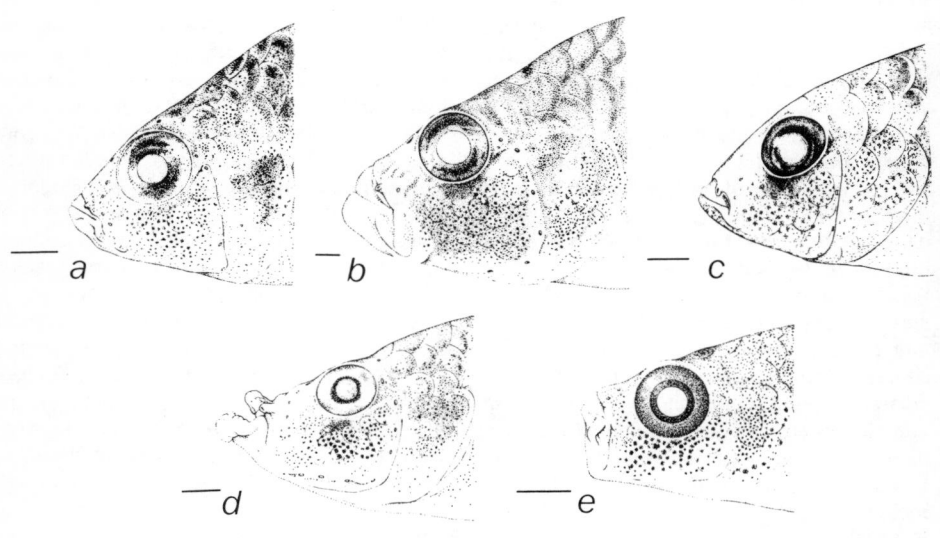

Figure 4. Head morphologies of the Chichancanab pupfishes.
Bar represents 2 mm. a) *C. beltrani,* male, 35.0 mm SL, UMMZ 201759. b) *C. maya,* female, 53.2 mm SL, UMMZ 201761. c) *C. verecundus,* male, 32.3 mm SL, uncataloged. d) *C. labiosus,* female, 36.0 SL, UMMZ 203908. e) *C. simus,* male, 24.7 mm SL, uncataloged.

The morphological changes seen in the Chichancanab pupfishes could, then, be expressions of only minor genetic reorganization during speciation. This would reconcile the lack of agreement between morphological and allozymic data. If simple gene changes are indeed responsible for the observed morphological changes, then the evolutionary processes giving rise to species or morphotypes in species flocks may not be equivalent to cladogenesis in the formation of allopatric species.

At least three alternative proposals have been made to explain the origin of species flocks, their apparent lack of allozymic differentiation, and the extensive evolution of trophic features.

1) If the rate of speciation has been much higher in species flocks (Kornfield 1978), then the minimal genetic evolution could be explained by recency of speciation. This alone, however, does not explain the continual recurrence of specialized feeding habits and morphologies.

2) A classic allopatric model of speciation would require extensive speciation, followed by range extensions into sympatry and extinctions of the original allopatric populations outside the basin. The monophyly of the endemic forms (Humphries unpublished) makes this an implausible sequence of events. Thus, intralacustrine speciation would seem more parsimoniously in accord with the data. For example Smith and Todd (1984) proposed a mechanism for speciation based on fluctuating lake levels. Laguna Chichancanab might seem ideal for this process as low water levels temporarily divide the basin into a number of individual lakes. There are two difficulties with this scenario. First, these individual lakes are very shallow, generally less than two meters, and the barriers between the lakes are low. As a result the peripheral lakes are very short-lived—in low water periods they dry up, and in high water periods they merge with the main basin. Second, any allopatric scheme needs to account for the development of reproductive isolation. Given the relative homogeneity of the Chichancanab basin and the probable short periods of isolation produced by fluctuating water levels, it seems likely that only weakly isolated forms would be produced. Pupfish species likely to have originated in

allopatry commonly produce fertile hybrids both in the laboratory (Turner, Liu 1977) and in the field (Miller 1968; Stevenson, Buchanan 1973; Kennedy 1977). Thus, unusually strong reproductive isolation would have to be produced in very short periods of time. Dominey (1984) has proposed sexual selection as a mechanism for producing such rapid divergence. However, the Chichancanab pupfishes do not show the expected sexual dimorphism or dichromatism usually attributed to sexual selection. Behavioral differences, as yet unknown for these species, might provide the basis for such reproductive isolation. Indeed, pupfishes generally show lek behavior, and at least one species is sexually selected (Kodric-Brown 1977).

3) Based on the foundations laid by Wright (1932), Maynard Smith (1966), Hutchinson (1968), and Thoday (1970), Rosenzweig (1978) proposed the concept of *competitive speciation*. In this model, niche space functions in the same manner as geographic space; that is, it can appear to the organism as a discontinuous series of suitable habitats. The first step is for a phenotype to "invade" an adjacent fitness peak. This new form need show only minor morphological divergence from the original phenotype. Assuming density dependent selection, the spread of this new phenotype will depress the fitness of intermediate forms. Two possible results of competitive speciation are discrete polymorphism, as in the Cuatro Cienegas cichlid, or assortative mating leading to reproductive isolation.

Several attributes of the Chichancanab system are in accord with competitive speciation. Assuming that the current absence of competitors or predators has always been true for Laguna Chichancanab, there would have been numerous ecological opportunities when the original pupfish population invaded the basin. The breeding conditions in the lake are ideal for pupfish and would allow the population to become sufficiently large that density dependent selection would have occurred.

Given that the most heterogeneous niche axis in a shallow tropical lake would be trophic in nature, phenotypes which could utilize the typically non-pupfish food resources would gain an advantage. At least currently, these food preferences appear relatively discrete between Chichancanab pupfishes. An assumption of the model is that they initiated that way. Selection against intermediates would ensue once phenotypes had crossed these adaptive gaps. If the evidence of monophyly for the Chichancanab pupfishes is correct, competitive speciation is a plausible hypothesis for the evolution of this species flock. This model explains both the low levels of genetic differentiation between closest relatives and the high degree of morphological divergence in trophic related characters.

SUMMARY

The five species of pupfish from Laguna Chichancanab exhibit striking contrasts between morphological and genetic differentiation. Genetic similarities among the species are extremely high for interspecific comparisons (Rogers index = .92-.99) while intraspecific variation is in normal ranges. Despite the extreme similarity of genotypes, analysis of gene frequencies indicates that the four sampled species are reproductively isolated.

The supposed uncoupling of genetics and morphology is interpreted as reflecting the effects of only minor genetic reorganization during speciation. The morphological differences among the species—head and mouth shape, tooth morphology, gut length, and body size—are hypothesized to be the result of simple genetic changes. Competitive speciation (Rosenzweig 1978) may have produced the intralacustrine differentiation seen in the Chichancanab pupfishes.

ACKNOWLEDGMENTS

Support for field work was provided by the Horace H Rackham School of Graduate Studies, the Division of Biological Sciences and the Museum of Zoology, all of the University of Michigan, and by National Science Foundation grants (GB-6272x, GB-35943x) to Robert R Miller. Thanks are also due the Government of Mexico for permission to collect specimens. Mark Orsen and Carrie DeLorenzo prepared Figure 4. Robert and Francis Miller, Jeffery Taylor, Michael Stevenson, Michael Smith, and Tom Grimshaw assisted with collecting or obtaining material. Wallace Dominey and Tony Echelle provided helpful comments on the manuscript.

REFERENCES

Avise J C 1976 Genetic differentiation during speciation. 106-122 F J Ayala ed. *Molecular Evolution.* Sinauer. New York

Avise J C, J J Smith, F Ayala 1975 Adaptive differentiation with little genetic change between two native California minnows. *Evolution* 29:411-426

Bailey R M, G R Smith 1981 Origin and geography of the fish fauna of the Laurentian Great Lakes basin. *Can J Fish Aquat Sci* 38:1539-1561

Berg L S 1949 Freshwater fishes of the U.S.S.R. and adjacent countries. Vol III. *Guide to the Fauna of the U.S.S.R.* No 30. Moscow. Israel Program for Scientific Translations. Jerusalem, 1965

Carson H L 1976 The unit of genetic change in adaptation and speciation. *Ann Missouri Bot Gard* 63:210-223

Darling J D S 1976 Electrophoretic variation in *Cyprinodon variegatus* and systematics of some fishes of the subfamily Cyprinodontinae. PhD thesis. Yale Univ. New Haven, Conn

Dobzhansky T 1970 *Genetics of the Evolutionary Process.* Columbia U Press. New York

Dominey W J 1984 Effects of sexual selection and life history on speciation: Species flocks in African cichlids and Hawaiian *Drosophila.* 231-250 A A Echelle, I Kornfield eds. *Evolution of Fish Species Flocks.* Univ Maine at Orono Press

Gould S J 1977 *Ontogeny and Phylogeny.* Cambridge and Belknap Press. London

Greenwood P H 1974 The cichlid fishes of Lake Victoria, East Africa: The biology and evolution of a species flock. *Bull Brit Mus Nat Hist (Zool)* Suppl 6:1-134

Humphries J M 1981 The evolution of a species flock in the genus *Cyprinodon* (Pisces Cyprinodontidae). PhD thesis Univ Michigan. Ann Arbor

―――― 1984 *Cyprinodon verecundus*, n. sp., a fifth species of pupfish from Laguna Chichancanab. *Copeia* 1984 In press.

Humphries J M, R R Miller 1981 A remarkable species flock of *Cyprinodon* from Lake Chichancanab, Yucatan, Mexico. *Copeia* 1981:52-64

Hutchinson G E 1968 When are species necessary? 177-186 R C Lewontin ed. *Population Biology and Evolution.* Syracuse Univ Press. New York

Kennedy S E 1977 Life history of the Leon Springs pupfish, *Cyprinodon bovinus. Copeia* 1977:93-103

Kirkpatrick M, R K Selander 1979 Genetics of speciation in lake whitefishes in the Allegash Basin. *Evolution* 33:478-485

Kodric-Brown A 1977 Reproductive success and the evolution of breeding territories in pupfish (*Cyprinodon*). *Evolution* 31:750-765

Kornfield I L 1978 Evidence for rapid speciation in African cichlid fishes. *Experientia* 34:335-336

Kornfield I L, D C Smith, P S Gagnon, J N Taylor 1982 The cichlid fish of Cuatro Cienegas, Mexico: direct evidence of conspecificity among distinct trophic morphs. *Evolution* 36:658-664

Kornfield I L, J N Taylor 1983 A new species of polymorphic fish, *Cichlasoma minckleyi*, from Cuatro Cienegas, Mexico (Teleostei: Cichlidae). *Proc Biol Soc Wash 96:253-269*

Maynard Smith J 1966 Sympatric speciation. *Amer Natur* 100:637-650

Mayr E 1963 *Animal Species and Evolution.* Belknap. London

Miller R R 1968 Two new fishes of the genus *Cyprinodon* from the Cuatro Cienegas basin, Coahuila, Mexico. *Occ Pap Mus Zool Univ Mich* 659:1-15

———— 1981 Coevolution of deserts and pupfishes (genus *Cyprinodon*) in the American southwest. 39-94 R S Naiman, D L Soltz eds. *Fishes in North American Deserts.* Wiley. New York

Nei M 1972 Genetic distance between populations. *Amer Natur* 106:283-292

———— 1977 F-statistics and analysis of gene diversity in subdivided populations. *Ann Hum Genet, Lond* 41:225-233

———— 1978 Estimation of average heterozygosity and genetic distance from a small number of individuals. *Genetics* 89:583-590

Nei M, A Chakravarti 1977 Drift variances of F_{st} and G_{st} statistics obtained from a finite number of isolated populations. *Theor Pop Biol* 11:307-325

Nevo E 1978 Genetic variation in natural populations: patterns and theory. *Theor Pop Biol* 13:121-177

Rogers J S 1972 Measures of genetic similarity and genetic distance. *Studies in Genetics VII. Univ Texas Publ.* 7213:145-153

Rosenzweig M L 1978 Competitive speciation. *Biol J Linn Soc* 10:275-289

Sage R D, R K Selander 1975 Trophic radiation through polymorphism in cichlid fishes. *Proc Nat Acad Sci* 72:4669-4673

Selander R K 1982 Phylogeny. 32-59 R Milkman ed. *Perspectives in Evolution.* Sinauer. Sunderland, Mass

Selander R K, M H Smith, S Y Yang, W E Johnson, J B Gentry 1971 Biochemical polymorphism and systematics in the genus *Peromyscus.* I. Variation in the old-field mouse (*Peromyscus polionotus*) *Stud Genet* 6 *Univ Texas Publ.* 1703:49-90

Shaw C R, R Prasad 1970 Starch gel electrophoresis—a compilation of recipes. *Biochem Genet* 4:297-320

Smith G R, T N Todd 1984 Evolution of species flocks of fishes in north temperate lakes. 47-68 A A Echelle, I Kornfield eds. *Evolution of Fish Species Flocks.* Univ Maine Press at Orono

Sokal R R, F J Rohlf 1969 *Biometry.* Freeman. San Francisco

Stevenson M M, T M Buchanan 1973 An analysis of hybridization between the cyprinodont fishes *Cyprinodon variegatus* and *C. elegans. Copeia* 1973:682-692

Templeton A R 1980 Modes of speciation and inferences based on genetic distances. *Evolution* 34:719-729

Thoday J M 1970 The probability of isolation by disruptive selection. *Amer Natur* 104:219-230

Todd T N 1981 Allelic variability in species and stocks of Lake Superior ciscoes (Coregoninae). *Can J Fish Aquat Sci* 38:1808-1813

Turner B J 1974 Genetic divergence of Death Valley pupfish species: biochemical versus morphological evidence. *Evolution* 28:281-294

Turner B J, D J Grosse 1980 Trophic differentiation in *Ilyodon,* a genus of stream-dwelling goodeid fishes: speciation versus ecological polymorphism. *Evolution* 34:259-270

Turner B J, R K Liu 1977 Extensive interspecific genetic compatibility in the New World killifish genus *Cyprinodon. Copeia* 1977:259-269

Turner J R G, M S Johnson, W F Eanes 1979 Contrasted modes of evolution in the same genome: Allozymes and adaptive change in *Heliconius. Proc Natl Acad Sci* 76:1924-1928

White M J D 1978 *Modes of Speciation.* Freeman. San Franciso

Wilson A C, S S Carlson, T J White 1977 Biochemical evolution. *Ann Rev Biochem* 46:573-639

Workman P L, J D Niswander 1970 Population studies on southwestern Indian tribes. II. Local genetic differentiation in the Papago. *Am J Hum Genet* 22:24-29

Wright S 1932 The roles of mutation, inbreeding, crossbreeding, and selection in evolution. *Proc VI Intl Cong Genetics* 1:356-366

———— 1965 The interpretation of population structure by F-statistics with special regard to systems of mating. *Evolution* 19:395-420

AFRICAN CICHLIDS AND EVOLUTIONARY THEORIES

P H GREENWOOD

INTRODUCTION

Fishes of the African Great Lakes have long featured in theories and arguments about evolutionary processes (Greenwood 1984). Current debate concerns the role of natural selection in species formation and the theory of punctuated equilibrium. Both areas of controversy are also basic to our interpretation of so-called macroevolutionary events, another disputed area (Greenwood 1979). Once again, fuel for these arguments can be found in the African lake fishes, in particular, the cichlid species flocks. In the new round of discussion, these flocks yield empirical evidence unavailable to paleontologists, and also harder data to stiffen the sometimes nebulous speculation engendered by arguments over evolutionary processes.

HISTORY AND DIVERSITY OF AFRICAN LAKE CICHLIDS

The large African lakes, Victoria, Edward, Malawi, and Tanganyika, and some smaller ones, are outstanding for the way in which their fish faunas are dominated, both in species numbers and in the extent of their ecological radiations, by members of a single family, the Cichlidae. Cichlid radiation is far more muted in the other African lakes, such as Albert and Turkana (Greenwood 1974b), and the continent's riverine cichlids show less diversity than those of South America.

This is a large and essentially freshwater family of about 800 species, with a wide geographical range embracing South and Central America, Africa, Madagascar, Sri Lanka and India. Only in Africa, however, have cichlids undergone explosive speciation. Apparently, since detailed information is lacking for the American species, only in Africa have they developed so many specializations in feeding habits and feeding mechanisms.

It is of interest that the African riverine species, as compared with the lake species, have relatively muted morphological and ecological diversity, since the primary ancestors of the lake faunas must have been derived from the rivers. Much of the diversity so characteristic of the present-day lake cichlids evolved within the lakes. Perhaps one should say "within and with the lakes" because the lakes themselves have had checkered histories (Fryer, Iles 1972; Greenwood 1974a). Regrettably, the fossil record for African cichlids is extremely poor. What few fossils there are certainly do not suggest that the pre-lake species were anatomically more diverse than the riverine species of today (Van Couvering 1982).

None of the "cichlid lakes" can be considered truly ancient. The origins of the oldest, Lake Tanganyika, are dated at about 2 my before present, and those of the youngest, Lakes Victoria and Edward, at about 750,000 years ago. Yet, in that short time, some 140 species have evolved in Lake Tanganyika, and over 200 in Lake Victoria, and probably about the same number in Lake Malawi, aged at about 1.5 my — Cromie (1982) suggested that the lacustrine environment may be a mere 100,000 years old. Lake Edward has far fewer cichlid species, about 45, but even this small number reflects an extensive diversity of anatomical specialization and trophic habits (Greenwood 1973, 1980). Lake Edward, however, should almost certainly be treated as a remnant of an earlier and more expansive lake (Greenwood 1974a, 1980) and its cichlid fauna, with that of Kivu, included with the cichlid fauna of Lake Victoria. The great majority of Lake Edward species, nevertheless, is endemic to that lake, as are the species of Lake Kivu.

Cichlid dominance is invariably correlated

with a reduction in the number of species belonging to other families, at least when compared with the numbers of non-cichlids occurring in rivers (Table 1). The biological or historical basis for this correlation is not clear since even in lakes where endemic cichlids are relatively few, the number of non-cichlid species is less than in the geographically related river systems, e.g., compare Lakes Albert and Turkana with the Nile (Table 1).

Unfortunately, we can get no evidence of how the initial cichlid and non-cichlid invaders of the developing lakes interacted on one another's evolutionary histories. That both cichlids and non-cichlids must have entered the lakes contemporaneously seems reasonable, as would the assumption that both groups could be potential competitors for the developing lacustrine niches.

Darwin (1898:327) comments on a similar problem in relation to oceanic islands where some taxa have undergone more marked evolutionary changes than others. His solution, that the unmodified "... species immigrated in a body so that their mutual relations have not been much disturbed," and there was "... frequent arrival of unmodified immigrants ..." would not seem applicable to the disparity between cichlids and non-cichlids in the African lakes. There is no reason to suppose that either group arrived separately, nor as long as migration routes were open, that both did not continue to arrive together. Moreover, Darwin's solution was applied to a phylogenetically less complex situation. The non-cichlids of African lakes represent a variety of families and orders, whereas Darwin's model applied to a single taxon or at most to a few closely related taxa. Various suggestions have been offered to explain the phenomenon of cichlid dominance and diversity (e.g., Fryer, Iles 1972; Liem 1973; Barbour, Brown 1974; Greenwood 1974a; Sage et al. 1984). Some border on the metaphysical by introducing concepts of "evolutionary potentiality" or lack of it, and others suggest unusual evolutionary mechanisms peculiar to the Cichlidae. None has been satisfactorily substantiated (see discussion in Greenwood 1974a,1979).

Among the "cichlid lakes" the Cichlidae of Lakes Tanganyika and Malawi are more diversified, particularly in morphological features,

Table 1. Approximate number of species in principal river systems and lakes of Africa. Those for Lake Edward include Lake George; those for Lakes Malawi and Victoria include species not yet formally described. Zambezi system data from Bell-Cross (1972); subscript identifies presumed faunal relationships of a river with specific lakes.

	Species			Families
	All	Cichlid	Non-Cichlid	Non-Cichlid
Rivers				
Nile$_1$	115	10	105	16
Niger	134	10	124	25
Zaire$_2$	690	40	650	23
Zambezi$_3$	110	20	90	17
Lakes				
Turkana$_1$	39	7	32	14
Victoria$_1$	238	200	38	11
Edward$_1$	57	40	17	7
Albert$_1$	46	9	37	14
Tanganyika$_2$	247	136	111	18
Malawi$_3$	242	200	42	8

than those of Lake Victoria. Until recently, this diversification and differentiation was reflected in the 35 endemic genera described from Lake Tanganyika and 20 from Lake Malawi, compared with 4 from Lake Victoria. The number of endemic genera in Lake Victoria has been increased considerably, however, in an attempt to delimit more clearly the phylogenetic relationships of its 200 + species (Greenwood 1980). Previously, most of these species were referred to Haplochromis, a "genus" widely distributed in tropical Africa, and one to which the bulk of Lake Malawi's endemic cichlid species was also assigned. This revisionary work has indicated that the Malawi "Haplochromis" also represent a number of lineages, and that these should be allocated to genera distinct from those in Lake Victoria, with which many seem to be but distantly related (Greenwood 1983a,b).

But despite such taxonomic upheavals, the fact remains that morphological diversity is highest, and interspecific differences most trenchant, among the Lake Tanganyika species, less so among those of Lake Malawi, and least among the Lake Victoria species. Various notions have been advanced to explain this pattern of diversity, the most favored being the different ages of the three lakes (Fryer, Iles 1972).

From an ecological view there are fewer

striking differences between the cichlid faunas of the three major lakes, although lakes Tanganyika and Malawi are noteworthy for their large numbers of rock-frequenting species, and Victoria for its bathybenthic species. The two former lakes have long stretches of rocky coastlines (see McKaye, Gray 1984), while Victoria has few. The deep bottom areas of Malawi and Tanganyika are permanently deoxygenated, but those of the much shallower Lake Victoria are almost permanently oxygenated (Greenwood 1974a).

Although the level of morphological diversity is relatively low in Lake Victoria, the species there are, in many respects, the most interesting. This is because their lack of clearcut morphological differentiation is due in large part to the persistence of species intermediate in form, which bridge gaps that make the difference between taxa in other lakes so much more trenchantly defined.

Since these gaps are filled in Lake Victoria, it is possible to identify, in living species, the developmental stages of an anatomical trait or set of correlated traits. At the same time, it is possible to study the ecological consequences of what is effectively a morphocline in presumed adaptive features. This is not to say that the stages in a morphocline represent actual ancestral and descendant species. It is possible that they do, but there is no way to establish that relationship.

The 200 + species of Lake Victoria at first seemed to be closely interrelated, and were generally considered to be of monophyletic origin, and a particularly good example of a species flock (Greenwood 1974a, 1984). But close analysis failed either to refute or substantiate the hypothesis of monophyletic origin. We can only recognize several apparently monophyletic lineages, with each lineage comprising from one to over 20 species. The interrelationships of the different lineages cannot be established, mainly because of the close similarity between their constituent species in all but the derived features characterizing a particular lineage. The remaining characters are so widely shared by the majority of the African cichlids that they are valueless for determining relationships at any but higher hierarchical levels. This difficulty does not reduce the value of the assemblage as an evolutionary model.

Each lineage can be treated as a phylogenetic entity, with the different entities having interacted as parts of an evolutionary whole during their respective histories in a developing lake. To that field of interaction must also be added the non-cichlid fishes, many of which share habitats and feeding requirements with the cichlids.

TROPHIC DIVERSITY IN LAKE VICTORIA CICHLIDS

The Lake Victoria cichlids, previously referred to the genus *Haplochromis* and its close relatives (all monotypic and endemic genera), have undergone an outstanding trophic radiation (Greenwood 1974a). There is apparently no food source in the lake which is not tapped by members of several haplochromine species, either as adults or as juveniles. Generally, more than one, and often several species seem to use the same food organisms in a particular habitat, and there are no truly lacustrine habitats which are not occupied by one or several haplochromine species (see Witte 1984).

This picture contrasts strongly with that of the species of other families. Not only are there fewer non-cichlid species in a particular habitat, especially in the deeper offshore waters of the lake, but there are fewer trophic specialists tapping any one food source and occupying any particular habitat. I use the term *specialist* to denote modal feeding habits, or food, of a species, particularly those of its adult members. Of course, there are ontogenetically correlated changes in feeding habits, and food organisms other than the modal one often are ingested.

Since trophic specialization is the hallmark of the haplochromines, it is not surprising that the different lineages in Lake Victoria are characterized principally by anatomical and functional features in the jaws, teeth, and skull, and in the pharyngeal jaws and their dentition (Liem 1973; Greenwood 1974a; Liem, Osse 1975). Contrasting with these cranial and dental differences, there is remarkably little diversity in general body form (Figure 1). Only the piscivorous predators and some predators on zooplankton depart from a body form shared with members of other trophic groups, as well as the majority of riverine haplochromine species. The

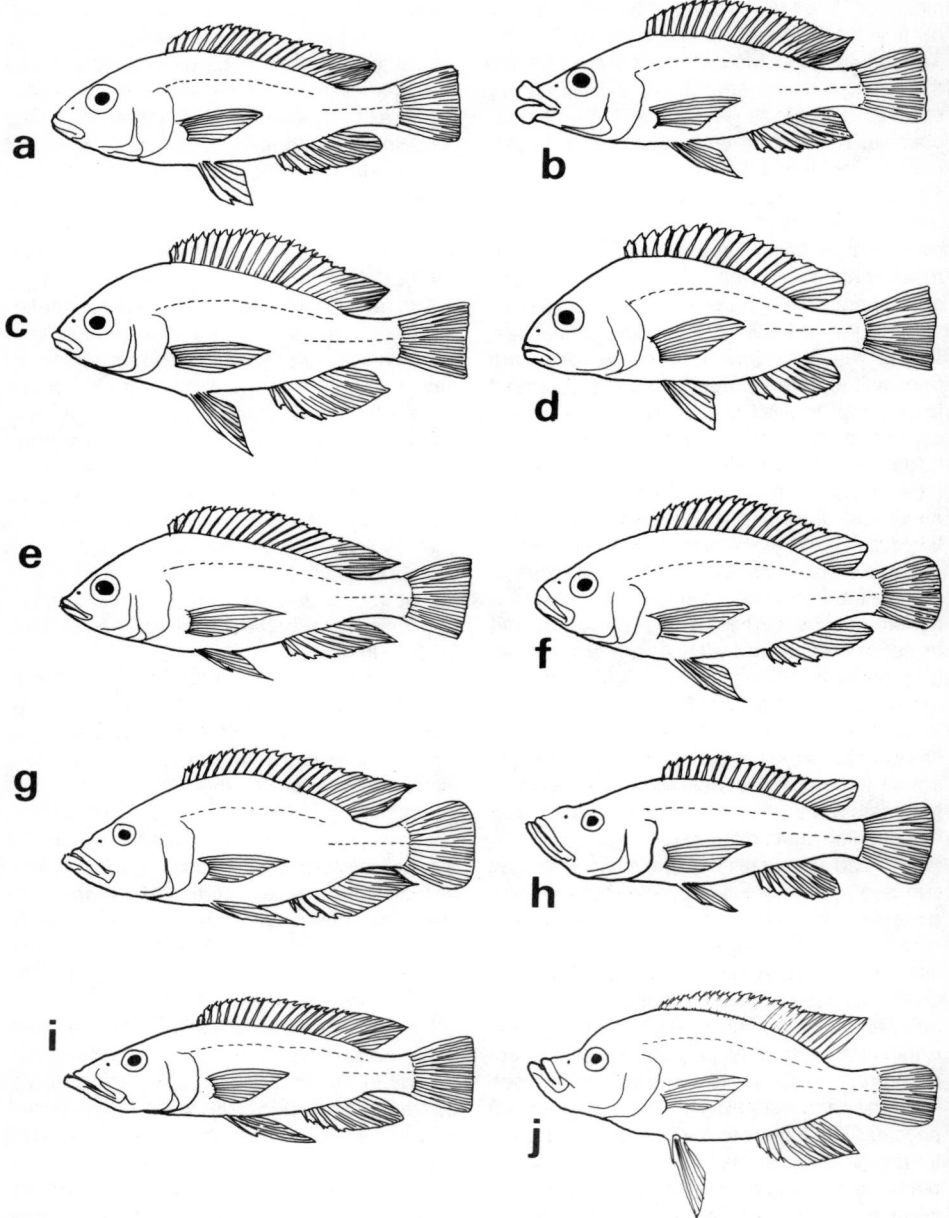

Figure 1. Body form in various Lake Victoria haplochromines showing the entire range of head and body shape represented amongst the endemic species. a. *Astatotilapia pallida* (generalized bottom feeder). b. *Paralabidochromis chilotes* (specialized insectivore). c. *Macropleurodus bicolor* (specialized mollusc eater, oral sheller). d. *Gaurochromis empodisma* (detritus eater). e. *Labrochromis ishmaeli* (specialized mollusc eater, pharyngeal sheller). f. *Lipochromis obesus* (paedophage). g. *Harpagochromis spekii* (piscivore). h. *Harpagochromis cavifrons* (piscivore). i. *Prognathochromis macrognathus* (piscivore). j. *Pyxichromis parorthostoma* (diet unknown). Figures are drawn to different scales.

predators on fish and zooplankton have a more elongate and streamlined shape, and the piscivores generally reach a large adult size, 120-220 mm standard length (SL). Members of other groups, including the predators on zooplankton, rarely exceed 130 mm SL, and several do not exceed 100 mm SL.

What is taken, on the basis of intrafamilial outgroup comparisons, to be the generalized oral dentition among haplochromines in Lake Victoria and elsewhere comprises an outer row of unequally bicuspid teeth backed by two or three rows of smaller, tricuspid teeth (Figure 2, Figure 3). This dental type occurs in most riverine species, and is the characteristic dentition in juveniles of all species, irrespective of their adult dental and oral specializations. It is a dental type not correlated with any particular feeding habit, being found in detritus feeders, insectivores, some mollusc eaters and even in certain piscivores. However, except in the insectivores, it is generally associated with modified pharyngeal dentition and/or form and proportions of the jaws.

The presumed basic pharyngeal dentition also is composed of bicuspid elements, albeit ones of a different shape from those in the jaws (illustrated in Barel et al. 1977). Individual teeth are moderately slender and arranged in close-set rows on the upper and lower pharyngeal bones. Teeth in the median two to four rows on the lower bone are somewhat stouter than the others, but retain essentially the same cusp pattern (Figure 4). Departures from these generalized features of the oral and pharyngeal dentition are manifest principally by changed relative proportions, in shape, in number (usually increased), or in differentiation of cusp patterns from the basic unequally bicuspid type (see illustrations in Barel et al. 1977).

Changes in relative proportions, variously correlated with one another and with dental changes, also occur in skull shape, shape of upper and lower jaw, and in various skeletal elements of the suspensorium (Barel et al. 1977). See Figure 5. There are correlated changes too in size and proportions of certain bones of the gill-arch skeleton, especially those associated with the pharyngeal jaws. In general, changes in the gill-arch skeleton are less obvious than those associated with the

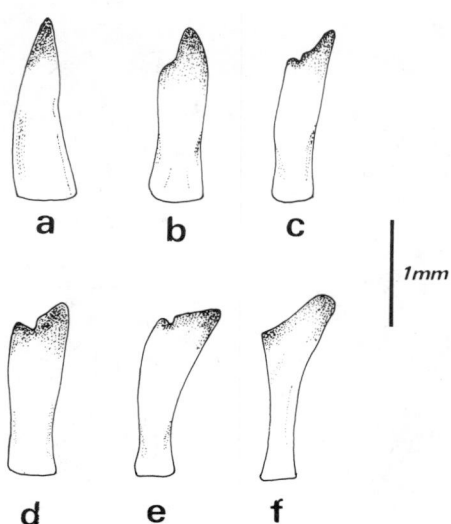

Figure 2. Outer row jaw teeth in various Lake Victoria haplochromines. The tooth shown in d is taken to represent the generalized bicuspid type. a. Unisuspid from a piscivore (*Prognathochromis estor*). b. Reduced bicuspid from an insectivore-piscivore (*Harpagochromis artaxerxes*). c. Reduced bicuspid from an insectivore-detritus eater (*Astatotilapia lacrimosa*). d. Generalized bicuspid from insectivore (*Astatotilapia macrops*). e. Bicuspid with somewhat protracted major cusp from a grazer on epiphytic algae (*Haplochromis lividus*). f. Extreme development of type e, from another epiphytic algal grazer (*Haplochromis obliquidens*).

teeth and osteology of the skull and oral jaws. But even those latter changes could not be described as especially outstanding, and many can be detected only after detailed comparative study.

The presumed generalized dentition and cranial anatomy occur in a large number, probably more than 50, Lake Victoria haplochromine species, but principally in those feeding on benthic insect larvae and pupae. Slight deviation from this anatomico-functional type occurs in at least five species which feed on the moribund phytoplankton and other organic debris which form extensive deposits over much of the lake floor. In these species the oral and pharyngeal dentition is finer, the crowns of the pharyngeal teeth are more hooked and the teeth are more densely arranged on their respective bones, and the orobranchial cavity is increased in volume;

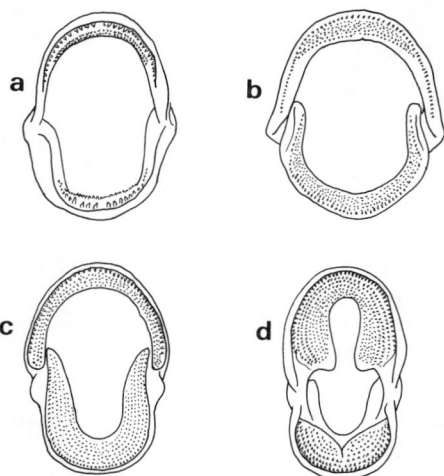

Figure 3. Tooth pattern in the jaws.
a. *Astatotilapia elegans* (an insectivore) to show the generalized condition. b. *Neochromis nigricans*, an epilithic algal grazer. c. *Hoplotilapia retrodens*, a molluscivore (oral sheller). d. *Platytaeniodus degeni*, detritus feeder on sandy substrates.

all are features associated with how food is gathered and swallowed (Greenwood 1953). The intestine in these fishes, as in other vegetarian species, is considerably longer than in insectivores.

Truly phytophagous haplochromines in the lake also have elongate guts, but show more pronounced changes in their oral dentition, and in some species, changes in skull shape and jaw morphology as well. In those which scrape algae from the stems and leaves of aquatic plants, the dental battery forms a brushlike surface. Species which graze algae from rocks and other hard surfaces also have broad bands of teeth, but their outer teeth have a different cusp form, and are stouter. Their jaws too are stouter than those of the epiphytic grazers, and the upper can be protruded downwards. This change in the direction of protrusion is effected mainly by the steeply inclined slope of the anterior skull region over which the ascending process of the premaxilla travels. The dentition in a third group of phytophages, those browsing on rooted plants, departs less markedly from the generalized insectivore type.

A scraping dentition involving greatly increased numbers of tooth rows but a

somewhat different tooth form, has evolved in the single lepidophagous species so far described from Lake Victoria (Greenwood 1980:57). The lepidophage, unlike the phytophages, has a body form closely resembling that of most piscivorous predators, and a relatively short gut. Molluscs are the main food source for more than 20 species. In one subgroup of mollusc eaters, the prey, usually snails, is crushed between the upper and lower pharyngeal jaws. These bones and their musculature are noticeably hypertrophied, and their dentition is variously modified towards replacement of the generalized pharyngeal dentition by stout, often molariform teeth. Apart from some strengthening of certain skull bones, particularly those articulating with the upper pharyngeal bones, the skull, jaws and oral dentition in these species do not depart appreciably from the generalized condition.

Members of the other mollusc-eating subgroup deal with prey very differently, either by crushing the shell between the oral jaws, or by wrenching a snail from its shell and swallowing the soft parts. In these species, the jaw bones are strengthened, the skull is strongly decurved anteriorly, enabling the protruding upper jaw to be directed down rather than forward, as it is in the generalized skull-jaw functional complex. The jaw teeth are strongly recurved, piercing unicuspids. The pharyngeal bones and dentition, however, are essentially of the generalized type.

In addition to the bottom-feeding insectivore mentioned earlier, there is another, more specialized, bottom-feeding insectivore lineage (Greenwood 1980:67). Members of this group remove prey from burrows or crevices in rocks and wood. Derived features which permit this specialization involve slight changes in skull form, relative elongation of certain parts of the upper jaw, and the presence of strong elongate unicuspid teeth anteriorly in both jaws. In combination, these features provide what are effectively forceps which can be thrust into narrow spaces to grasp prey.

All the Lake Victoria haplochromines whose breeding habits are known practice female buccal incubation, the young leaving the mouth on attaining a length of 10-15 mm. This breeding habit is exploited as a food source

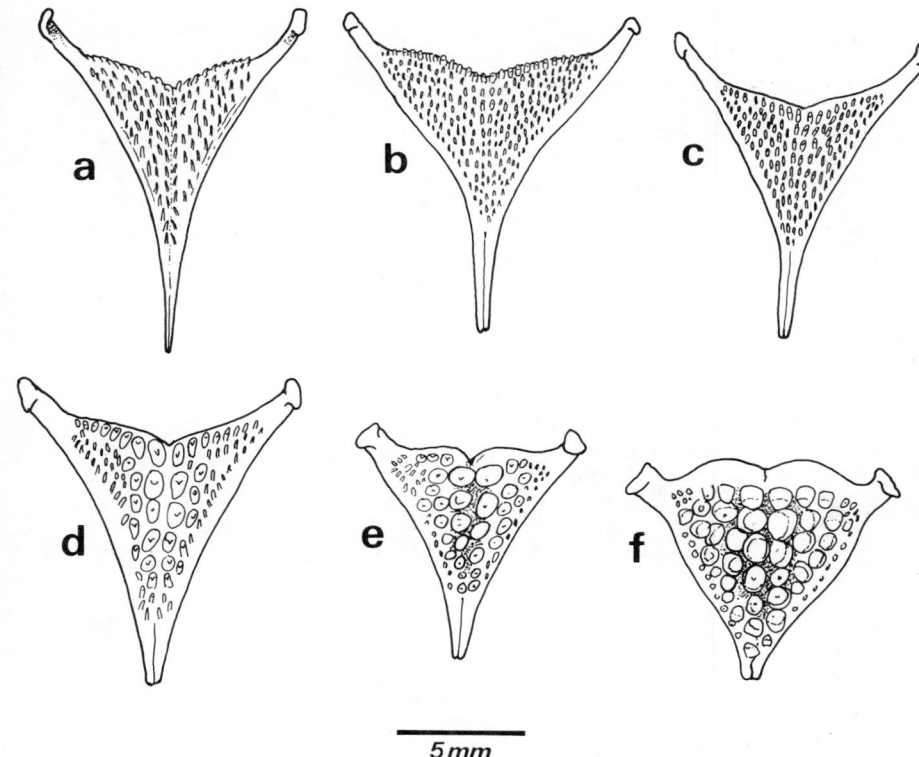

Figure 4. Lower pharyngeal bone and dentition, in occlusal view. a. Piscivore. b. Insectivore. c. Detritus feeder. d. Mixed mollusc and insect eater. e. Another mixed mollusc-insect eater. f. Specialized mollusc eater. All three mollusc eaters are pharyngeal crushers. A bone and dentition like that in b are taken to represent the generalized condition.

by at least 10 species, whose diet consists mainly of embryos and larvae obtained from brooding female fishes. Various feeding strategies are employed by these paedophagous haplochromines, including engulfing the snout and mouth of a brooding fish, or harassing the female by butting and chasing until she jettisons the brood (Greenwood 1974a; Wilhelm 1980). Most paedophages have small jaw teeth deeply embedded in the oral mucosa, and they have jaws which are markedly protrusible and distensible. This latter feature is effected mainly by the relative elongation of the ethmovomerine skull region over which the premaxilla slides when the mouth opens.

Finally, there are numerous predatory piscivorous species which feed on fishes larger than those taken by the paedophages. Piscivores may comprise 30-40 percent of the known haplochromine species in Lake Victoria (Greenwood 1974a). A few piscivores are virtually indistinguishable in gross morphology, pharyngeal and oral dentition, and cranial architecture from the larger generalized insectivores. Many, in fact, regularly include insects and other invertebrates in their diet. Most piscivores, however, appear to subsist as adults or near-adults on other haplochromines. The overall appearance of these species shows the greatest departure from the generalized body form, and adults have the largest body sizes among haplochromines, 120-220 mm SL. The body and head are slender and elongate, though in some species, neither feature differs

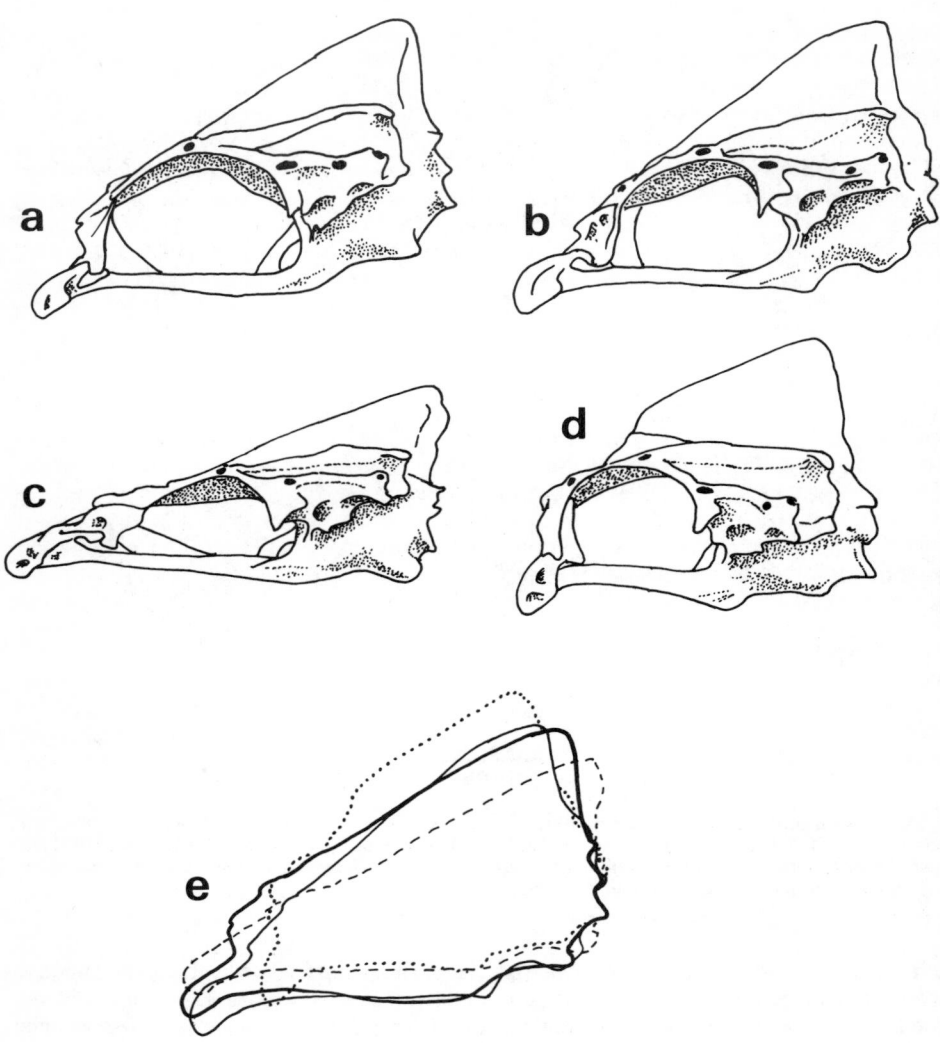

Figure 5. Skull form in various Lake Victoria haplochromines. a *Astatotilapia macrops*, an insectivore. b. *Harpagochromis maculipinna*, a piscivore. c. *Prognathochromis gowersi*, a piscivore. d. *Macropleurodus bicolor*, an oral shelling mollusc eater. e. Four skull lines superimposed show their proportional changes: dotted = *M. bicolor*, dashed = *P. gowersi*, thick line = *A. macrops*, thin line = *H. maculipinna*. Skull form of *A. macrops* is taken to represent the generalized condition. The skulls are drawn at different scales, but to the same occupital-lateral ethmoid length.

greatly from the generalized type. The jaws are large, very protrusible and expansible, the oral teeth are strong, curved, piercing unicuspids, and the skull has a protracted and shallow preorbital region. At least superficially, the pharyngeal jaws do not seem to differ from the generalized condition, but the teeth

have a more clearly defined cutting edge on the major cusp.

These features of the oral and pharyngeal dentition, and the enhanced protrusibility and expansibility of the jaws are associated with the piscivore manner of feeding. They capture and hold the prey fish in the jaws and gradually macerate it with the pharyngeal dentition as it passes through the pharynx into the esophagus (Greenwood 1974a). Several species whose adult members feed on zooplankton, a trophic group commonest in the deeper waters of the lake, have a close superficial resemblance to the more streamlined piscivores. Skull and jaw form resemble those of the piscivores, but the oral dentition is finer and usually consists of slender tricuspid teeth, of which the functional significance is unknown.

This survey of feeding habits and associated anatomical features is, perforce, an inadequate account of the often subtle morphological features characterizing members of a major trophic group (for more details see Greenwood 1974a,1981; Witte 1984). Its intention is chiefly to show that the morphological features correlated with different feeding habits are very slight indeed, as are their levels of departure from the basic and trophically unspecialized haplochromine condition. It is also intended to show that most morphological changes from that basic condition seem to involve simple alterations in relative proportions of various skeletal parts, an increase in body size, changes in proportion of cusp size in the teeth, and changes in pattern of tooth distribution in the jaws. None would appear to involve macromutations or other extraordinary evolutionary events (also, see Strauss 1984). In terms of overall food resource partitioning through trophic specialization, the consequences of such minor changes are profound (Liem, Osse 1975). We seem to have in these fishes evidence for an ecological revolution brought about without a corresponding morphological one.

MORPHOCLINES IN LAKE VICTORIA

It is of particular interest that one can trace among still existing species those gradual changes which underlie the development of any one specialized endpoint. In the group that crushes molluscs intrapharyngeally, there are some species with only a few enlarged pharyngeal teeth carried on slightly hypertrophied pharyngeal bones. Others have many more enlarged and molariform teeth, noticeable enlargement of the supporting bones, and some hypertrophy of pharyngeal

musculature. Finally, there are species in which few pharyngeal teeth remain non-molariform, the pharyngeal bones are greatly enlarged, and there is an associated increase in hypertrophy of the pharyngeal musculature. There is some suggestion that in some of these species, the ultimate development of extreme hypertrophy may indeed be environmentally induced, with only the basic steps in that direction being under genetic control (Greenwood 1965).

In the complex of epiphytic algal-grazers one can trace a step by step and species by species change in the oral dentition, from one in which the jaw teeth barely differ from the generalized bicuspid type to one in which the upper cusp is greatly protracted and lies obliquely to the vertical axis of the tooth, and the minor cusp is suppressed (Greenwood 1974a:63). Associated with this progression is an increase in the inner tooth rows and a tendency for the outermost tricuspid teeth of the inner series to be replaced by teeth closely resembling those of the outer rows.

Among the piscivorous predators it is possible to detect series showing a gradual, species by species changeover from a bicuspid biting dentition to unicuspid grasping dentition. Likewise one can trace changes in relative skull proportions, leading by differential growth of particular skull regions, from the generalized shape to one which is elongate and shallow, which has a relatively attenuated and gently sloping ethmo-vomerine region over which the premaxillary slides when the mouth is protruded. Associated with the skull elongation is a gradual refinement in body form, change in shape of jaw elements, and enhanced jaw protrusibility.

Such trends can be seen in every trophic group and many phyletic lineages, generally with more than one, and often several species representing a particular point in a morphocline of character development. Since we are dealing with coexisting biological species, the pattern cannot be interpreted as one of phyletic gradualism, nor can it be intraspecific polymorphism of the kind seen in the remarkable cichlid of Cuatro Cienegas, Mexico (Sage, Selander 1975; Kornfield et al. 1982; Kornfield, Taylor 1983; Liem, Kaufman 1984).

Stanley (1979) argued that in a punctuational model, trends are predominantly the outcome of species selection. The trends with which he was concerned have an extended time range, and none of his examples involves the contemporaneous existence or coexistence of steps in the trend seen in the Lake Victoria haplochromines. Certainly the situation among the latter species would not seem to be an obvious example of what Gould (1982) described as the outcome of ".. a higher order process that operates upon the differential birth and death of species .."

Lake Victoria is not unique in this respect. Similar patterns are detectable in several of the Lake Malawi lineages, and less clearly, since the gaps are greater, in lineages from Lake Tanganyika. But the phenomenon is not apparent in the small haplochromine assemblages of Lakes Albert and Turkana (Greenwood 1974b), nor in the riverine species.

The ecological and evolutionary aspects of this gradualism, expressed through cladogenesis, raise numerous questions. If natural selection has played some past role in the evolution of different characters, why have so many different levels of anatomical specialization evolved when, at least over a large part of a morphocline, the various species seem equally well equipped to exploit the same trophic niche? As a corollary, how is it that the species still coexist if each has similar demands on a particular food resource and on other environmental factors? From another level, can interspecific selection, *sensu* Stanley (1979), have been operating?

I find no answer to that question in the Lake Victoria haplochromines. One could speculate that the swarm is young, and that, as yet, there has been no concatenation of circumstances which might invoke the necessary pressures to effect selection at the species level. Such an argument contributes nothing to our understanding of the situation. To me, the whole concept of species selection suffers from all the weaknesses inherent in the concept of natural selection —weaknesses which are becoming more apparent as further data are accumulated about the "classical" examples of natural selection (Jones 1982; Leith 1982).

One paradox may contribute to the resolution, or nonexistence, of this problem in nature; namely that specialized species are potentially able to exploit a wider range of food than can their structurally generalized relatives (Greenwood 1981; Liem 1981). Thus, piscivorous predators sometimes feed on insect larvae, as do the paedophages. Mollusc-eaters can be effective insectivores. Even the algal-grazers ingest and apparently digest small animals. Generalized species, on the other hand, are unable to exploit trophic niches for which these other species have the necessary anatomical specializations.

The question of whether natural selection has played a major part in the evolution of the various derived or specialized characters still cannot be answered, nor can that of species selection in the evolution of the different morphoclinal trends. That the anatomical characters involved in trophic specialization are derived ones seems beyond dispute. That they are used in exploiting certain food sources also seems indisputable. What can be contested is their identification as specific adaptations that have evolved through the deterministic action of selection.

When considering the role of natural or species selection in the evolution of these fishes, we must note that, in many cases, the same "adaptive" characters are present in several species. Often more than one of those species occur together in the same habitat, and each species apparently exploits the same environmental resources. If selection operated in the origin and early establishment of a character —and we cannot determine that— and if it be assumed that the character spread through several species as the result of repeated speciation events involving related ancestors, then selection would not seem to be effective now that the different but "adaptively" similar species are syntopic. More information on the ecology and behavior of such trophically and anatomically similar species might throw light on some aspects of the problem, but we would still be unable to establish what part selection played in the origin and initial establishment of the derived characters or of the species themselves.

Likewise, it is difficult to imagine how selection can be invoked to explain the evolution of highly derived features, such as dental specialization, when functionally and morphologically intermediate stages in their development are still present in species occurring syntopically with the more derived forms. Any explanation is made more difficult when, as in the Lake Victoria haplochromines, the intermediates have seemingly the same "adaptive" status as the more derived species, and when the several replicas of the more specialized conditions coexist with each other and with the less specialized forms. That this pattern of evolution has occurred, and that it has been achieved through a process of descent with modification, seems well substantiated. To go beyond this now would be to carry speculation far beyond its utility as an intellectual stimulant.

When considering the Lake Victoria haplochromines in the broader context of evolutionary theory, it should be emphasized that the various taxa are discrete, noninterbreeding units: they are biological species. It is also important to note that the development of a character or suite of characters, as manifest in a morphocline of species, often involves recognition of very slight changes in the form of those characters. Without the biological information we have which indicates that the different stages are represented by distinct species, it would be easy to dismiss them as intraspecific variants, as was often done when fishes were studied by museum-bound workers. Similarly, it would often be impossible to recognize that any one level of specialization is represented by two or more species, especially when those taxa are sibling species (Marsh et al. 1981; McKaye et al. 1982).

In other words, if the taxa were fossils, it would be impossible to derive from them the information which is available to neontologists. It is therefore not feasible to compare the Lake Victoria haplochromines with any fossil assemblage, and thus to assess whether similar situations existed in the past. The value of the haplochromine flocks lies in demonstrating what can happen and did happen, within the geologically short period of 750,000 years. Similar restraints are involved when considering the fossil record for the so-called periods of stasis that are postulated to follow punctuation events. The number of species surviving the effects of "species selection" could well be higher than the fossil

remains might seem to indicate.

The recently discussed Pleistocene mollusc fauna of Lake Turkana (Williamson 1981) invites comparison with these extant fishes, bearing in mind the restraining factors discussed above. The Lake Turkana molluscan fauna was cited (Williamson 1981) as a particularly clear example of a evolutionary pattern conforming to the "punctuated equilibrium" model (Eldredge, Gould 1972; Gould, Eldredge 1977; Gould 1980, 1982). Williamson cited one particular section of the Turkana sequence as a good example of the punctuational phase of evolution. But arguments can be leveled against his claims that the distinct morphotypes which he recognizes are in fact species rather than ecophenotypic variants.

I believe that the Lake Victoria haplochromines would provide a far better example of a punctuational phase, because the specific biological identity of the fishes is far better established than that of the fossil molluscs. Admittedly, the Victoria fish species are no longer allopatric, but the geological history of the lake strongly suggests that they originated allopatrically within the area of the present lake basin (Fryer, Iles 1972; Greenwood 1974a). Even if some sympatric speciation were involved (Greenwood 1974a, 1981), the haplochromines would still serve as a good example of rapid speciation associated in an overall sense with marked phenotypic changes occuring in a geographically circumscribed area.

In the preceding sentence I say "overall" because the extant haplochromine flock, unlike the fossil molluscs, exhibits gradual changes in the development of particular characters. But since various stages of character development are represented by distinct and contemporary species, the process should be identified as *cladistic* rather than phyletic gradualism (Greenwood 1981).

To my knowledge, this pattern of contemporaneous cladistic gradualism has not been recognized in any of the fossil assemblages used to exemplify punctuational phases in evolution. But it is unlikely that any fossil assemblage could provide the type of fine-scale data given by the Lake Victoria haplochromines. I would estimate that even if the present haplochromine fauna were to be fossilized under optimal conditions, the number of species which future investigators might recognize is unlikely to exceed 50. This maximum would be reduced through the inevitable imperfections of the fossil record. From these losses, future investigators of the hypothetical fossil "flock" would see larger morphological gaps between taxa, and since they would be unlikely to recognize that intermediate stages in character development were present and represented in contemporaneous species, the phenomenon of cladistic gradualism would not be apparent to them.

The punctuational or explosive phase of the punctuated equilibrium model clearly upsets some Darwinians. In reviewing Stanley's (1979) book, *Macroevolution, Pattern and Process*, Lande (1980) criticizes the way punctuated equilibrists interpret both the paleontological evidence, particularly the absence of transitional forms, and the genetic background to speciation and morphological changes, especially their implications for macromutation and "hopeful monsters". He also queries their scaling of the time elements. The Lake Victoria haplochromines would provide counters to most of Lande's arguments, particularly regarding morphological features and the problem of time. The genetic factors remain unknown, but their phenotypic manifestation in the haplochromines would not suggest genetic revolutions or even large scale systematic mutations. If anything, they suggest the involvement of genes controlling developmental rates (see Sage et al. 1984; Strauss 1984).

On the problem of intermediate forms, the living and contemporaneous haplochromine species of Lake Victoria clearly demonstrate an abundance of intermediates, both morphologically and ecologically. Whether these intermediates would be recognizable as fossils seems unlikely, but their existence in this living model is beyond doubt.

Ruse (1981:218) wrote critically of what he considered to be the punctuated equilibrists implied "drastic change .. totally new forms of animals .. [and] .. fundamentally modified ground plans." Neither drastic changes nor fundamentally modified ground plans are involved in the production of a cichlid species flock, especially one at the stage represented in Lake Victoria. Yet in evolutionary terms, or in terms of overall morphological change and ecological differentiation, the result is outstanding and the time span is remarkably short.

So far, only the haplochromines of Lake Victoria have been considered in relation to the punctuated equilibrium model. I have noted the greater diversity of the cichlids in the older Lake Tanganyika, a diversity which encompasses more extreme levels of anatomical and trophic specialization than that of the Lake Victoria cichlids (Fryer, Iles 1972; Myers 1960). The Lake Tanganyika cichlids also differ from those of Lake Victoria in that their endemic lineages, i.e., genera, are more sharply demarcated morphologically. The phenomenon of cladistic gradualism is far less obvious and replaced by a more punctuated one. Yet in most cases, it is still possible to identify the prior condition from which the more extreme form could have evolved. However, the Tanganyika cichlids are like those of Victoria in that the morphological differences stem from changes in differential growth, and hence, changes in relative proportions (e.g., Strauss 1984). As in other lakes, the anatomical features are mainly those concerned with feeding.

The readily differentiable lineages of Lake Tanganyika, coupled with the fact that interspecific differences are also more trenchant than in Lake Victoria might suggest that the Tanganyika cichlids, after a period of species selection, are in a phase of stasis postulated in the punctuated equilibrium model. Unfortunately, this phase and that of species selection can only be recognized from the fossil record, and none is available.

Lake Malawi's cichlid fauna occupies a position intermediate between those of Victoria and Tanganyika (Fryer, Iles 1972). It has fewer examples of extreme anatomical specialization than has Tanganyika, rather more than has Victoria, and shows many examples of cladistic gradualism among the 100 + species which, like those in Victoria, were previously placed in the genus *Haplochromis*. Since there is virtually no information about the genetics of African cichlids, and since we are still dealing only with crude ecological data, we cannot draw meaningful conclusions about the evolutionary processes involved in producing these different forms. I suspect that this is true for most attempts to explain evolutionary patterns and processes above the intraspecific level.

Possibly the most interesting conclusion to be drawn from the African cichlids, and from all known examples of island and lake species flocks is the relatively low level of morphological differentiation necessary to initiate and produce "adaptive radiations" of considerable magnitude. The cichlids also demonstrate how, by a process of cladistic gradualism and its correlate, multiple and rapid speciation, anatomical specializations can evolve rapidly by what seem to be ontogenetically simple means. It may be that "hopeful monsters" did and do exist, and that we have failed to recognize them because, to taxonomists, they are not monstrous enough. Seen in that light, many aspects of the punctuated equilibrium model gain credibility, and, from a biological viewpoint, feasibility too. How widespread the role of punctuated or "explosive evolution" has been in the history of life remains an unanswered, and perhaps unanswerable question, as does the role of natural selection in evolution above the intrapopulation level.

SUMMARY

The species flock or flocks of haplochromine fishes in Lake Victoria, East Africa, appear to be an outstanding example of an extant punctuational evolutionary phase. During the 750,000-year history of the lake, its endemic haplochromine species have come to occupy a dominant position in the lake's ecology as a result of their extensive and diverse trophic specializations. Clearcut trends in the development of different anatomical specializations associated with the various feeding habits are discernible. The stages of derivation in any one morphocline are represented by distinct species, often with more than one species at a particular level of specialization still extant. Individual lineages in the flock thus seem to represent a form of cladistic gradualism rather than one of phyletic gradualism. The principal anatomical features involved in the radiation seem to have been brought about through simple changes in relative growth, and are manifest chiefly in syncranial characters and in oral and pharyngeal dentition.

It is suggested that this flock provides little evidence for the effects of natural selection or of species selection in its origin and development. On the other hand, it seemingly provides better evidence for the probability of punctuational evolutionary phases than do most of the paleontological examples cited by the proponents of the punctuated equilibrium theory. However, it does not provide any evidence to support a thesis that extraordinary genetic or epigenetic processes were involved in producing its pattern of speciation and morphological diversity.

REFERENCES

Barel C D N, M J P Van Oijen, F Witte, E Witte-Maas 1977 An introduction to the taxonomy and morphology of the haplochromine cichlids from Lake Victoria. *Neth J Zool* 27:381-389

Barbour C D, J H Brown 1974 Fish species diversity in lakes. *Amer Natur* 108:473-489

Bell-Cross G 1972 *Arnoldia* 5:1-19

Cromie W J 1982 Beneath the African Lakes. *Mosaic* 13:2-6

Darwin C 1898 *The Origin of Species*. 6th ed. Murray, London

Eldredge N, S J Gould 1972 Punctuated equilibria: an alternative to phyletic gradualism. 82-115 T Schopf ed. *Models of Paleobiology*. Freeman, Cooper

Fryer G, T D Iles 1972 *The Cichlid Fishes of the Great Lakes of Africa. Their Biology and Evolution*. Oliver Boyd. Edinburgh

Gould S J 1980 Is a new and general theory of evolution emerging? *Paleobiol* 6:119-130

——— 1982 Punctuated equilibrium—a different way of seeing. 26-30 J Cherfas ed. *Darwin Up to Date*. New Scientist. London

Gould S J, N Eldredge 1977 Punctuated equilibrium: the tempo and mode of evolution reconsidered. *Paleobiol* 3:115-151

Greenwood P H 1953 Feeding mechanism of the cichlid fish *Tilapia esculenta* Graham. *Nature* 172:207-208

——— 1965 Environmental effects on the pharyngeal mill of the cichlid fish *Astatoreochromis alluaudi*. *Proc Linn Soc.* 176:1-10

——— 1973 A revision of the *Haplochromis* and related species (Pisces, Cichlidae) from Lake George, Uganda. *Bull Brit Mus Natur Hist (Zool)* 25:139-242

——— 1974a Cichlid fishes of Lake Victoria, east Africa: The biology and evolution of a species flock. *Ibid* Suppl 6:1-134

——— 1974b The *Haplochromis* species (Pisces: Cichlidae) of Lake Rudolf, east Africa. *Ibid* 27:139-165

——— 1979 Macroevolution—myth or reality? *Biol J Linn Soc* 12:293-304

——— 1980 Towards a phyletic classification of the "genus" *Haplochromis* (Pisces: Cichlidae)

and related genera. Part II. *Bull Brit Mus Nat Hist (Zool)* 39:1-101

_____ 1981 Species flocks and explosive evolution. 61-74 P H Greenwood, P L Forey eds. *Chance, Change and Challenge — The Evolving Biosphere*. Cambridge Univ Press and British Museum (Natural History). London.

_____ 1983a The *Ophthalmotilapia* assemblage of cichlid fishes reconsidered. *Bull Brit Mus Natur Hist (Zool)* 44:249-290

_____ 1983b On *Macropleurodus, Chilotilapia* (Teleostei, Cichlidae), and the interrelationships of African cichlid species flocks. *Ibid* 45:209-231

_____ 1984 What *is* a species flock? 13-20 A A Echelle, I Kornfield eds. *Evolution of Fish Species Flocks*. Univ Maine Press. Orono

Jones J S 1982 More to melanism than meets the eye. *Nature* 300:109-110

Kornfield I, D C Smith, P S Gagnon, J N Taylor 1982 The cichlid fish of Cuatro Cienegas, Mexico: direct evidence of conspecificity among trophic morphs. *Evolution* 36:658-664

Kornfield I, J N Taylor 1983 A new species of polymorphic fish, *Cichlasoma minckleyi*, from Cuatro Cienegas, Mexico (Teleostei: Cichlidae). *Proc Biol Soc Wash* 96:253-269

Lande R 1980 Microevolution in relation to macroevolution. *Paleobiol* 6:235-238

Leith B 1982 *The Descent of Darwin*. Collins. London

Liem K F 1973 Evolutionary strategies and morphological innovations: cichlid pharyngeal jaws. *Syst Zool* 22:425-441

_____ 1981 A phyletic study of the Lake Tanganyika cichlid genera *Asprotilapia, Ectodus, Lestradea, Cunningtonia, Ophthalmochromis* and *Ophthalmotilapia*. *Bull Mus Comp Zool* 149:191-214

Liem K F 1973 Evolutionary strategies and morphological innovations: cichlid pharyngeal jaws. *Syst Zool* 22:425-441

_____ 1981 A phyletic study of the Lake Tanganyika cichlid genera *Astrotilapia, Ectodus, Lestradea, Cunningtonia, Ophthalmochromis* and *Ophthalmotilapia*. *Bull Mus Comp Zool* 149:191-214

Liem K F, L Kaufman 1984 Intraspecific macroevolution: functional biology of the polymorphic cichlid species *Cichlasoma minckleyi*. 203-216 A A Echelle, I Kornfield eds. *Evolution of Fish Species Flocks*. Univ Maine Press at Orono

Liem K F, J W M Osse 1975 Biological versatility, evolution and food resource exploitation in African cichlid fishes. *Amer Zool* 15:427-454

McKaye K R, W N Gray 1984 Extrinsic barriers to gene flow in rock-dwelling cichlids of Lake Malawi: macrohabitat heterogeneity and reef colonization. 169-184 A Echelle, I Kornfield eds. *Evolution of Fish Species Flocks*. Univ Maine Press at Orono

McKaye K R, T Kocher, P Reinthal, I Kornfield 1982 A sympatric sibling species complex of *Petrotilapia* Trewavas from Lake Malawi analysed by enzyme electrophoresis (Pisces: Cichlidae). *Zool J Linn Soc* 76:91-96

Marsh A C, A J Ribbink, B A Marsh 1981 Sibling species complexes in sympatric populations of *Petrotilapia* Trewavas (Cichlidae, Lake Malawi). *Zool J Linn Soc* 71:253-264

Myers G S 1960 The endemic fish fauna of Lake Lanao and the evolution of higher taxonomic categories. *Evolution* 14:323-333

Ruse M 1981 *Darwinism Defended*. Addison Wesley. Reading Mass

Sage R D, R K Selander 1975 Trophic radiation through polymorphism in cichlid fishes. *Proc Nat Acad Sci* 72:4669-4673

Sage R D, P V Loiselle, P Basasibwaki, A C Wilson 1984 Molecular versus morphological change among cichlid fishes (Pisces: Cichlidae) of Lake Victoria. 185-202 A A Echelle, I Kornfield eds. *Evolution of Fish Species Flocks*. Univ Maine Press at Orono

Stanley S M 1979 *Macroevolution, Pattern and Process*. Freeman. San Francisco

Strauss R E 1984 Allometry and functional feeding morphology in haplochromine cichlids. 217-230 A Echelle, I Kornfield eds. *Evolution of Fish Species Flocks*. Univ Maine Press at Orono

Van Couvering J A 1982 Fossil cichlid fish of Africa. *Spec Pap Paleont* 19:1-103

Wilhelm W 1980 The disputed feeding behaviour of a paedophageous haplochromine cichlid (Pisces) observed and discussed. *Behaviour* 74:310-323

Williamson P G 1981 Paleontological documentation of speciation in Cenozoic molluscs from Turkana Basin. *Nature* 293:437-443

Witte F 1984 Ecological differentiation in Lake Victoria haplochromines: comparison of cichlid species flocks in African lakes. 155-168 A A Echelle, I Kornfield eds. *Evolution of Fish Species Flocks*. Univ Maine Press at Orono

ECOLOGICAL DIFFERENTIATION IN LAKE VICTORIA HAPLOCHROMINES: COMPARISON OF CICHLID SPECIES FLOCKS IN AFRICAN LAKES

F WITTE

INTRODUCTION

Comparison of the cichlid species flocks of the three Great Lakes of Africa, lakes Victoria, Malawi and Tanganyika, has been extensively discussed. In this paper, the term *flock* denotes a group of similar species in a topographically well defined area; monophyletic origin is not implied. Based on the available ecological data, it has been concluded that the Lake Victoria haplochromines are ecologically less specialized and consequently form less close knit communities than the cichlids of the other two lakes (Greenwood 1974, 1981; Fryer 1977; Fryer, Iles 1972). However, recent work by the Haplochromis Ecology Survey Team on the haplochromines in the Mwanza Gulf of Lake Victoria (Figure 1) shows that ecological specialization and habitat restriction of Victorian haplochromines are more strongly developed than originally thought. This brings into question a number of previous conclusions regarding the evolutionary dynamics of the Lake Victoria flock.

NOMENCLATURE

I use, herein, the generic name *Haplochromis* for all Lake Victoria haplochromines except those belonging to the monotypic genera as defined by Greenwood (1956, 1959). Thus, I do not follow the recent classification of the genus *Haplochromis* (Greenwood l979, l980). For a critical analysis of some of Greenwood's proposals, see Hoogerhoud (1984). Informal names of the many new, as yet undescribed species of Lake Victoria haplochromines are placed in quotation marks. A reference collection of these species has been deposited at the Rijksmuseum van Natuurlijke Historie, Leiden, The Netherlands.

RESULTS AND DISCUSSION

The Species Concept. Our research group has distinguished more than 250 different forms of haplochromines in the Mwanza Gulf alone. These forms, especially those belonging to the same trophic group, are often separated by small morphological differences. The taxonomic status of these forms is questionable. Are they all true species, or does the diversity include a number of polymorphic species? The problem is exemplified by the works of Sage and Selander (1975), Kornfield et al. (1982), Kornfield and Taylor (1983), and Liem and Kaufman (1984) which demonstrate that *Cichlasoma minckleyi*, a cichlid from Cuatro Cienegas, Mexico, includes two morphologically and functionally distinct trophic types. In addition, *Cichlasoma citrinellum* of Middle America includes two color morphs in both males and females, which interbreed despite a certain degree of habitat partitioning during spawning periods (McKaye 1980). In contrast, underwater observations by Holzberg (1978) and Marsh et al. (1981) demonstrated assortative mating among Lake Malawi color morphs of *Pseudotropheus zebra* and *Petrotilapia tridentiger*. Enzyme electrophoresis revealed genetic discontinuity among color morphs and supported the conclusion that they represent sibling species (McKaye et al. 1982). Underwater observations are difficult in the murky water of Lake Victoria. However, the following examples indicate that at least in the three cases examined, closely similar coexisting forms in Lake Victoria are genetically isolated: 1) *H. iris* and *H. hiatus* are isolated by differences in depth preference and in spawning period, although in both cases there is overlap (Hoogerhoud et al. 1983). 2) The two zooplanktivores *H.*

"argens" and *H*. "heusinkveldi" differ in depth preference and seem to spawn in different areas. The example of *Cichlasoma citrinellum* indicates that even differences in spawning areas, when not complete, may not prevent interbreeding, which requires care in interpreting these data. 3) Three closely similar forms of epilithic algal grazers kept together in an aquarium for more than two years were not observed interbreeding although spawning occurred regularly (Witte-Maas, Witte pers obs). Thus far, studies on African lake cichlids provide no examples of polymorphism comparable to that in *Cichlasoma minckleyi*. Only preliminary genetic data are available for African cichlids (Kornfield 1978; Sage et al. 1984), and the possibility remains that such polymorphisms are present.

Trophic Groups. Subdivision into trophic groups was the first attempt at ecological classification of Lake Victoria haplochromines (Greenwood 1974). A trophic group is defined as a group of species feeding predominantly, but not exclusively, on a certain food type (Greenwood 1981; Witte 1981). Van Oijen et al. (1981) previously noted that the definition of food types and consequently of trophic groups is not consistent, and is based on a "... mixture of criteria ... on the systematic position of food items (molluscivores), on the size (ontogentic stage) of the food (paedophages), or on the location of the food, e.g., periphyton versus phytoplankton eaters ..." Yet our research reveals that these trophic groups can be useful in ecological studies, as the included species generally share a number of ecological parameters such as distribution patterns, feeding and migration patterns, and breeding patterns, which are related in some way to their major food type (Figure 2).

Morphologically, most trophic groups are recognizable by a particular facies. A facies and its constituent structures is related more to the way food is collected than to the food type itself, as with oral shelling or pharyngeal crushing of molluscs (see Barel 1983).

Formerly it was thought that the number of trophic groups in the species flocks of Lake Victoria was smaller than that of lakes Tanganyika and Malawi (Fryer, Iles 1972). However, although the morphological range is different in the Lake Victoria flock, the

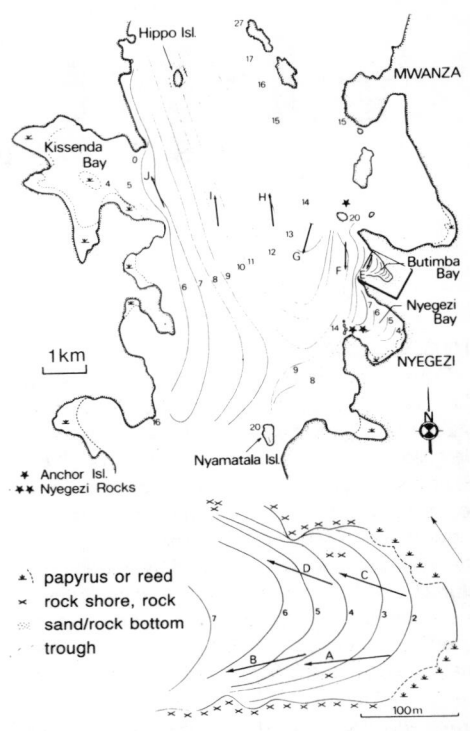

* Anchor Isl.
** Nyegezi Rocks

⌐) papyrus or reed
× rock shore, rock
▒ sand/rock bottom
⋯ trough

Figure 1. Primary sampling area of the Haplochromis Ecology Survey Team, Mwanza Gulf, Lake Victoria. Depth contours in meters. Butimba Bay inset, below, with 100 m scale.

number of trophic groups is similar in the flocks of the three lakes (Barel et al. 1977; Greenwood 1974). At present 11 main trophic groups are recognized in Lake Victoria. These include 8 groups described by Greenwood (1974): molluscivorous oral shellers and crushers and pharyngeal crushers, epiphytic and epilithic algal scrapers, insectivores, macrophytivores, detritivores, phytoplanktivores, prawn eaters and piscivores, comprising piscivores sensu stricto, paedophages and scale eaters, and, more recently discovered, zooplanktivores and parasite eaters (van Oijen et al. 1981; Witte 1981; Witte, Witte-Maas 1981) and a crab eater (van Oijen pers comm). Because there are far more than 250 haplochromine species in Lake Victoria and only 11 main trophic groups, many species use similar food sources.

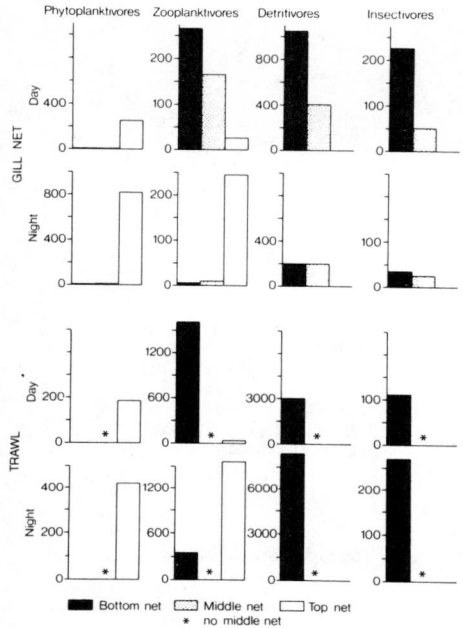

Bottom net ■   Middle net □···   Top net □
* no middle net

Figure 2. Total fishes of four trophic groups indicating habitat preference and diel migration. Taken in 8 different day and night catches through 1981-1982 at 3 levels: bottom, mid-water and surface at station G, Figure 1.

Ecological restriction and segregation.

Fryer and Iles (1972) emphasized that there were marked differences in levels of ecological restriction and specialization between the cichlid flock of Lake Victoria and those of Lakes Malawi and Tanganyika. They say: "... communities in Lake Victoria appear to be less close-knit than those in Lake Malawi. This reflects both the nature of the available habitats and the smaller degree of specialization of the Lake Victoria species ... many ... cichlids [of Lake Victoria] are certainly much less rigorously restricted to particular habitats than are those of Lake Tanganyika and Lake Malawi." Nevertheless, Fryer and Iles do not deny well-defined habitat preferences in *some* Lake Victoria species, and in this context, the epilithic algal grazer, *H. nigricans* is mentioned. Illustrating the "wide ecological tolerance" they refer particularly to four piscivorous species. Fryer and Iles conclude that the Lake Victoria haplochromines probably can "move

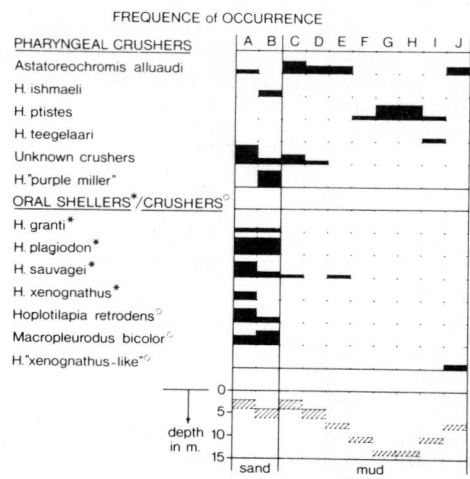

Figure 3. Percentage occurrence of molluscivores. Monthly trawl catches from sampling stations A-J, February 1979-February 1980. Percentage = percent of total number of catches which included certain species. Each square = 100 percent.

relatively easily from one community to another."

Although more cautious, Greenwood (1974: 48,113) arrived at similar conclusions, "...that in Lake Victoria there is, as compared to Lake Malawi, less obvious stenotopy amongst members of the cichlid species flock," but he would "be chary of suggesting" that in Lake Victoria, *Haplochromis* sometimes shift habitat. He states further that "The role of habitudinal segregation in the speciation of Lake Victoria *Haplochromis* is difficult to establish. Few of the contemporary species show such close association with a particular habitat as do several Lake Malawi cichlids." Although far from complete, our studies in the past six years reveal stronger ecological restriction and specialization than originally ascribed to Lake Victoria cichlids. The following examples, based on studies of four trophic groups illustrate this point.

Molluscivores. Restricted depth distribution and substrate preferences for sand or mud occur among molluscivores (Greenwood 1974; Witte 1981). The degree of restriction becomes particularly clear from the results of Witte (1981; Fig 3) when one realizes that the centers of stations A, B, C and D are no more than 100 m apart, and that the size of

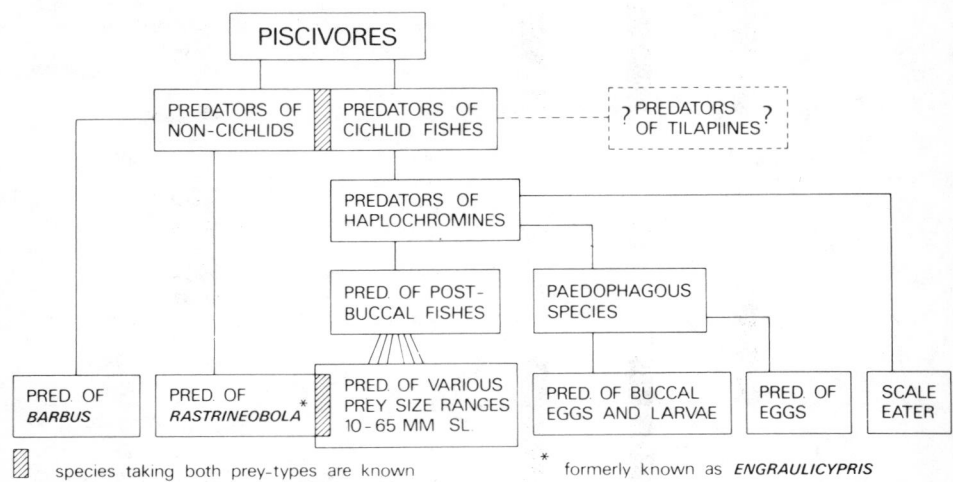

Figure 4. Subtrophic categories dividing piscivores by food preference. (van Oijen 1982)

Butimba Bay is less than 0.5 km² (Figure 1). The initial results suggest that the pharyngeal crushers feed on smaller molluscs than do the oral shellers (Hoogerhoud pers comm). To some extent, ecological segregation is realized by habitat restrictions and food preferences, but considerable overlap remains, especially among the oral shellers which occur almost exclusively over sandy substrates (Greenwood 1974; Witte 1981)
Piscivores. Piscivores represent the most speciose trophic group in the lake, and they comprise approximately 40 percent of the species (Greenwood 1974; van Oijen et al. 1981). This, taken with their apparently ubiquitous distributions, suggested considerable niche overlap among these forms (Fryer, Iles 1972; Greenwood 1974). However, van Oijen (1982) recently indicated that upon closer examination this overlap is greatly reduced. Segregation along the food axis is one of the important factors in ecological separation (Figure 4). The piscivores specialized on different prey taxa (*Haplochromis* spp.; *Barbus* spp. and the cyprinid *Rastrineobola argentea*), different prey sizes, different ontogenetic stages (eggs, larvae, juveniles, adults), and even different parts of the prey, such as scales, though the specific identity of many prey has not been resolved. In addition they use different predation techniques, which

include ambush and pursuit hunting (Galis, Barel 1980; van Oijen 1982), and among peadophages, include snout engulfing versus egg snatching (Wilhelm 1980; Witte-Maas 1981).

Habitat partitioning is also important among the piscivores. Substrate types, depth ranges (Figure 5) and exposure to wind are major factors involved. Although van Oijen (1982) did not discuss piscivores exclusively coexisting in the same area, he concluded that: "... there are clear indications that each species has its own unique combination of food preferences, habitat and feeding behaviour."
Insectivores. Two morphologically similar insect-eating species, *H. hiatus* and *H. iris* are considered here. Both species feed mainly on *Chaoborus* and chironomid larvae and pupae which are collected from the bottom mud during daylight periods. Molluscs and bottom debris are taken as well. With the exception of male coloration, these two species are so similar that at first it was not certain whether they should be considered color morphs or distinct species (Hoogerhoud, Witte 1981). Closer examination has subsequently revealed partial segregation in spawning period, and in depth of occurrence (Hoogerhoud et al. 1983). *H. hiatus* occurs mainly between depths of 3-9 m, while *H. iris* is more frequent between 8-15 m. This segregation

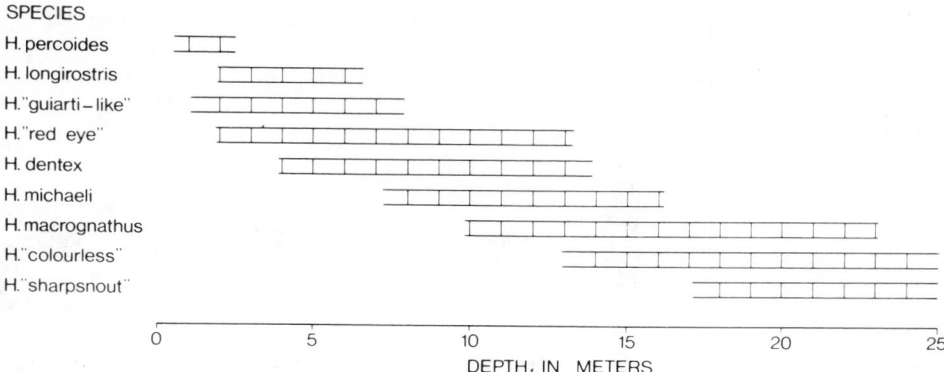

Figure 5. Depth distribution of adult piscivores of Mwanza Gulf, Lake Victoria. (van Oijen 1982)

is apparently correlated with oxygen concentrations. Hypoxic conditions occasionally occur in water deeper than 8 m (van Oijen et al. 1981). The deeper-living *H. iris* has a significantly larger (1.6 x) gill surface than *H. hiatus* (Hoogerhoud et al. 1983). A similar correlation of depth segregation with gill surface area occurs in two morphologically similar detritivores, *H.* "75" and *H.* "cinctuslike" (van Oijen et al. 1981; Galis pers comm). These examples show that species which are virtually identical in general aspect can, at a finer level, exhibit structural differences which are related to differences in ecological demands. *Zooplanktivores.* The zooplanktivores are mainly slender-bodied fishes resembling *H. laparogramma* and *H. fusiformis* described by Greenwood and Gee (1969). Of the 8 species regularly encountered in our research transect (Figure 1) two were restricted to hard substrates. One of them, *H.* "crimson" is confined to rocks and is remarkably similar to *Cynotilapia afra*, a zooplanktivore member of the Mbuna group, the rock-frequenting cichlids of Lake Malawi (Witte-Maas pers. comm). The other, *H.* "double stripe," is caught over sand as well as near rocks. The remaining 6 species occur over mud or both sand and mud, and bottom trawling reveals some segregation along the bottom profile. Exposure to wind may play a role in distributions of zooplanktivores as some species occur predominantly in bays. Despite evidence of segregation along the bottom profile, considerable overlap occurs at station G (Figure 1)

where 4 species, *H.* "argens," *H.* "orangehead," *H.* "heusinkveldi" and *H.* "reginus" apparently coexist. Detailed investigations of the vertical distribution in the water column, however, reveals partial segregation (Figure 6). *H.* "argens" occurs mainly in the surface layers; *H.* "reginus" lives closer to the bottom; *H.* "orangehead" and *H.* "heusinkveldi" apparently cover a wider range, but they are more frequent in the lower half of the water column during the day. However, vertical distribution is not the same during day and night. All species migrate upward during the night, although to different degrees (Figure 6).

Gut content analysis (Goldschmidt pers comm) indicates that niche overlap is further reduced by differences in food composition (Figure 7). During the day, *H.* "reginus," which in body shape resembles a phytoplankton-detritus feeder such as *H. erythrocephalus* (Greenwood, Gee 1969) takes more bottom material (detritus and chironomid larvae) than do *H.* "orangehead" and *H.* "heusinkveldi." *H.* "heusinkveldi" and *H.* "orangehead" show considerably more overlap, both in vertical distribution and in food composition. At night these two species again have virtually the same distribution pattern, but food overlap decreases (Figure 7). At night, *H.* "reginus," which does not migrate as far upward as *H.* "heusinkveldi" and *H.* "orangehead," shifts almost completely to diatoms (*Melosira* spp.). *H.* "argens," *H.* "heusinkveldi" and *H.* "orangehead" also show a remarkable diel

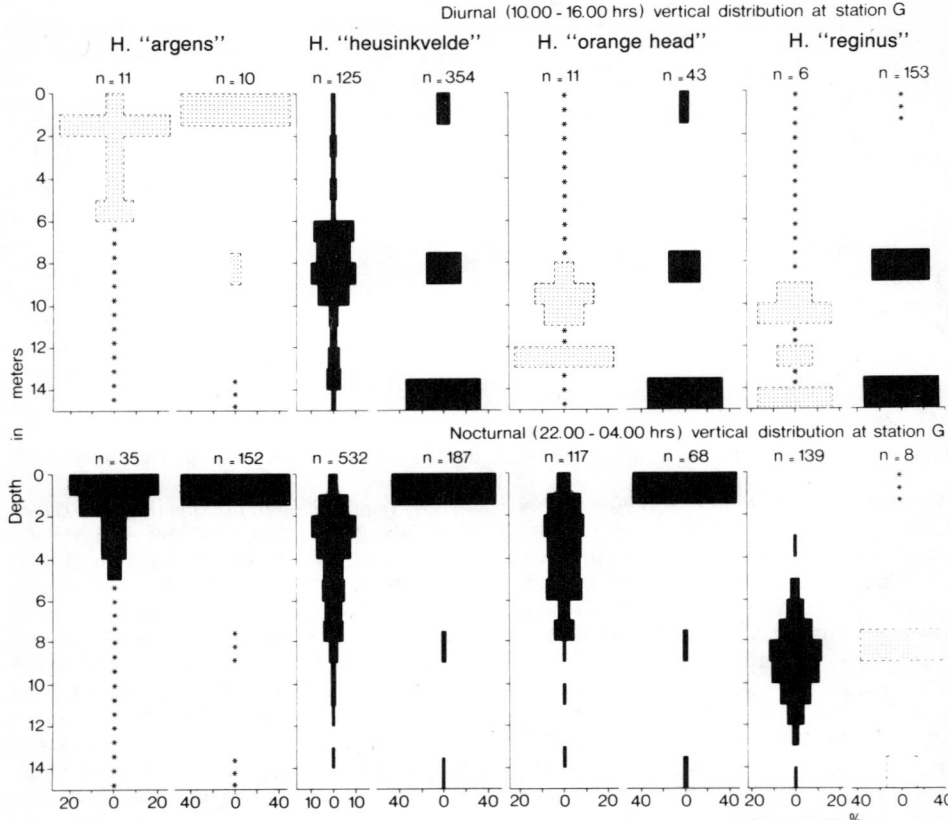

Figure 6. Diurnal and nocturnal vertical distribution of four coexisting zooplanktivores at station G. Two types of gill nets used: a net of 1'' mesh, 15 m deep, 5 m wide covering the whole water depth; three nets, each with panels of 1'', 1½'', & 2'' mesh, each net 60 m long, 1½ m deep; set at the bottom, in midwater, and at the top. Data lumped for 8 catches through 1981 and 1982 and for data from one catch by the set of three nets in 1980. N = total number caught. Dotted areas indicate numbers too low to calculate a reliable distribution. Asterisks indicate empty nets.

change in feeding pattern. *Chaoborus* larvae (3rd and 4th instars) and pupae become a very important food source during the night (Figure 7). By day, these *Chaoborus* larvae apparently live in the bottom mud and are not accessible to zooplanktivores (McDonald 1956). However at night, these larvae migrate towards the water surface to feed on zooplankton, and they become available to the zooplanktivorous haplochromines. As noted above, benthic insectivores like *H. iris* and *H. hiatus* collect *Chaoborus* larvae by day from the bottom debris. This is another example of different feeding strategies applied to the same prey. Pelagic zooplanktivores and benthic insectivores show resource partitioning

by temporal and spatial segregation. However, if availability of *Chaoborus* is a limiting factor, competition may exist in spite of this temporal and spatial segregation. The switching by the zooplanktivores from small zooplankton prey such as cladocerans and copepods to the larger *Chaoborus* larvae at night seems in accordance with optimal foraging theory (Krebs 1978). It remains to be demonstrated, however, whether *Chaoborus* is relatively more profitable in terms of search time, handling time, or food value.

Data on breeding are known for only two of the coexisting species. Brooding females of *H.* "argens" are frequently caught with beach seines in bays, and apparently are restricted

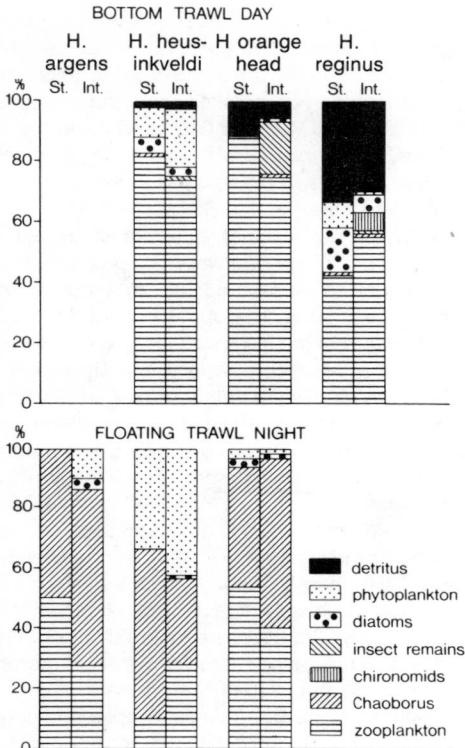

BOTTOM TRAWL DAY

FLOATING TRAWL NIGHT

■ detritus
▦ phytoplankton
•:•: diatoms
▨ insect remains
▥ chironomids
▨ Chaoborus
☰ zooplankton

Figure 7. Mean volume percentages of food types in stomachs (St.) and intestines (Int.) of 4 coexisting zooplankivores. Caught at station G at approximately midday and midnight. Data of 8 catches through 1981, 1982 lumped. Each bar represents foods from 10-29 fish. Phytoplankton is primarily *Microcystes* sp.; diatoms include mainly *Melosira* sp. Insect remains are unidentifed pupae. Chironomids are larvae; *Chaoborus* includes larvae and pupae; zooplankton includes cladocerans and copepods. The nocturnal and diurnal foods of, respectively, *H. reginus* and *H. argens* are not shown.

to the shallow littoral habitat. Brooding females of *H.* "heusinkveldi" occur in open and deeper water. Thus, these two species probably are also partitioning spawning areas.

In summary, we have observed the following different modes of ecological segregation in Lake Victoria haplochromines of the Mwanza Gulf area: bottom type preference; distribution along the bottom profile; vertical distribution in the water column; qualitative differences in food composition, including food-size partitioning; quantitative differences in

food composition; differences in food collecting strategy; and partitioning of spawning areas. The various combinations of these segregations make niche differentiation very complex.

Discreteness of Littoral Habitats. According to Fryer and Iles (1972) and Greenwood (1974) the littoral habitats in Lake Victoria tend to merge imperceptibly with one another and are less conducive to ecological isolation of fishes than are the littoral habitats of lakes Malawi and Tanganyika. These observations were made primarily in the northern part of Lake Victoria and do not hold for the Mwanza area (Figure 1). Here, rocky shores may change abruptly to sandy or muddy shores. Papyrus fringes are mainly restricted to inshore ends of bays. Isolated rocky islands often surrounded by relatively deep troughs (Figure 1, shaded area) emerge from the soft mud bottoms in the Mwanza Gulf and the Speke Gulf. Distances between similar habitats may be many kilometers, but even a small embayment like Butimba Bay (◄ 0.5 km²) includes distinct rocky, muddy, and sandy habitats with entirely different haplochromine communities. I should reemphasize that substrate type is only one of several factors involved in habitat restriction. Further, some of the simple factors include more than one parameter. Increased depth usually means decreased oxygen concentration and light intensity and increased pressure. Oxygen concentration (van Oijen et al 1981; Hoogerhoud et al. 1983) and water pressure (Hill, Ribbink 1978; Ribbink, Hill 1979) may form important barriers for movement of shallow water species into deeper water. In conclusion, the littoral habitats in the Mwanza area are topographically well defined. Excepting the gently sloping shores, these habitats closely resemble those of the southern part of Lake Malawi.

Isolated Populations and Geographical Restriction. Fryer and Iles (1972) commented that, in Lake Victoria "... there is no evidence today of isolated *Haplochromis* populations which are recognizably distinct in different parts of the lake," and that this suggests "a certain lack of spatial isolation within the present-day basin." This absence of geographic restriction, they say, is in contrast to the situation in lakes Malawi and Tanganyika. Similarly, Greenwood (1974)

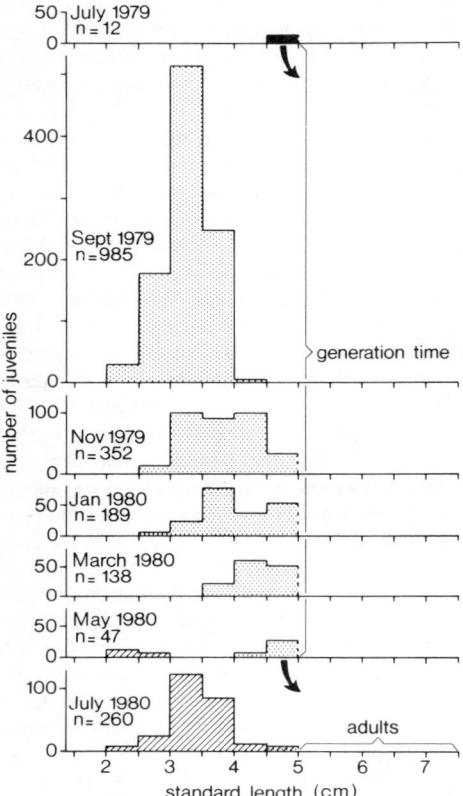

Figure 8. Length frequencies of all juveniles of H. "nigrofasciatus", one of the abundant phytoplankton/detritivores occurring at stations C,D,E,F and G. The cohorts appear after the spawning season, which is approximately from March through July. The generation time approximates one year. (From Witte 1981.)

stated that "none of the Lake Victoria *Haplochromis* or related species shows any geographical restriction within the lake, nor have any morphologically distinguishable populations been discovered. In both these respects the Victoria species differ from those of Lake Malawi, and especially, Lake Tanganyika."

There are, however, several recent examples of intraspecific geographic differences within Lake Victoria (van Oijen et al. 1981). A summary, amplified by some new arguments follows:

1) There are apparent differences in morphometrics, although overlapping, and

coloration between conspecifics from the northern, Jinja, and southern, Mwanza, parts of the lake. *H. macrognathus* and *H. michaeli* have larger eyes, smaller heads and deeper bodies in the Mwanza Gulf than in northern areas, and *H. plagiostoma* and *H. cryptogramma* from the two areas differ in male coloration (van Oijen et al. 1981).

2) Small but distinct differences in male coloration occur between populations of conspecifics separated by much smaller distances, such as populations of *H.* "red tridens" from the Mwanza Gulf and the Speke Gulf at a distance of approximately 75 km.

3) In a rock-dwelling algal grazer, *H.* "velvet black," there are small differences in morphometric measurements between populations from different islands separated by distances of 2.5-7 km in the Mwanza Gulf (Witte-Maas unpubl). The island populations differ in the frequency of the Orange Blotched (OB) morph among females. At most islands, such as Nyegezi rocks, Hippo Island and Rocky Island (Figure 1), the OB morph is rare, whereas at Anchor Island, the OB morph is more common than the brown morph.

Although the foregoing examples indicate spatially isolated populations, more detailed studies are needed before any detailed conclusions can be drawn. One problem concerns the taxonomic status of these isolated populations. In some cases the morphometric and color differences are as great as seen between highly similar species such as *H. iris* and *H. hiatus*. Our decision to call them populations of the same species is based on the observation that these forms apparently replace each other in the same habitat in different areas. If some of these prove to be distinct species, then we are dealing with geographically restricted taxa. The many new species that one encounters when collecting fishes in an unexplored area of Lake Victoria indicates that geographical restriction indeed occurs in the lake. However, one must be careful, because if an unexplored area comprises new habitat types, then the differences in species composition do not prove geographical restriction. In fact, a lake-wide survey of the specific habitat of a species is necessary to confirm its lake-wide distribution.

Comparison of fishes drawn from the north and the south end of Lake Victoria revealed that at least 60 of the 250 + species occur in both areas and may have a lake-wide distribution in specific habitats (van Oijen et al. 1981).

Differences in Origins of the Species Flocks. An important difference in the geographical history of Lake Victoria compared to the two rift valley lakes, Malawi and Tanganyika, is its supposed origin from a number of small isolated lakes in which allopatric speciation took place (Fryer, Iles 1972). This and the supposed reduced ecological restriction of species, lack of isolated populations, and ill-defined habitats in Lake Victoria led Fryer and Iles (1972) to the conclusion that "unlike Malawi and Tanganyika, Lake Victoria is not an outstanding example of intralacustrine speciation; in fact, it may not exhibit the phenomenon at all." Accordingly, Greenwood (1974) states: "... there is sufficient evidence both historical and contemporary (see Fryer and Iles 1972) indicating that the cichlid flocks of Lakes Malawi and Victoria did have somewhat different evolutionary histories." Somewhat later, Fryer (1977) criticized the last statement, saying that "... if the evidence for these two lakes has been correctly interpreted ... the histories of these species flocks were not 'somewhat' but fundamentally different. Evidence of intralacustrine, but allopatric speciation in Lake Malawi (and Lake Tanganyika) are extremely strong; in Lake Victoria it is nonexistent." Greenwood (1981) repeats: "Although ultimately, the cichlid flock in Lake Victoria came to resemble those of Lakes Tanganyika and Malawi, the history of the lakes, and the phylogeny of their cichlid flocks differ in several respects."

As demonstrated in earlier paragraphs, these interpretations of Fryer and Iles and Greenwood are based to some extent, not on real differences, but on disparate knowledge of the species flocks of the three lakes, that of the Lake Victoria cichlids being least known. Moreover, the fishes that were compared were often inconsistent. In lakes Malawi and Tanganyika most ecological studies were carried out on the rock-frequenting cichlids, a group which, excepting H. nigricans, had not yet been discovered in Lake Victoria. To illustrate the wide ecological range exhibited by the Lake Victoria haplochromines, Fryer

and Iles considered among others the substrate preferences of a number of piscivores. As van Oijen (1982) has noted, piscivores occurring over different substrates probably have less strict benthic habits, but this does not imply that they have a wide ecological range. It is this inconsistency to which Greenwood (1974) refers when he states: "I do not find all their [Fryer and Iles] arguments equally convincing, particularly since they are based essentially on a detailed study of only certain elements in the cichlid fauna: the strength of habitat restriction is unknown for most of the Haplochromis species in Lake Malawi." Fryer (1977) found Greenwood's comment on the lack of information about Lake Malawi Cyrtocara "too sweeping," and he refers to the Utaka (zooplanktivorous Cyrtocara of Lake Malawi) which, in spite of their open water habits, are restricted to particular shore types where breeding "... seems to demand a particular type of substratum." However, at least one Utaka species has recently been observed to be a free spawner (Eccles, Lewis 1981). As described in an earlier section, indications of such segregation in spawning habitat also occur in two partially overlapping zooplanktivores, H. "argens" and H. "heusinkveldi."

Failure to recognize different species is another serious source of misinterpretation concerning habitat restriction (van Oijen 1982): in such cases, "... the habitats seem larger and more varied than in fact they are," because the habitats of two or more species are compounded. There are several examples of such misinterpretations for Lake Victoria haplochromines (van Oijen 1982; pers obs).

From recent knowledge of Lake Victoria haplochromines a sounder comparison can be made between these fishes and the cichlids of the two rift valley lakes. Comparison of habitat restriction in the rock-dwelling fishes of lakes Malawi, Tanganyika and Victoria reveals similar magnitudes. Just as in the other two lakes, the rock-fishes in Lake Victoria are rarely caught more than a few meters from rocky substrate, though one of the presumed algal grazers feeds at night about 100 m offshore on surface phytoplankton (Witte, Goldschmidt unpubl). On the other hand, in spite of close association with rocks, migration of rock-frequenting fishes toward artificial rock habitats has been observed in all three lakes. In Lakes Tanganyika and Victoria, these fishes appeared at jetties that were built on sandy beaches (Fryer, Iles 1972; Marlier 1959; van Oijen et al. 1981) and in Lake Malawi, Mbuna species migrated to artificial reefs (McKaye, Gray 1984). Apparently, some rock-frequenting cichlids do leave their substrate under certain circumstances. The degree of isolation of a population of these rock-fishes depends on the frequency of migrations and migration distance as compared to the mean distance between adjacent rock habitats (McKaye, Gray 1984).

That habitat restriction of the Lake Malawi species was generally somewhat exaggerated is further indicated by Lewis' (1981) comments: "... bottom type appears to be an important factor in governing distribution patterns in Lake Malawi cichlids though the influence is greater in some species than in others. ... Amongst the Lake Malawi species, members of the mbuna group probably display the greatest degree of selection of substrate type." A recent study of microdistribution of cichlids on a rocky shore of Lake Tanganyika (Hori et al. 1983) also revealed that algal grazers show strong substrate preferences while most piscivores and omnivores are ubiquitous.

The foregoing appears to refute or at least temper many of the arguments used by Fryer and Iles (1972), Fryer (1977) and Greenwood (1974, 1981) to demonstrate the likelihood of different speciation patterns in the species flocks of lakes Victoria, Malawi and Tanganyika. A remaining argument is the origin of the lakes. There are indications that, like Lake Victoria, the history of Lake Malawi may also have involved a number of sub-basins, although it is unknown whether they existed as separate lakes (Yairi 1977). Moreover, small lake cutoffs like those which occur in Lake Victoria (Greenwood 1965) may have played a role in speciation in Lake Malawi (McKaye, Gray 1984)

Niche Segregation, Coexistence, and Competition. Species that exploit the same resource may coexist in the same habitat (Fryer, Iles 1972; Greenwood 1974, 1981). This was illustrated by the example of the oral shellers/crushers of which 7 species occur in the same sandy habitat. Fryer and Iles (1972) explained such coexistence by using Wynne-Edwards' (1962) condominium principle. This principle states that "... nearly related species of very similar habitats can be united ecologically, for a part of their annual cycles at least, and in some of the more extreme cases the evidence suggests that they can be united throughout the whole year." The members of such an association are not in competition with one another as species, but as "... individuals belonging to a mixed indivisible society." Greenwood (1974) referred to the same principle, but because there were no *Haplochromis* species known to him with absolutely identical ecological requirements, he preferred the term *commonweal* for such groups in Lake Victoria.

These examples are regarded as important exceptions to Gause's principle of competitive exclusion (Fryer, Iles 1972; Greenwood 1974). Fryer and Iles (1972) say that it "emphatically refutes" this principle, and Greenwood (1974) talks about Wynne-Edwards' demonstration of the principle being "logically and ecologically unsound."

A crucial point is that virtually nothing is known about competition among cichlids in African lakes. Thus far, there has been no attempt to trace and measure competition experimentally (e.g., Connell 1983). The only available data deal with habitat and food overlap, and these data are generally not sufficiently detailed. Besides, competition need not be present in every place and moment and at the same intensity. During lean periods it probably intensifies (Schoener 1982). Laboratory experiments on food preferences of two trophic morphs of *Cichlasoma minckleyi* indicate that food segregation takes place only during periods of low food abundance (Liem, Kaufman 1984). Further, species apparently specialized for certain food sources may switch food preferences when other foods become superabundant. In Lake Victoria, zooplanktivores, detritivores, and insectivores switch to *Melosira* spp. when there is a bloom of these diatoms (pers obs), and in Lake Malawi specialized algal grazers switch to zooplankton when this food source is abundant (cf. Barel 1983; McKaye, Marsh 1983; McKaye, Gray 1984). These examples demonstrate that incidental observations at one time or place are generally not sufficient to decide whether competitive exclusion exists or not. Much work remains to unravel the ecological dynamics of species flocks. I therefore cannot agree with Greenwood (1981) who states: "Frequently this [lacustrine explosive evolution] has led to the coexistence in one niche of two or more species (often close relatives) with what appear to be identical demands on that niche. In other words, this appears to be an apparent negation of the competitive exclusion principle."

Although Greenwood later admits that "Such close coexistence may, of course, be more apparent than real, the consequence of insufficiently detailed research," he concludes that "even apparent total interspecific overlap in such fundamental requirements as food, breeding times, and spawning sites raises the question of whether interspecific competition is inevitably, or even commonly, a major factor in evolution." However, as seen in the zooplanktivores of Lake Victoria, each case of apparent total interspecific overlap that has been studied in detail reveals niche segregation. Whether current competition for resources plays a role in segregation of species is as yet unknown because it has not been shown that the fishes would expand their niches in the absence of the other species.

Breeding. Based on observations made in northern Lake Victoria, Greenwood (1974, 1981) refers to two aspects of reproductive biology that have contributed to the explosive evolution and adaptive radiation of African cichlids:

1) year-long breeding, thus implying that fishes can produce several broods per year; and

2) short generation times compared with noncichlids. Our research in the Mwanza Gulf area has shown, however, that many species, especially detritus phytoplanktivores and zooplanktivores, have distinct breeding seasons (Figure 8; Witte 1981): the total breeding period of these seasonal spawners may extend over half a year, but peak reproduction occurred only during two or three months of the year. Because brooding and brood care take 4 to 6 weeks (Fryer, Iles 1972) and females must build up new fat reserves after brooding, only 1-2 broods can be raised during such a peak period. Species that breed throughtout the year, such as the insectivore H. "thickskin" (Witte 1981) may produce several broods per year, but this is not yet documented. Instead of repeated breeding by the same individuals, the image of continuous spawning may reflect spawning by different segments of the population at different periods. No data on generation times of Lake Victoria cichlids were available until recently. Greenwood (1974) referred to the closely related Lake George fishes which we have have been breeding in aquaria. These fishes had a generation time of approximately half a year (Barel pers comm). However, in the field, generation times of the smaller species, detritus-phytoplanktivores and zooplanktivores, are approximately one year (Figure 8). For the large species such as molluscivores and piscivores, the generation time may be even longer. Consequently, the differences in spawning frequency and generation time between many of the extant haplochromine cichlids and other small-size fishes such as cyprinids in Lake Victoria may be less striking than supposed. Additional research on reproductive biology is essential.

SUMMARY

Recent work on the cichlids of Lake Victoria has demonstrated that differences between the ecology of the cichlid flocks of Lake Victoria and those of lakes Malawi and Tanganyika are weaker than supposed by Fryer (1977), Fryer and Iles (1972) and Greenwood (1974, 1981). Consequently, the process of explosive evolution of the flocks in these lakes may have been more similar than previously conceived. Arguments for allopatric speciation in a single body of water ("truly intralacustrine speciation," Fryer, Iles 1972) apply as well to the flock in Lake Victoria as to those of lakes Malawi and Tanganyika. Furthermore, like Lake Victoria, Lake Malawi, at least, may encompass a number of smaller, once-isolated bodies of water (McKaye, Gray 1984; Yairi 1977), and extralacustrine speciation associated with absolute physical barriers (Fryer, Iles 1972) may have played a role in both lakes. Consequently, intra- and extralacustrine allopatric speciation processes, as well as sympatric speciation (cf. Holzberg 1978; Hoogerhoud et al. 1983; McKaye et al. 1982) may have been important in forming the species flocks in all three lakes.

Although the work of Greenwood (1974, 1981), Fryer (1959), and Fryer and Iles (1972) on cichlids of the Great African Lakes was pioneering, many of the conclusions concerning differences between the Lake Victoria flock and those of lakes Malawi and Tanganyika have been refuted or weakened by recent information. I would hope that the recent studies will be as stimulating as the earlier work, and that they will generate further comparative research on these cichlid species flocks. This is clearly essential to a more complete understanding of the biology and evolution of these interesting fish assemblages.

ACKNOWLEDGMENTS For critical reading and helpful discussion I thank C D N Barel, M J P van Oijen, F Galis, J J M van Alphen, G Fryer, I L Kornfield, M L Staissny, R J C Hoogerhoud, E L M Witte-Maas, Prof. P Dullemeijer, and Prof. K Bakker. P T Goldschmidt provided data on gut content of planktivores. M K I Vogelsang-Liem typed various drafts. M L Brittijn drew the figures. The Mwanza Freshwater Fisheries Research Centre and the Freshwater Fisheries Training Institute assisted the Haplochromis Ecology Survey Team (HEST). HEST is supported by the Netherlands Foundation for Advancement of Tropical Research, by the Education and Research Programmes Department of the Directorate of International Cooperation, and by the State University of Leiden.

REFERENCES

Barel C D N 1983 A constructional morphology of cichlid fishes (Teleostei: Perciformes) *Neth J Zool* 23:357-424

Barel C D N, M J P van Oijen, F Witte, E L M Witte-Maas 1977 An introduction to the taxonomy and morphology of the haplochromine Cichlidae from Lake Victoria. *Neth J Zool* 27:333-389

Connell J H 1983 On the prevalence and relative importance of interspecific competition: evidence from field experiments. *Amer Nat* 122:661-696

Eccles D H, D S C Lewis 1981 Midwater spawning in *Haplochromis chrysonotus* (Boulenger) (Teleostei: Cichlidae) in Lake Malawi. *Env Biol Fish* 6:201-202

Fryer G 1959 The trophic interrelationships and ecology of some littoral communities of Lake Nyasa with especial reference to the fishes, and a discussion of the evolution of a group of rock-frequenting Cichlidae. *Proc Zool Soc London* 132:153-281

Fryer G 1977 Evolution of species flocks of cichlid fishes in African lakes. *Zeit Zool Syst Evolut-forsch* 15:141-165

Fryer G, T D Iles 1972 *The Cichlid Fishes of the Great Lakes of Africa.* Oliver Boyd, Edinburgh

Galis F, C D N Barel 1980 Comparative functional morphology of the gills of African lacustrine Cichlidae (Pisces, Teleostei): an eco-morphological approach. *Neth J Zool* 30:392-430

Greenwood P H 1956 The monotypic genera of cichlid fishes in Lake Victoria. *Bull Br Mus Nat Hist (Zool)* 3:295-333

_____ 1959 The monotypic genera of cichlid fishes in Lake Victoria, Part 2. *Bull Br Mus Nat Hist (Zool)* 5:163-177

_____ 1965 The cichlid fishes of Lake Nabugabo, Uganda. *Bull Br Mus Nat Hist (Zool)* 12:315-357

_____ 1974 The cichlid fishes of Lake Victoria, East Africa: the biology and evolution of a species flock. *Bull Br Mus Nat Hist (Zool)* Suppl 6

_____ 1979, 1980 Towards a phyletic classification of the 'genus' *Haplochromis* (Pisces: Cichlidae) and related taxa. *Bull Br Mus Nat Hist (Zool)* Part 1, 35:265-322; Part 2 39:1-99

_____ 1981 *The Haplochromine Fishes of the East African Lakes.* Kraus Int'l, Muenchen

Greenwood P H, J M Gee 1969 A revision of the Lake Victoria *Haplochromis* species (Pisces: Cichlidae) Part 7. *Bull Br Mus Nat Hist (Zool)* 18:1-65

Hill B J, A J Ribbink 1978 Depth equilibrium of a shallow-water cichlid fish. *J Fish Biol* 13:741-745

Holzberg S 1978 A field and laboratory study of the behaviour and ecology of *Pseudotropheus zebra* (Boulenger), an endemic cichlid of Lake Malawi (Pisces: Cichlidae). *Zeit Zool Syst Evolut-forsch* 16:171-187

Hoogerhoud R J C 1983 A critical analysis of Greenwood's definitions of the genera *Gaurochromis* Greenwood, 1981, and *Labrochromis* Regan, 1920. *Neth J Zool* In press

Hoogerhoud R J C, F Witte 1981 Revision of species from the "*Haplochromis*" empodisma group. *Neth J Zool* 31:232-272

Hoogerhoud R J C, F Witte, C D N Barel 1983 The ecological differentiation of two closely resembling *Haplochromis* species from Lake Victoria (*H. iris* and *H. hiatus*, Pisces, Cichlidae) *Neth J Zool* 33:283-305

Hori M, K Yamaoka, K Takamura 1983 Abundance and microdistribution of cichlid fishes on a rocky shore of Lake Tanganyika. *African Study Monogr* 3:25-38

Kornfield I, D C Smith, P S Gagnon, J N Taylor 1982 The cichlid fish of Cuatro Cienegas, Mexico. Direct evidence of conspecificity among distinct trophic morphs. *Evolution* 36:658-664

Kornfield I L, J N Taylor 1983 A new species of polymorphic fish *Cichlasoma minckleyi* from Cuatro Cienegas, Mexico (Teleostei: Cichlidae). *Proc Biol Soc Wash* 96:253-269

Krebs J R, N B Davies eds. 1978 *Behavioral Ecology, an Evolutionary Approach.* Blackwell Scientific Publ

Lewis D S C 1981 Preliminary comparisons between the ecology of the haplochromine cichlid fishes of Lake Victoria and Lake Malawi. *Neth J Zool* 31:746-761

Liem K F, L S Kaufman 1984 Intraspecific macroevolution: functional biology of the polymorphic cichlid species *Cichlasoma minckleyi.* 203-216 A A Echelle, I Kornfield eds. *Evolution of Fish Species Flocks.* Univ Maine Press at Orono

MacDonald W W 1956 Observations on the biology of chaoborids and chironomids in Lake Victoria and on the feeding habits of the "elephant snout fish" (*Mormyrus kannume* Forsh.) *J Anim Ecol*25:36-53

Marlier G 1959 Observations sur la biologie littorale du lac Tanganyika. *Rev Zool Bot Afr.* 59:164-183

Marsh A C, A J Ribbink, B A Marsh 1981 Sibling species complexes in sympatric populations of *Petrotilapia trewavas* (Cichlidae, Lake Malawi). *Zool J Linn Soc* 71:253-264

McKaye K R 1980 Seasonality in habitat selection by the gold color morph of *Cichlasoma citrinellum* and its relevance to sympatric speciation in the family Cichlidae. *Env Biol Fish* 5:75-78

McKaye K R, W N Gray 1984 Extrinsic barriers to gene flow in rock-dwelling cichlids of Lake Malawi: macrohabitat heterogeneity in reef colonization patterns. 169-184 A A Echelle, I Kornfield eds. *Evolution of Fish Species Flocks.* Univ Maine Press at Orono

McKaye K R, T Kocher, P Reinthal, I Kornfield 1982 A sympatric sibling species complex of *Petrotilapia trewavas* from Malawi analyzed by enzyme electrophoresis (Pisces: Cichlidae). *Zool J Linn Soc* 76:91-96

McKaye K R, A Marsh 1983 Food switching by two specialized algae-scraping cichlid fishes in Lake Malawi, Africa. *Oecologia* 56:245-248

Oijen M J P van 1982 Ecological differentiation among piscivorous haplochromine cichlids of Lake Victoria (East Africa). *Neth J Zool* 32:336-362

Oijen M J P van, F Witte, E L M Witte-Maas 1981 An introduction to ecological and taxonomic investigations on the haplochromine cichlids from the Mwanza Gulf of Lake Victoria. *Neth J Zool* 31:149-174

Ribbink A J, B J Hill 1979 Depth equilibrium by two predatory fish from Lake Malawi. *J Fish Biol* 14:507-570

Sage R D, P V Loiselle, P Basasibwaki, A C Wilson 1984 Molecular versus morphological change among cichlid fishes (Pisces: Cichlidae) of Lake Victoria. 185-202 A A Echelle, I Kornfield eds. *Evolution of Fish Species Flocks.* Univ Maine Press at Orono

Sage R D, R K Selander 1975 Trophic radiation through polymorphism in cichlid fishes. *Proc Nat Acad Sci* 72:4669-4673

Schoener T W 1982 The controversy over ecological competition. *Amer Sci* 70:586-595

Wilhelm W 1980 The disputed feeding behaviour of a paedophagous haplochromine cichlid (Pisces) observed and discussed. *Behaviour* 74:310-323

Witte F 1981 Initial results of the ecological survey of the haplochromine cichlid fishes from the Mwanza Gulf of Lake Victoria (Tanzania): breeding patterns, trophic and species distribution. *Neth J Zool* 31:175-202

Witte F, E L M Witte-Maas 1981 Haplochromine cleaner fishes. A taxonomic and eco-morphological description of two new species. Revision of the haplochromine species (Teleostei: Cichlidae) from Lake Victoria, Part I. *Neth J Zool* 31:203-222

Witte-Maas E L M 1981 Egg-snatching: an observation on the feeding behaviour of *Haplochromis barbarae* Greenwood, 1967 (Pisces: Cichlidae). *Neth J Zool* 31:786-789

Wynne-Edwards V C 1962 *Animal Dispersion in Relation to Social Behaviour.* Oliver Boyd. Edinburgh

Yairi K 1977 Prelim. acc't of lake-floor topography of Lake Malawi in relation to the formation of Malawi Rift Valley. *2nd Prelim Rep Afr Stds* Nagoya Univ 51-69

EXTRINSIC BARRIERS TO GENE FLOW IN ROCK-DWELLING CICHLIDS OF LAKE MALAWI: MACROHABITAT HETEROGENEITY AND REEF COLONIZATION

KENNETH R MCKAYE AND W NOEL GRAY

INTRODUCTION

Lake Malawi, southernmost of the rift valley lakes of East Africa, is the world's ninth largest lake at 22,480 km^2, 570 km long, 80 km wide, and 704 m maximum depth. Malawi, along with two other East African great lakes, Tanganyika and Victoria, has been the site of explosive speciation of cichlid fishes (Fryer, Iles 1972; Greenwood 1974; Barel et al. 1977). Icthyologists estimate that there are over 500 endemic species in the lake (Lewis, Ribbink pers. comm.; pers. obs.) . The mechanisms by which so many species have evolved and by which they coexist have evoked interest and controversy (Kosswig 1947, 1963; Lowe-McConnell 1959; Greenwood 1974; Fryer 1977; Fryer, Iles 1972; McKaye 1980).

Fryer's (1959; Fryer, Iles 1972) microallopatric model of speciation for the Lake Malawi cichlids hypothesizes that intralacustine speciation was due to habitat isolation followed by divergence of populations. The emphasis of the earliest ecological studies was on a group of rock-dwelling cichlids known by the local name *mbuna*. This group includes the genera *Pseudotropheus*, *Labeotropheus*, *Melanochromis*, *Cyathochromis*, *Genyochromis*, *Gephyrochromis*, and *Iodotropheus* which are probably of monophyletic origin (Fryer 1959; Lewis 1982, pers. comm.). Mbuna species are usually small, less than 20 cm long, and over 95 percent are confined to rock habitats. Fryer's (1959) research concentrated on the mbuna, which has turned out to be even more speciose than he realized (Lewis 1982; Marsh 1981; Ribbink et al. 1983). Fryer suggested that the lake shore is a mosaic of habitats, with alternating stretches of rock, sand, and weed environments. Significant habitat or patch discontinuity could serve to cluster and isolate fish populations such as the mbuna from each other. He also speculated that fluctuating lake levels would serve, 1) to isolate, and 2) to recombine fish populations from these patches (Figure 1).

However, neither the number, size and distribution of habitat patches in the lake, nor the effect of changing lake levels on the Lake Malawi shoreline have been examined. Some assume that sand habitat is scarce in Lake Malawi (Ostro 1979), and the idea that Lake Malawi is dominated by rocky shoreline may reflect the popularity of the mbuna in the aquarium trade. Our data suggest that habitat variation along Lake Malawi's western shoreline is significant, and that habitat isolation may be possible for some species.

We asked the following questions about the significance of isolated rock patches for cichlid occurrence: 1) Do rock-dwelling cichlids cross barriers and colonize new habitats? 2) Are the first rock-dwelling cichlids to colonize new habitats those that feed on zooplankton and are thus less dependent on rock habitats for food? 3) Does the size, structure or depth of a new habitat patch affect the species number or total number of fishes on the habitat? 4) How does species number and population density change over time? Movement of fishes between patches and colonization of new patches were examined by regular surveys of the fish on eight experimental reefs placed over sand 1 km from the nearest rock outcropping.

Our evidence shows that: 1) the Malawi lake shore is a mosaic of habitats with many rock patches isolated by large stretches of weed and sand areas, and 2) some of the rock-dwelling cichlid species colonize new habitats

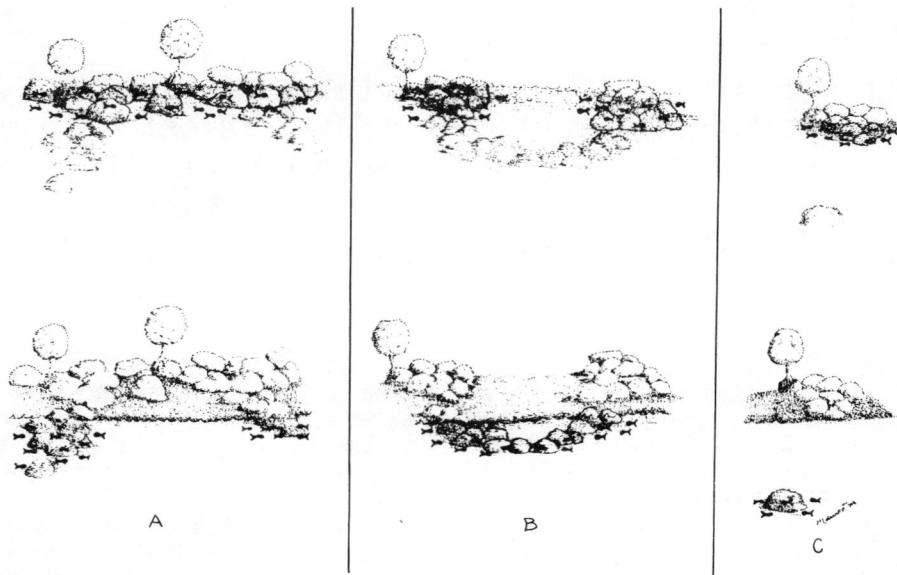

Figure 1: Effects of fluctuation in lake level and habitat utilization by cichlid fish. At A, a rise in lake level unites two populations. At B, a rise in lake level splits a single population into two. At C, a rise in lake level places a previously occupied rock habitat into water that is too deep, but opens up a new rocky shore habitat for colonization (Fryer 1977 modified).

within a short time. This assessment of the role of microhabitats isolating cichlid populations provides important baseline information for future studies on speciation and gene flow among cichlid populations.

METHODS AND MATERIALS
Aerial Shoreline Analysis.

To quantify the relative proportion of rock, sand, and weed habitat, we examined a series of 1972 aerial photos of the western shoreline of Lake Malawi. Four parameters were used to detail the pattern of habitat availability: 1) relative proportion of rock, sand and weed; 2) frequency of alteration of rock and sand-weed habitats along the shore; 3) patch size of each type of habitat; 4) distribution of rock and sand-weed habitats for four main regions along the western lake shore. We examined additional aerial photographs taken between 1966 and 1980 for the effect of increasing water level on shoreline configuration.

We divided the western lake shore into four main geographic regions from north to south: 1) Karonga to Chilumba (9°40'S to 10°15'S); 2) Chilumba to Nkhata Bay (10°15'S to 12°00'S); 3) Nkhata Bay to the Nankumba

Peninsula; 4) from the Nankumba Peninsula to the Shire River (Figure 2). Habitat type was determined from 1972 aerial photographs and scored on survey maps 0933D2; 1034A3, A4, C1, C3; 1134A1, A3, A4, C1, C2, C3; 1234A3, C1, C3, C4; 1334A3, A4, C2, D1, D3; 1434B1, B3, B4; 1435A3 (Malawi Survey Dept).

The effect of rising lake level on shoreline topography was analyzed for two relatively flat regions, Sungu Point and Sungu Spit, for which we located a series of photographs that spanned 16 years. For these sites, which border the southwest edge of the lake, three sets of aerial photos were available and used (Figures 3 and 4). The scale of photographs taken in 1964 and 1966 for Sungu Point (13°35'S, 34°32'E) and Sungu Spit (12°55'S, 34°18'E) was 1/40,000, and the scale of photographs taken in these areas in 1972 and 1980 was 1/25,000. All figures were reduced to a common scale of 1/50,000, and area measurements were taken from the reduced diagrams with a Haff-315 planimeter.

Experimental Reefs.

Eight experimental reefs were placed offshore of the Cape Maclear Fisheries Research Station in May 1978 (McKaye 1981). Reefs

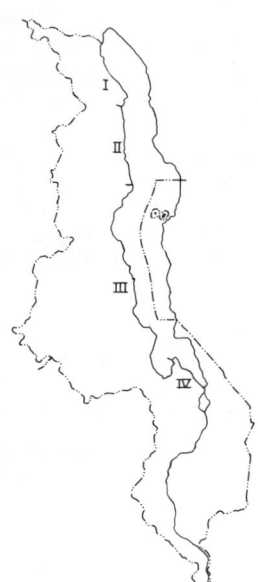

Figure 2: Four main regions of Lake Malawi

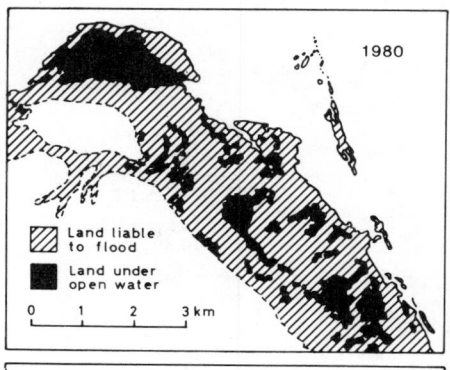

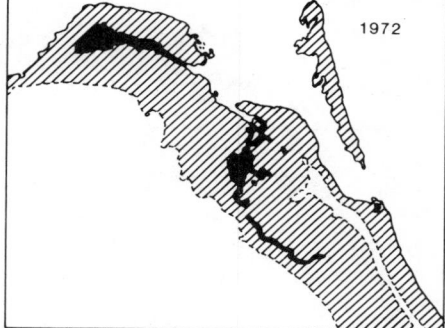

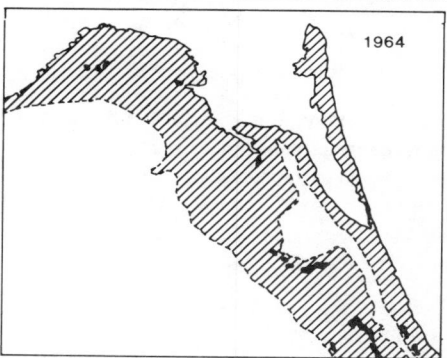

Figure 3: Sungu Point from 1964, 1972, and 1980 aerial photographs. White area on landward side, lower left, is not liable to flooding.

were structured from cement blocks in two basic designs and two sizes (Figure 5). Four "homogeneous" reefs, two large and two small, were constructed with large interspaces. Four "heterogeneous" reefs, two large and two small, were constructed with blocks of medium square and round interspaces. The large (10 m) reefs consisted of 26 cement blocks. The small reefs consisted of 13 blocks. The height of all reefs was 80 cm. Four reefs, one of each type were placed at 6 m depth; actual depth varied between 5.5 and 7 m as lake levels changed during the course of this study. Four other reefs were placed at 9 m depth; actual depth varied between 8.5 and 10 m. The reefs were 50 m apart, and the ordering of the reefs at each depth, from west to east, was small homogeneous, large homogeneous, large heterogeneous, and small homogeneous. Otter Point, the nearest rock outcropping, was 1 km to the west (McKaye 1981).

Census data exists for both the dry (June-November) and the wet (December-May) seasons for the 5-year period 1978-1983, except the 1982 wet season. The reefs were censused every 2-10 days. Sampling periods

and number of censuses were: Dry-1978, 18; Wet-1979, 8; Dry-1979 13; Wet-1980, 7; Dry-1980, 21; Wet-1981, 10; Dry-1981, 18; Dry-1982, 12; Wet-1983, 12. The reefs were designed to allow divers to see completely through all holes. Therefore, all fish were easily counted. All fish which were in reef holes, feeding on the reef, or orienting to the reef within 2 m were included in the census.

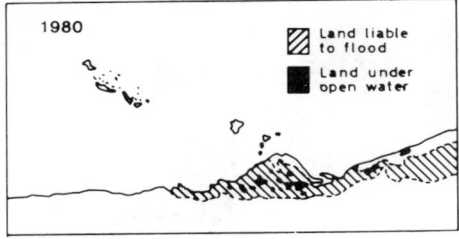

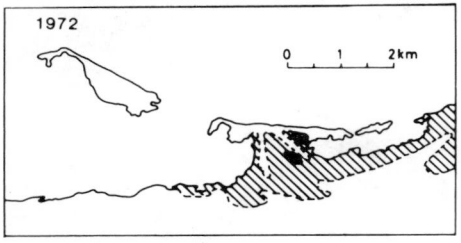

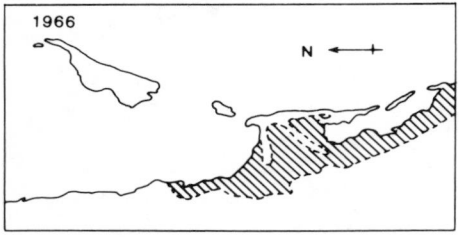

Figure 4: Sungu Spit from 1964, 1972, and 1978 aerial photographs. The shore of the lake is on the left side of the continuous line from left to right.

Excluded were cichlid species which occurred and fed only on the sand.

Water current velocities over the reefs were estimated in the dry season of 1980 by measuring the time suspended particulates in the water moved 30 cm. On the 21 reef censusing dates from June through August, 7 days had no current; 8 days had a slight current less than 2 cm/sec; and 6 days had a moderately strong current of 2-15 cm/sec, from east to west.

Zooplankton Availability.
Zooplankton samples were taken weekly from March 1978 to August 1980, 20 m off-shore of the deeper set of experimental reefs, in 17 m water depth. Three samples were taken at depths of 5 and 10 m, which corresponded closely to the depths of nearby reefs. Samples were collected with a 9-liter Van Dorn bottle and concentrated through a 64 μm net. The three samples from each depth were combined for analysis. Mechanical breakdowns and bad weather caused the departure from weekly sampling. All plankters were identified to species (McKaye 1983), but we only report data for diaptomids and

Diphanosoma excisum. These are the primary plankters eaten by the rock-dwelling cichlids (McKaye, Marsh 1983) and by most other cichlids (McKaye 1983, unpubl. data).

RESULTS

Habitat Distribution and Availability.
As Fryer predicted (1959), Lake Malawi's shoreline was a distinct mosaic of alternating patches of sand, rock and weed habitats. In 1972, 61.0 percent of the western shoreline was sand, 20.9 percent weed, and 18.1 percent rock (Table 1). The lake level then was 474 m above sea level (Drayton 1979). The Region I shoreline of 86.6 km was a stretch of sand and weed, with no observable (►0.05 km) rock habitats (Table 1). The 222-km shoreline of Region II had the highest proportion, 44 percent, of rock habitats, among the four regions. Isolated patches of rock in this section ranged from 0.05 km to 6.3 km in length (Table 1, Figure 6). Region III was predominantly sand, with rock patches comprising less than 4 percent of this 369 km shoreline (Table 1). The longest continuous stretch of sand-weed habitat we observed, 99.5 km, was in this region between 13°02'S and 13°40'S. Region IV, which included the Nankumba Peninsula and Cape Maclear to the Shire River outlet of the lake, had about 27 percent rock along its shoreline. This southernmost region had the longest continuous stretch of rock we observed, 8.5 km.

Figure 7 shows the frequency distribution of patch sizes of rock and of sand habitats in the four regions. Median patch length for rock was 300 m. Median patch length for combined sand-weed habitats was 400m. Though median patch size was similar, the rock habitats occur frequently as small islands, since rock predominates in patch sizes under 100 m. Sand and weed form the longest continuous stretches of habitat. Region II, with the greatest absolute amount of rock, had the highest number of rock patches: 43.2 per 100 km, vs. 9.2 for Region III and 23.6 for Region IV.

Shoreline Configuration over Time

Analysis of photographs taken between 1966-1980 demonstrated the creation and dissolution of isolated habitats over time. In 1966, Sungu Point consisted of a spit 4.2 km long pointing NNW that sheltered Domira bay to the west. On the landward side of the shoreline, about 17.5 ha were under water, and 1530 ha more were potentially flooded during the high rainy season (Figure 3).

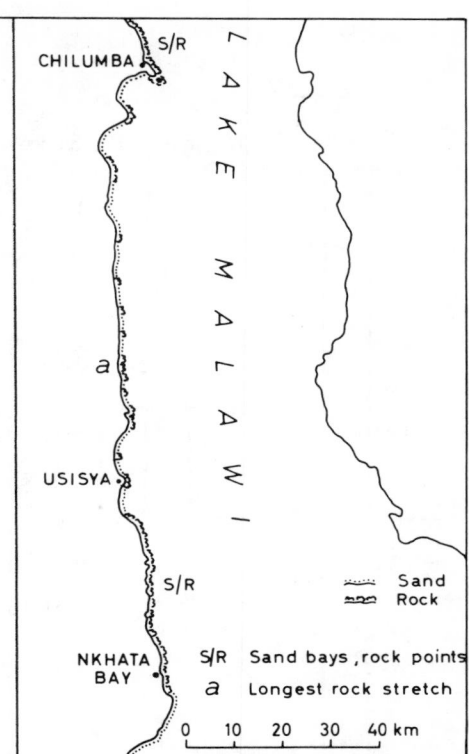

Figure 6: Region II from Chilumba to Nkhata Bay showing longest stretch of uninterrupted rock in this region.

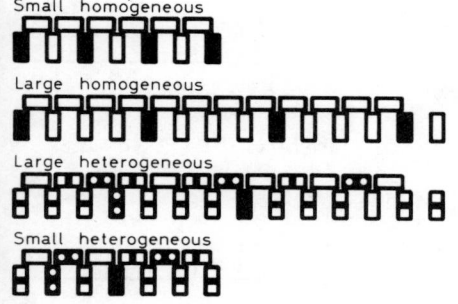

Figure 5: Structure of experimental reefs.

Table 1: Summed lengths (km) of three kinds of shoreline in four regions of western Lake Malawi, 1972 (See Figure 1)

	Sand	Marsh	Rock
Region I	76.9	9.7	0.1?
Region II	115.5	9.5	97.0
Region III	235.7	119.1	14.5
Region IV	63.3	29.8	34.7
Total	491.4	168.1	146.2

By 1972, following a slight increase of about 0.5 m in the lake level, the spit length shrank to 3.5 km. The landward extremity was separated from the mainland. An estimated 135 ha were under water with about 1500 ha liable to flood, and the shoreline had encroached noticeably onto the spit, to include some areas previously under water and usually isolated from the lake.

By 1980 the lake level increased 2 m and the spit had fragmented into a series of small islands separated by 1-2 km from the shore. Land under water had increased to an estimated 252 ha. Land not liable to flood had retreated, giving 1535 ha of flood plain (Figure 3).

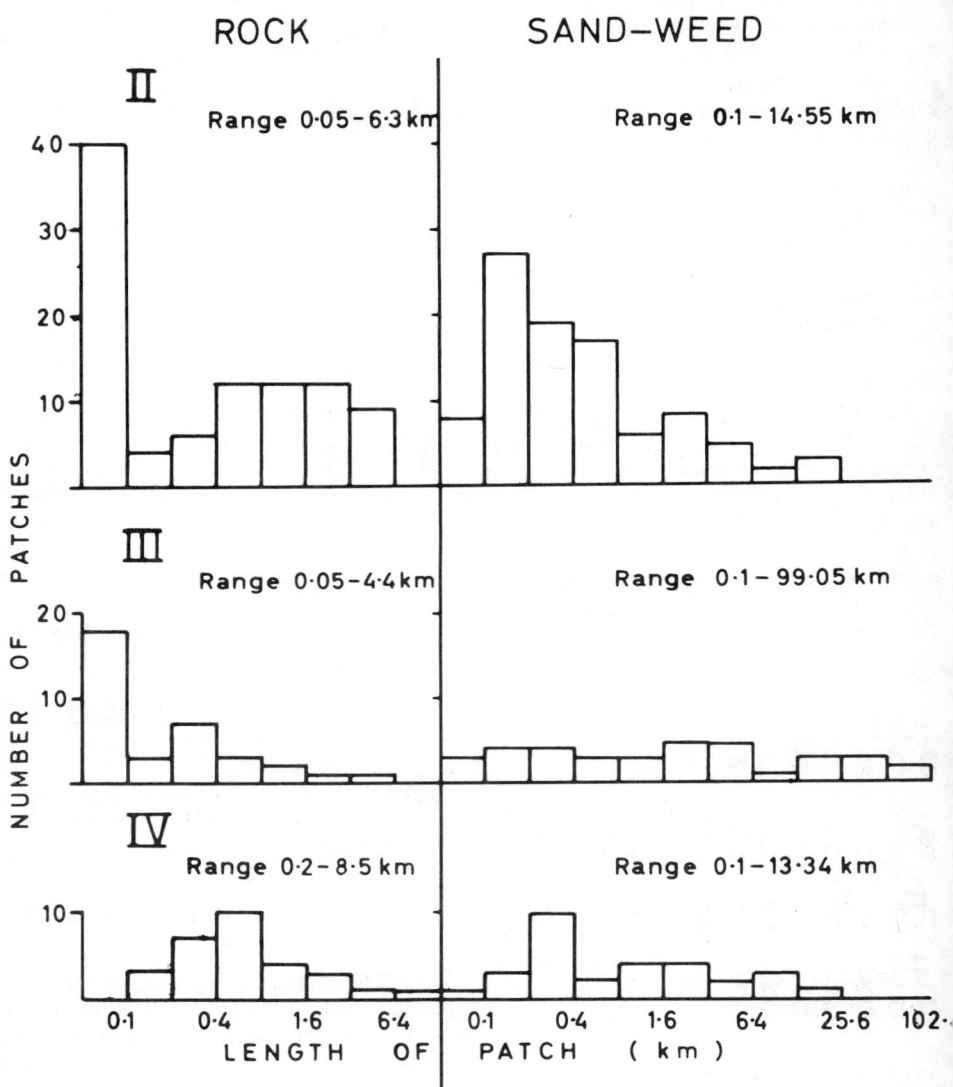

Figure 7: Distribution of habitat patch length in Lake Malawi: Regions II-IV. Horizontal axis is on log scale.

Another pattern of change is illustrated by Sungu Spit (Figure 4). In 1966 it was a detached spit 2.5 km long and 1 km from shore, pointing NNE. Probably it was attached earlier when the lake level was lower (Drayton 1979). To the south of the spit the shoreline contained a sheltered lagoon, 2.5 km x 0.5 km, or about 77.5 ha that was not completely isolated from the lake. Besides the lagoon, there was no open water on the landward side; however, 320 ha was near lake level and liable to flooding.

By 1972 there was a change in shape and size of the spit. On the landward side, sand had been deposited towards the spit. To the south, the lagoon remained about the same

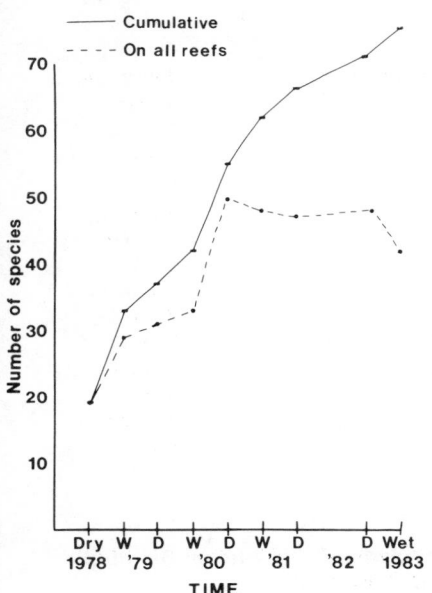

Figure 8: Pooled species numbers for all reefs,
versus time. Solid line marks cumulative number
of different species which ever visited the 8 reefs;
dashed line marks the total number of different
species on all eight reefs during any one season.

area, 77 ha, but an extra 5 ha of open water
appeared to the north. The area liable to flood
was reduced to 292.5 ha.

By 1980 a dramatic change had occurred,
as with Sungu Point. The spit had been
reduced to a few fragments. The landward
extension had disappeared. The shoreline had
retreated inland and eliminated the southern
lagoon. Within the new shoreline, 12 ha was
under open water, and only 200 ha remained
liable to flood. Thus in only 16 years, land
forms changed, islands appeared and disappeared, and shoreline continuity was
significantly modified.

Reef Colonization: Rock-Dwelling Species

Data presented here for the five-year period,
late 1978-early 1983, show the colonizing
abilities of the rock-dwelling cichlids and the
pattern in species diversity on the experimental reefs. Seventy-five different fish species
have been recorded on the habitats since the

reefs were established. New species which
had never before been seen on the reefs continued to appear at a rate of about 5 per six
months after 4.5 years. The rate of immigration of new species onto the reefs does not
as yet appear to be diminishing (Figure 8).
During the dry season of 1980 the maximum
of 50 different species occurred on the eight
reefs. At the wet season 1983 census, the
number of different species on all habitats had
dropped to 42. During this study, 18 species
which have never been recorded over sand
in the Cape Maclear region, either in 155 transects of 100 m^2 or 600 transects of 200 m^2
(McKaye 1983; McKaye, Reinthal unpubl.
data), have occurred on the reefs (Table 2).
Ten of these 18 species belong to the rock-
dwelling mbuna complex (Fryer 1959; Lewis
1982) which includes the genera
Pseudotropheus and *Petrotilapia*. All of these
mbuna species except one undescribed
Pseudotropheus sp. ("Yellow dorsal fin") were
seen on the reefs every season, after their
initial appearance. These nine resident mbuna
species included individual territorial males or
brooding females with young in their mouths.
The other rock-dwelling cichlids did not remain
consistently on the reefs from one season to
the next.

**Colonization Related to Reef
Characteristics.**

1. Large vs. Small. The four large reefs had
both more species and greater numbers of
fishes on them at all seasons than the comparable small reefs (Figures 9, 10). At 5 years
the average large reef had 1.6 times as many
species as did the average small reef.
2. Deep vs. Shallow. The four deeper reefs
had both more species and a greater number
of fishes on them during all seasons than the
comparable shallower reefs (Figures 9, 10).
At 5 years the average deep reef had 1.6 times
as many species as did the average shallow
reef.
3. Heterogeneous vs. Homogeneous. The four
heterogeneous reefs had more species of
fishes on them in 92 percent of the 36 censuses than the comparable homogeneous
reefs (Figure 9, Sign test, p ◄ .01); and 81
percent of the 36 censuses had greater
numbers of fishes on the heterogeneous reefs
(Figure 10, Sign test, p ◄ .05). At 5 years the
average heterogeneous reef had 1.3 times as

Table 2: Rock dwelling species appearing on experimental reefs.

Species Identification	1st Observation
Pseudotropheus "macrophthalmus" (Blue "tropheops")	1st season
Pseudotropheus tropheops (White with yellow fins)	1st season
Pseudotropheus tropheops (Orange head)	1st season
Petrotilapia spp. (Three species, females not distinguished)	1st season
Cyrtocara linni	½ year
Pseudotropheus elongatus (Species complex)	½ year
Pseudotropheus tropheops (Yellow)	2 years
Cyrtocara fenestratus / taeniolatus	2 years
Cyrtocara euchilis	2 years
Pseudotropheus zebra (Blue-black)	2½ years
Pseudotropheus sp. (Yellow dorsal fin)	2½ years
Alanocara sp. (nyassae)	2½ years
Labeo cylindricus	2½ years
Docimodus johnstoni	3 years
Trematocranus sp. (jacobfreibergi?)	4 years
Cyrtocara heterodon	4 years
Pseudotropheus zebra (Cobalt)	4½ years
Pseudotropheus tropheops (Dark)	4½ years

*Members of the mbuna complex.

many species as the average homogeneous reef.

Species and Total Fish Numbers through Time.

Each habitat has had a different pattern of species accumulation and population densities. Nevertheless, certain patterns emerge when the reefs are examined as a unit. The mean number of species per habitat peaked at 9 after 2 years, and has declined to 7 at 5 years. The mean number of individuals occurring on the reefs reached a peak within the first year, at 50 per reef, but declined steadily since the 1980 dry season. The last census in the 1983 wet season revealed an average of 22 fish.

Effect of Currents on Species Number and Population Densities

When currents were greater than 2 cm/sec, mean number of species and individuals on all eight reefs were greater than when currents were absent or less than 2 cm/sec (Figure 11). On average, there were 1.2 times more species, and 2.0 times more individuals on the reefs when the currents were running (Figures 11, 12). The increase in numbers of fish was 177 percent for the shallower reefs, and 43 percent for the deeper reefs.

Zooplankton Resources: Distribution and Seasonality

For the two years of sampling which corresponded with the beginning of the experiment, there was consistently less zooplankton in 5 m of water than in 10 m (Figure 13). Near the habitats, there appeared to be a bimodal peak of *Diaphanosoma* and diaptomids, one at the beginning of the rains in December-January, and one at the beginning of the *mwera*, or southerly winds, in June-July of 1979 and 1980. These results are consistent with observations on frequency of zooplankton feeding by fishes on the reefs (McKaye unpubl.).

DISCUSSION

The western shore of Lake Malawi is a heterogeneous environment of sand-weed and rock. Eighteen percent of the western shore is composed of 158 rock patches isolated by stretches of sand or weed. Such isolated rock sections provide the necessary physical situation for microallopatric speciation. Isolation would be possible if the movements of rock-dwelling cichlids are restricted, relative to the distances between rocks (Fryer 1959, 1977; Fryer, Iles 1972). Wide separation would minimize gene flow

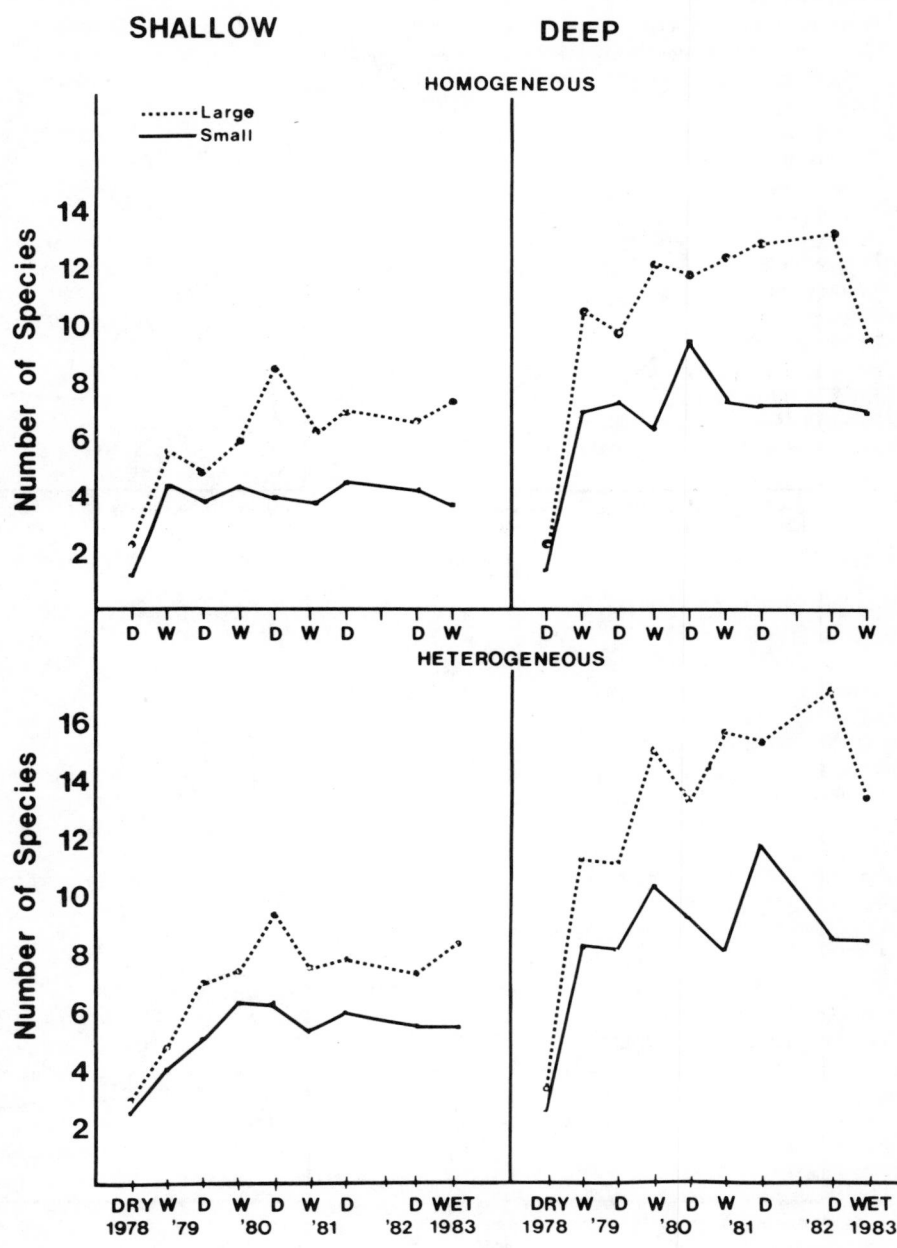

Figure 9: Mean number of species for each season on each of eight reefs.

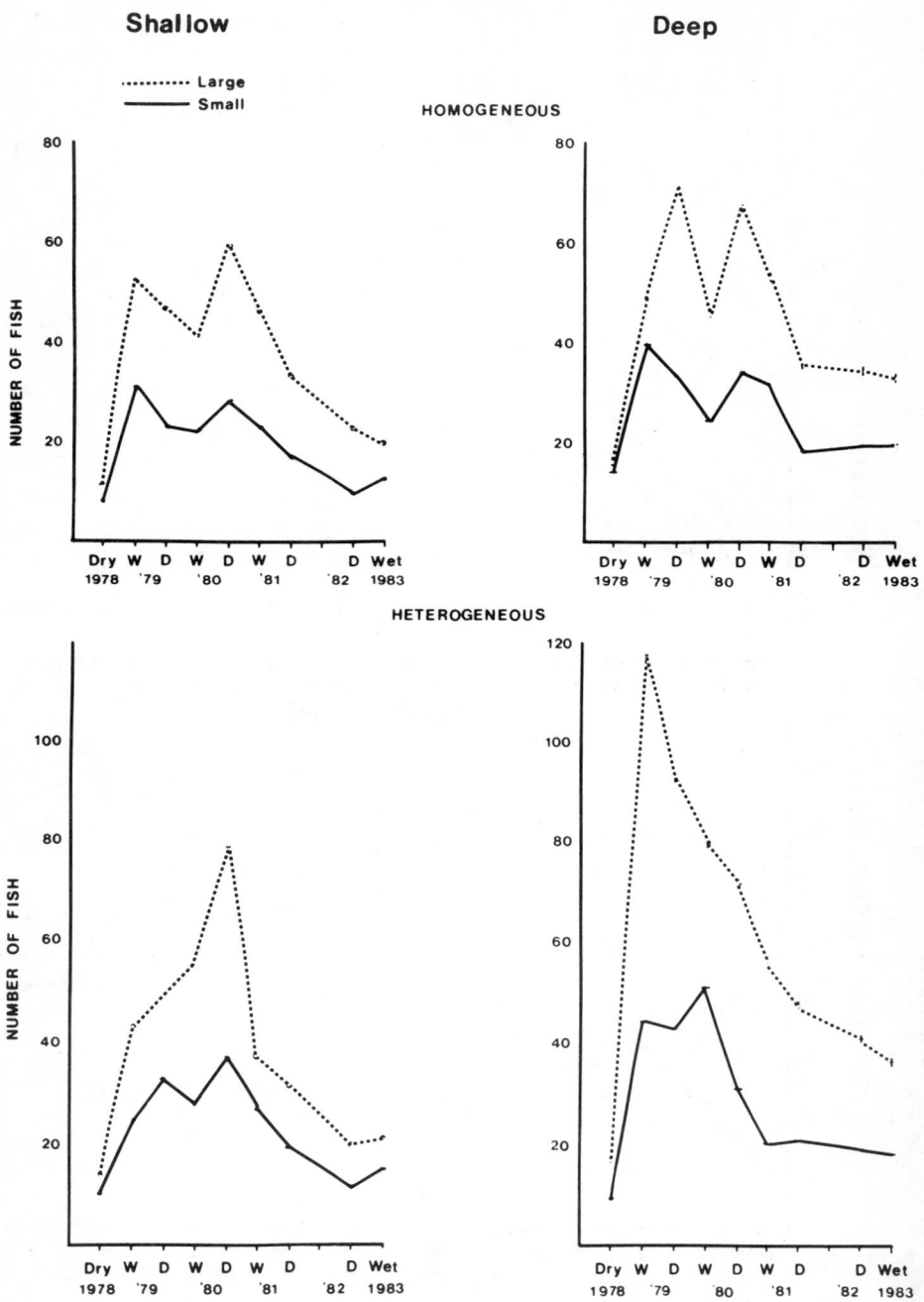

Figure 10: Mean number of individual fish for each season on each of the eight reefs.

Figure 11: Cichlid fishes feeding on zooplankton in back of reef when strong current is flowing.

among populations and consequently would maximize the possibility of speciation due to geographic isolation.

The dynamics of the shoreline indicate that fragmentation of the lake into varying habitat can and does occur over surprisingly short timespans (Figures 3, 4). The rapid modification of habitat continuity along the mainland shoreline, plus the existence of numerous isolated islands and rock habitats in the lake, provide evidence that the physical discontinuity required for microallopatric isolation exists, and thus gives some support to Fryer's (1959, 1977) model for the evolution of cichlid species flocks in Lake Malawi. However, data from the experimental reefs suggest the process may be complex and differential among species subsets. Different groups, depending in part on trophic ecology, appear to disperse at different rates.

The results of the reef experiment specifically demonstrate this key point: that a subset of rock-dwelling cichlids will disperse over sand and colonize newly formed habitats. Further, fish species diversity and zooplankton productivity were correlated. Thus, food diversity and availability may contribute to

determining species composition and successful colonization of new habitats. All of the species on the reefs were observed to feed on zooplankton at least some of the time. When currents were flowing, bringing more zooplankton onto the reefs, more species were counted. The most dramatic response in fish numbers with increased current flow occurred on the shallow reefs where plankton was at lower absolute density (Figures 11, 12, 13). The mbuna colonizers early in the 5-year research period belong to the four species complexes which feed extensively on zooplankton: the *Pseudotropheus zebra* complex (McKaye, Marsh 1983; Holzberg 1978), the *P. tropheops* complex (pers. obs.), the *P. elongatus* complex (pers. obs.), and the *Petrotilapia* complex (McKaye, Marsh 1983). These species are also ones that Liem (1978, 1980, pers. comm.) suggests are facultative in their feeding repertoire. It appears that the mbuna which fed extensively on zooplankton are the most vagile and thus, the most likely species to colonize a new habitat. These species groups may provide the founders for new populations and lead to rapid allopatric speciation (Mayr 1963). On

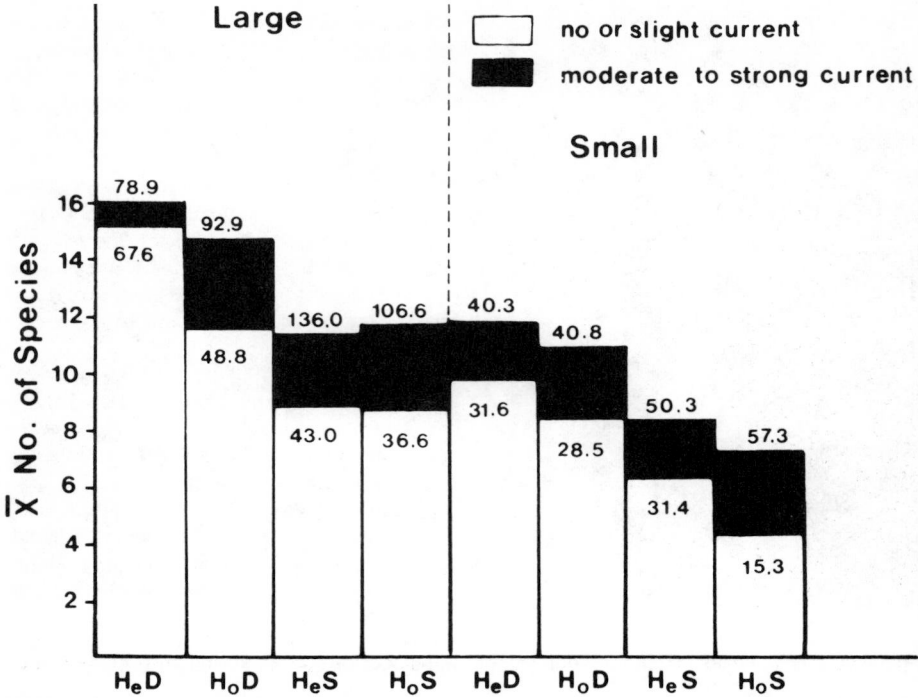

Figure 12: Effect of current on numbers of individual species occurring on reefs. On all eight reefs, numbers of individuals and species increased when currents were greater than 2 cm/sec. Lower numbers count fish on reef with little or no current; upper numbers count fish on reef with high current. Letters below bars signify heterogeneous (He) or homogeneous (Ho) reefs in deep (D) or shallow (S) water.

the other hand, increased gene flow between populations could retard allopatric speciation, and intrinsic isolating mechanisms could be more important than extrinsic ones in splitting species (McKaye 1980). More data are required to fully appreciate the implications of these observations.

The mapping data support the view that fragmentation of populations and subsequent reproductive isolation could occur both on a microgeographic scale, less than 20 km separation, for the less vagile species, and on a more macrogeographic scale. Along the western shore of the lake, the rock portions in the north were separated from those in the south by extensive sand-weed beaches. The longest single stretch was nearly 100 km. Northern mbuna species would have to traverse these large distances over sand in order to interbreed with southern conspecifics. Thus, the data lead to the prediction of distinctive northern and southern subsets of species

occurring in each region. The available genetic data are consistent with this hypothesis. Electrophoretic analysis of isoenzymes of *Pseudotropheus zebra* show that Chilumba and Nkhata Bay populations in the north have gene frequencies that are not statistically different from each other, but do differ statistically from those of *P. zebra* populations at Mumbo and Domwe islands in the southern Cape Maclear region (McKaye et al. 1984). The longest sand-weed stretch between Chilumba and Nkhata Bay is 15.5 km; the longest sand beach between Nkhata Bay and Cape Maclear is close to 100 km. Thus, though *P. zebra* may feed on zooplankton while migrating, 100 km of sand appears to be a significant barrier while 15.5 km may not be. More data on migratory abilities of these fish are required before a definitive conclusion can be made about what constitutes a major barrier to gene flow between populations.

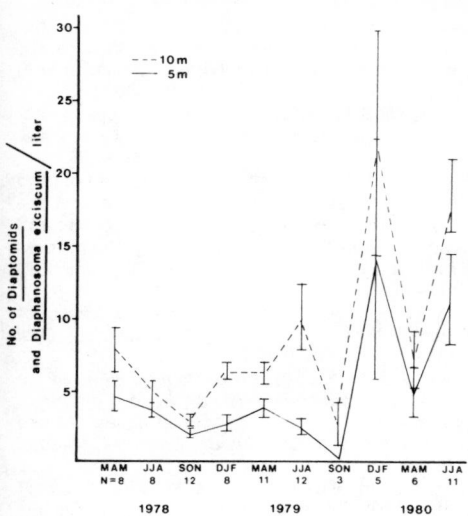

Figure 13: Density of zooplankton fed on by the majority of cichlids colonizing reefs at 5 and 10 m depths.

A related prediction to guide future research can be derived from the map data. We expect that the steep rocky shores in Region II should serve to isolate many of the obligate sand-dwelling cichlid populations in the northern Karonga area from those in the south. If so, the sand-dwelling fauna to the south of Nkhata Bay down to the outlet of the Shire River should be genetically differentiated from those in the northern sand portion of the lake.

Despite the limited extent of this survey regarding the history of the lake, these data should be useful in evaluating the contribution of island biogeography theory (MacArthur, Wilson 1967) to understanding cichlid species packing in communities. The isolated rock outcroppings in the lake, surrounded by sand, are islands of habitat. Species area curves, constructed for many groups of animals, predict the increase in number of species relative to increased habitat area. Barbour and Brown (1974) applied such analysis to fish faunas of many lakes throughout the world. In Lake Malawi we find rock outcroppings ranging in size from 0.05 to 8.5 km, with all intermediate sizes present (Figure 7). It should be possible

in this diverse vertebrate assemblage, to test the applicability of the established species area relationship (MacArthur, Wilson 1967) for fishes on habitat patches within a large lake. Deviations from expected species diversities on given patches would focus attention on the processes determining community structure in this diverse fish assemblage. Our experimental data suggest that not only the size, but also spatial heterogeneity of the available habitat is important in determining how many species may ultimately find and inhabit a new region (Figure 10).

Greenwood's (1965) study of the endemic cichlids of Lake Nabugabo may also provide a relevant paradigm for the evolution and speciation of shore-bound cichlids in Lake Malawi. In Lake Nabugabo, which was separated from Lake Victoria for about 4000 years, five endemic cichlid species coexist. These species are each most closely related to cichlids in Lake Victoria, suggesting that they evolved allopatrically from those parent populations. The appearance and disappearance of lagoons in Lake Malawi, as indicated by the 16-year series of photographs (Figures 3, 4) could be analogous to the Lake Nabugabo situation. More detailed geological studies of Lake Malawi should allow us to begin to determine the age and continuity of isolation among various habitat patches or regions. With such information, we will be in a better position to determine the rate of evolution of the cichlid fishes in this lake, and to assess the degree of isolation and time necessary for new species to arise.

Interestingly, the mbuna species that feed extensively on zooplankton and colonized the experimental reefs would appear to be the groups least likely to have restricted gene flow. However, they are among the most species-rich groups of mbuna (Marsh 1982; Marsh et al. 1981; McKaye et al. 1982, 1983; Lewis, Marsh, Ribbink pers. comm.). Much further work is required before any definitive conclusions can be drawn concerning the evolutionary processes which caused this species-rich vertebrate flock to arise. We hope these data will stimulate research on the geololgy, history and fauna of Lake Malawi. This inland sea, with its habitat heterogeneity and highly diverse fauna, has the potential to provide further insight into the processes which mold the evolution and structure of complex communities.

SUMMARY

1. Aerial photos of the western shoreline of Lake Malawi support Fryer's view that Lake Malawi consists of a mosaic of alternating patches of sand (61%), weed (21%) and rock (18%).

2. Analysis of photographs taken between 1966-1980 documented the creation and dissolution of isolated habitats due to rising lake levels.

3. Data from eight experimental reefs placed on sand demonstrated that a subset of rock dwelling cichlids will move one kilometer from the nearest rock outcropping and colonize new habitats. These early colonizing species are facultative in their feeding repertoire and primarily exploit zooplankton.

4. Doubling the size of a reef resulted in a 60 percent increase in species number, and increasing the reef's spatial heterogeneity resulted in a 30 percent increase in species numbers. Deep reefs (9 m) had approximately 60 percent more species than did identical shallow reefs (6 m). This increase was due to the greater zooplankton densities at the greater depth. Zooplankton is the primary food of most cichlids.

5. The species groups most likely to colonize newly formed rock outcroppings are those which are among the most species rich. This anomaly suggests that in some circumstances intrinsic isolating mechanisms could be more important than extrinsic ones in splitting species. Processes of both allopatric intralacustrine speciation and sympatric speciation are probably occurring in the cichlid species flocks of the Great Lakes of Africa.

ACKNOWLEDGMENTS

We thank the Malawian Fisheries and Survey Departments for continuing assistance, J. Trendall and M. Salmon for reviewing the manuscript, and S. Louda for constant encouragement and patient review of the manuscript. The research was sup- by the United States National Science ation.

REFERENCES

Barbour C D, J H Brown 1974 Fish species diversity in lakes. Amer Nat 108:473-489

Barel C D N, M J P van Oijen, F Witte, E L M Witte-Maas. 1977 An introduction to the taxonomy and morphology of the Haplochromine Cichlidae from Lake Victoria. Neth J Zool 27:381-389

Drayton R S 1979 A study of the causes of the abnormally high levels of Lake Malawi in 1979. Water Resources Division, Malawi WRD-NO. TP 5

Fryer G 1959 The trophic interrelationships and ecology of some littoral communities of Lake Nyasa with especial reference to the fishes and a discussion of the evolution of a group of rock-frequenting Cichlidae. Proc Zool Soc Lond 132: 153-281

_____ 1977 Evolution of species flocks of cichlid fishes in African lakes. Z Zool Syst Evolut-forsch 15:141-165

Fryer G, T D Iles 1972 The Cichlid Fishes of the Great Lakes of Africa. Oliver, Boyd. Edinburg

Greenwood P H 1965 The cichlid fishes of Lake Nabugabo, Uganda. Bull Br Mus Nat Hist (Zool) 12:315-357

_____ 1974 Cichlid fishes of Lake Victoria, East Africa: The biology and evolution of a species flock. Bull Br Mus Nat Hist (Zool) Suppl 6:1-134

Holzberg S 1978 A field and laboratory study of the behaviour and ecology of Pseudotropheus zebra (Boulenger): an endemic cichlid of Lake Malawi (Pisces: Cichlidae) Z Zool Syst Evolut-forsch 33:478-485

Kosswig C 1947 Selective mating as a factor for speciation in cichlid fish of East African lakes. Nature 159:604-605

_____ 1963 Ways of speciation in fishes Copeia 1963:238-244

Lewis D S C 1982 A revision of the genus Labidochromis (Teleostei: Cichlidae) from Lake Malawi. Zool Jour Linn Soc 75:189-265

Liem K F 1978 Modulatory multiplicity in the functional repertoire of the feeding mechanism in cichlid fishes. J Morphol 158:323-366

_____ 1980 Adaptive significance of intra-and interspecific differences in the feeding repertoires of cichlid fishes. Am Zool 20:295-314

Lowe-McConnell R H 1959 Breeding behavior patterns and ecological differences between Tilapia species and their significance for evolution within the genus Tilapia (Pisces: Cichlidae). Proc Zool Soc Lond 132:1-30

MacArthur R, E O Wilson 1967 The Theory of Island Biogeography. Princeton Univ Press. Princeton NJ

Marsh A C 1981 A contribution to the ecology and systematics of the genus Petrotilapia (Pisces: Cichlidae) in Lake Malawi. MS Thesis, Rhodes Univ

Marsh AC, A J Ribbink, B A Marsh 1981 Sibling species complexes in sympatric populations of Petrotilapia Trewavas (Cichlidae, Lake Malawi). Zool J Linn Soc 71:253-264

Mayr E M 1963 Animal Species and Evolution. Belknap. Cambridge Mass

McKaye K R 1980 Seasonality in habitat selection by the gold color morph of *Cichlasoma citrinellum* and its relevance to sympatric speciation in the family Cichlidae. *Env Biol Fish* 5:75-78

McKaye K R 1981 Death feigning: A unique hunting behavior by the predatory cichlid, *Haplochromis livingstonii* of Lake Malawi. *Env Biol Fish* 6:361-365

———— 1983 Ecology and breeding behavior of a cichlid fish. *Cyrtocara eucinostomus*, on a large lek in Lake Malawi, Africa. *Env Biol Fish* 8:81-96

McKaye K R, A Marsh 1983 Food switching by two specialized algae-scraping cichlid fishes in Lake Malawi, Africa. *Oecologia* 56:245-248

McKaye K R, T Kocher, P Reinthal, I Kornfield 1982 A sympatric sibling species complex of *Petrotilapia* Trewavas from Lake Malawi analyzed by enzyme electrophoresis (Pisces: Cichlidae). *Zool J Linn Soc* 76:91-96

McKaye K R, T Kocher, P Reinthal, R Harrison, I Kornfield 1984 Genetic evidence for allopatric and sympatric differentiation among color morphs of a Lake Malawi cichlid fish. *Evolution* 38:215-219

Ostro M E 1979 *Tropical Fish Hobbyist* 27:71

Ribbink A J, B A March, A C Marsh, A C Ribbink, B J Sharp 1983 A preliminary survey of the cichlid fishes of rocky habitats in Lake Malawi. *S Afr J Zool* 18:1-180

MOLECULAR VERSUS MORPHOLOGICAL CHANGE AMONG CICHLID FISHES OF LAKE VICTORIA

Richard D Sage, Paul V Loiselle, Pereti Basasibwaki, A C Wilson

INTRODUCTION

Although biologists have been intrigued for 70 years with the explosive evolution displayed by cichlid fishes of Lake Victoria (Fryer, Iles 1972; Fryer 1977; Greenwood 1974, 1981, 1984), attempts to quantify the degree and rate of this process are recent. Such quantification is needed for rigorous testing of ideas about factors determining the rate of evolution. Our goal is to present quantitative information on degrees of divergence among these fishes.

Thorough ecological studies of Lake Victoria cichlids are just beginning (Oijen et al. 1981; Oijen 1982; Witte 1984), and more field work is needed. New field data considered in relation to the molecular results reported for cichlids in Lake Malawi (Kornfield 1978; Marsh et al. 1981; McKaye et al. 1982, 1984; McKaye, Gray 1984) suggest that the evolutionary process in the Great Lakes of East Africa is even more dramatic than indicated by preserved specimens. The new studies show that cichlids previously considered as color morphs of one species probably do not interbreed, and, in Lake Victoria, the species show more ecological separation than previously believed.

Proposing that genetic rather than ecological constraints are the principal limits to community diversity, Felsenstein (1981:124) states that under a model of strong ecological competition, "one would expect to find nearly infinite numbers of species ... the number of species in nature is far smaller, and their size far larger, than such a model would predict." But do the African lakes suggest exceptions to the conclusion of a genetic limit on the speciation process? Perhaps the general theoretical models need to encompass new parameters to explain the explosive radiation of species flocks. It has been suggested that cultural drive, a potent nongenetic force, may be responsible for the rapid morphological evolution and speciation in some mammal and bird groups (Wyles et al. 1983; Larson et al. 1984). After presenting quantitative data on rates of evolution in the cichlid flock of Lake Victoria, we discuss whether cultural drive may be rushing the ecological process in the Great Lakes cichlids faster than indicated by the theoretical model.

DIVERGENCE: QUANTITATIVE ESTIMATE

This paper approaches the problem of measuring the extent of structural gene divergence among Lake Victoria cichlids and also their degree of overall anatomical divergence. These measurements considered in relation to the age of the lake allow approximate rates of evolution at these two levels of biological organization to be calculated and compared with those available for other groups of organisms.

Starch gel electrophoresis, which has been used extensively in estimates of structural gene divergence between populations and species of numerous vertebrates (Avise, Aquadro 1982), is applied here to estimate genetic distances within and among cichlid species from Lake Victoria. Only recently have evolutionary biologists shown much interest in quantitative comparative studies of morphological divergence and the question of differential rates of organismal evolution (King, Wilson 1975; Wilson 1975; Wake 1981; Cherry et al. 1982). Quantitative anatomical comparisons have had a major role in systematic biology, but the question of just *how much* anatomical divergence has occurred between taxa has not been so actively investigated. Recognizing it as a crude initial approach,

TABLE 1. DISTRIBUTION AND FEEDING ECOLOGY OF FISHES STUDIED FROM LAKE VICTORIA

Taxa	Distribution	Feeding habits
1. *Haplochromis lividus* (G)	Lake Victoria	periphyton
2. *Harpagochromis "pectoralis"*	Lake Victoria	fish
3. *Lipochromis maxillaris* (T)	Lake Victoria	fish eggs and fry
4. *Lipochromis "microdon"*	Lake Victoria	fish eggs and fry
5. *Lipochromis obesus* (B 1906)	Lake Victoria	fish eggs and fry
6. *Lipochromis parvidens* (B 1911)	Lake Victoria	fish eggs and fry
7. *Neochromis nigricans* (B 1911)	Lake Victoria, Victoria Nile	periphyton
8. *Paralabidochromis chilotes* (B 1911)	Lake Victoria, probably Victoria Nile	insects
9. *Prognathochromis longirostris* (H)	Lake Victoria, possibly Victoria Nile	fish
10. *Prognathochromis "macrognathus"*	Lake Victoria	fish, insects
11. *Hoplotilapia retrodens* (H)	Lake Victoria, probably Victoria Nile	bivalve molluscs
12. *Astatoreochromis alluaudi* (P)	Lake Victoria, adjacent lakes, Victoria Nile	gastropod molluscs
13. *Tilapia nigra*	Introduced to Lake Victoria	omnivore (?)

(B) Boulenger 1906, 1911; (G) Greenwood 1956; (H) Hilgendorf 1888; (P) Pellegrin 1903; (T) Trewavas 1928.

we have adapted the method of Cherry et al. (1982) to compare extent of morphological divergence among cichlids from Lake Victoria with that for a group of species from North America, the sunfishes (Centrarchidae).

The sunfish-cichlid comparison is appropriate because of their general ecological and morphological resemblance and because protein electrophoresis has already been used to estimate the genetic relationships of sunfish (Avise, Smith 1977). Thus, we regard the centrarchids as a control group that presumably has evolved in a fairly conventional allopatric manner, i.e., in separate drainages, rather than as a flock within one lake.

Nomenclature. Until recently, most of the endemic cichlid species of Lake Victoria were included in the genus *Haplochromis* (see Greenwood 1974). Among the endemic species we studied, only *Hoplotilapia retrodens* has always been treated as a separate genus. Greenwood (1979, 1980) revised the genus *Haplochromis*, and separated the species into many genera. Although we use his revised taxonomy in the tables and appendices, for simplicity in the text we refer to the first 11 species in Table 1 as the "Haplochromis" group.

MATERIALS AND METHODS

Electrophoresis. Electrophoresis was done in two phases. The first study was performed (by ACW) with freshly killed specimens in Nairobi, and the second was performed (by RDS and PVL) in Berkeley with frozen materials taken to Berkeley in 1973 and run in 1979. In 1973, one of us (PB) collected live animals of 12 species near Jinja, Uganda, and took them alive to Nairobi. Using horizontal starch gel electrophoresis, 11 specimens each of *L. obesus* and *P. "macrognathus"* were screened for electrophoretic variation at 13 loci. Extracts were made from eye, heart, and liver by homogenization in 0.25 M sucrose or (Karig, Wilson 1971) by soaking in a phenoxyethanol solution. Electrophoresis was done in 12-18 percent starch gels, (a) at pH6 in phosphate-citrate buffer (Karig, Wilson 1971) or (b) at pH 8.4 in Tris-glycine buffer (glycine 21.8 gm/l; Tris 4.5 gm/l).

Specimens for return to the United States were stored frozen at -10° C in Nairobi, transported to Berkeley in dry ice at -79° C, and stored at -10° C, and later at -76° C before fresh extracts were prepared. In 1979 a live specimen of *A. alluaudi* from Lake Victoria was purchased and one of *H. lividus* was obtained from an aquarist for use as fresh materials to compare with the old specimens.

At Berkeley, aqueous extracts were made from body muscle and from the eye of each specimen by mincing the tissue with a razor blade and mixing with an equal volume of distilled water. High speed centrifugation (27,000 g, 40 min) produced cleared extracts used in electrophoresis. Samples were stored at -76° C.

Horizontal starch gel electrophoresis, with starch from Connaught Laboratories, done in Berkeley, followed Sage's (1978) methods. Buffers and running conditions were: 1) Tris-hydrochloric acid (pH 8.5) 250 v, 1.5 hrs; 2) Continuous Tris-citrate (pH 8.0) 135 v, 3 hrs; 3) Discontinuous Tris-citrate (pH 8.2) 250 v, 3 hrs; 4) Phosphate-citrate (pH 7.0) 100 v, 4 hrs; 5) Tris-maleic-EDTA (pH 7.4) 100 v, 4 hrs, with NADP added to the starch just before cooling; 6) Continuous Tris-citrate (pH 7.0) 180 v, 3 hrs.

The following 13 enzymes were studied (parenthetic indicators refer to buffers and running conditions described above— a,b = Nairobi runs, 1-6 = Berkeley runs): alcohol dehydrogenase, Adh (1); glycerol-3-phosphate dehydrogenase, Gpd (2); L-iditol dehydrogenase, Iddh (b); lactate dehydrogenase, Ldh (a,b, 3); malate dehydrogenase, Mdh (a); isocitrate dehydrogenase, Icd (4); 6-phospho-gluconate dehydrogenase, Pgd (5); "nothing dehydrogenase," Ndh (b); superoxide dismutase, Sod (b); aspartate aminotransferase, Aat (a,b); phosphoglucomutase, Pgm (6); mannose phosphateisomerase, Mpi (2); glucosephosphate isomerase, Gpi (3). Two hemoglobin loci, Hb (b), were studied in Nairobi.

An initial survey of enzyme stains available in the Berkeley laboratory was performed on tissue extracts from old and new samples of A. alluaudi. This showed that the 1973 samples had lost much of their enzymatic activity. For systems retaining sufficient staining activity, comparison of the old with the new samples showed no mobility changes; this gave us confidence to proceed with the study of the aged materials.

The anodal alcohol dehydrogenase locus we scored is one that reacts with trans-2-hexen-1-ol and not with ethanol (Holmes 1977). On the gels stained for glycerol-3-phosphate dehydrogenase, two bands were present, but only one locus appeared to be represented because the observed variants covaried in the same manner. Three lactate dehydrogenase loci are present in cichlids. The fastest migrating lactate dehydrogenase, encoded by the Ldh-C locus, is found in the retinal tissue (Markert et al. 1975). The slowest migrating form (Ldh-A) predominates in the muscle extracts. The

form of intermediate mobility (Ldh-B) was present in both types of extracts, but stained so faintly that many specimens in the Berkeley study could not be scored. The isocitrate dehydrogenase system scored is the slower migrating of two known loci, and probably corresponds to the cytosol form of the enzyme. Two phosphoglucomutase systems were detected, but only the more anodal was scored. Two glucophosphate isomerase loci are present in cichlid fishes (Avise, Kitto 1973). Our Gpi-1 corresponds to the faster migrating of the two enzymes.

In the Berkeley study all 16 specimens of A. alluaudi and two specimens of each of the other taxa were examined at all ten loci: Adh, Gpd, Ldh-A, Ldh-C, Icd-2, Mpi, Pgd, Pgm, Gpi-1, and Gpi-2. We studied only one H. lividus because the available fishes were not wild-born, and we had no idea how many native genomes were present in this aquarium-reared material. All specimens were tested and scored for the polymorphic Gpi and Ldh-A loci (Appendix 2). Four specimens of N. nigricans were scored for Icd-2 variability.

Morphological Measurements and Calculations. The cichlid specimens from which tissues were taken for electrophoresis were preserved in neutral buffered formalin and stored in ethanol for morphological measurements. Voucher specimens deposited in the collection of the Museum of Comparative Zoology, Harvard University, were examined by P. H. Greenwood, who made identifications confirming and supplementing those made by one of us (PB) when the animals were collected. He made only tentative identifications of three of the species that we studied: the Harpagochromis specimens belong to the second of two species in the "pectoralis" group; exact determination of species in the Lipochromis "microdon" group requires knowledge of adult male coloration, which we lack; and the Prognathochromis specimens are possibly P. "macrognathus." We use these tentative names for the three taxa.

Measurements were made on our cichlid specimens and on sunfish specimens from the California Academy of Sciences, San Francisco. Eight traits were measured on the preserved carcasses: 1) total body length, tip of upper lip to caudal peduncle; 2) head length (HL) from upper lip to posterior edge of operculum; 3) body height (BH) at the point above end of operculum; 4) body width (BW) at this same level on trunk; 5) eye width (EW), horizontal distance across eye socket; 6) pectoral fin length (PF), longer of the two fins, from tip of longest fin ray to its attachment on radial bones; 7) eye to snout length (ES), chord distance from tip of upper lip to center point of eye; 8) premaxilla length (PL), chord distance of premaxillary bone from mid-dorsal line to its end at side of mouth. Trunk length (TL), total body length minus head length.

Relative trait lengths (Cherry et al. 1982) were computed. For each specimen each trait length was divided by the sum of the eight trait lengths from the same animal and expressed as parts per 10,000.

The proportional distance, Delta, between taxa was computed as follows:

$$\Delta = 200p^{-1} \sum_{i=1}^{p} | d_i | \, (x_i + y_i)^{-1}$$

where x_i and y_i are the mean values of the ith trait in species X and Y, d_i is the difference between x_i and y_i and p is the total number of traits. Cherry et al. (1982) justify this metric as an approximate measure of overall morphological difference. Though we consider it appropriate to compare Delta values obtained for cichlids with those for centrarchids, we caution against comparing them to the Delta values reported by Cherry et al. (1982) for land vertebrates because many traits are not homologous between studies.

RESULTS
Genetic Variability within Species

Ldh-C and *Mpi* were monomorphic and identical in all of the Berkeley specimens. The remaining 8 loci showed inter- or intraspecific variation. Frequencies of allelic variants found in Nairobi and Berkeley are given in Appendices 1 and 2. Both studies revealed loci segregating for more than one allele. The presumptive heterozygotes had the appropriate appearance and number of bands consistent with the quaternary structure of the enzymes. The two *Gpi* loci were the most variable ones encountered, both in terms of

TABLE 2. MEAN HETEROZYGOSITY: LAKE VICTORIA CICHLIDS, BASED ON TEN BIOCHEMICAL LOCI

Species	H̄
1. *Haplochromis lividus*	---
2. *Harpagochromis "pectoralis"*	.069
3. *Lipochromis maxillaris*	.034
4. *Lipochromis "microdon"*	.051
5. *Lipochromis obesus*	.012
6. *Lipochromis parvidens*	.014
7. *Neochromis nigricans*	.054
8. *Paralabidochromis chilotes*	.105
9. *Prognathochromis longirostris*	0
10. *Prognathochromis "macrognathus"*	.050
11. *Hoplotilapia retrodens*	.114
12. *Astatoreochromis alluaudi*	.057
13. *Tilapia nigra*	0
Average	**.047**

intra- and interspecific variability.

Mean heterozygosity (H) corrected for small sample size (Nei 1978) was calculated to measure intraspecific variability. In the Nairobi study of 11 individuals of each species, the H value for *L. obesus* was 0.062, and for *P. "macrognathus"* was 0.040. Estimates of heterozygosity for the Berkeley samples (Table 2) range from zero in *P. longirostris* and *T. nigra* to 11 percent in *H. retrodens*. The heterozygosity estimates for *L. obesus* and *P. "macrognathus"* in the two studies are not significantly different.

Genetic Distances Between Species.

Relationships among species were estimated by calculating genetic distances, D (Nei 1978), corrected for sample size (Table 3). *Tilapia nigra* is distantly associated with *Astatoreochromis* and the "Haplochromis" species, with D values always exceeding 1.0. *Astatoreochromis* is more similar to the "Haplochromis" species, genetic distances falling between 0.3 and 0.4. The 11 "Haplochromis" species, belonging to 7 genera, are extremely similar to one another, with genetic distances ranging from 0.00 to 0.04.

Shared Polymorphisms Between Trophic Groups

Besides exhibiting standard levels of heterozygosity (Avise, Aquadro 1982) and extremely small genetic distances, the "Haplochromis" group of species tends to share polymorphisms. The most striking cases involve species belonging to different trophic groups or genera. Consider the *Sod-S* locus, at which the same polymorphism occurs in both *L. obesus*, a paedophage, and *P. "macrognathus"*, a piscivore (Appendix 1). These two species, based on previous morphological analysis, were thought to be remotely related phylogenetically. Indeed, they are separated by at least 6 speciation events (Greenwood 1974). The *Gpi* loci provide similar examples of shared polymorphisms between species from different trophic groups (Appendix 2).

Morphological Distances

Figure 1 and Appendices 3 and 4 summarize results of comparisons of morphological differences found in the cichlids and centrarchids. The mean morphological distance between congeneric species of cichlids is about 8, compared to a value of about 11 for intergeneric comparisons within the "Haplochromis" flock. These means are statistically homogeneous. Nearly the entire range (Delta = 4-19) of shape difference found in the Lake Victoria cichlids is present among the genera of the "Haplochromis" group. There is almost no increase in the amount of change seen by including the more distantly related *Astatoreochromis* or *Tilapia*. Among the centrarchids the morphological distances are slightly greater. There is much morphological divergence among congeners, particularly among the *Lepomis* species (Range = 4-22 Delta units). The means and ranges increase, but not significantly, in the genetically more distant intergeneric comparisons (Figure 1). [1]

DISCUSSION

While recognizing existing explanations for the rapid speciation and morphological change displayed by East African cichlids (Liem 1973; Dominey 1984), we think that the continuing fascination with lacustrine flocks of fish species may signify a need for a new hypothesis as to the driving force for this sort of evolution. Our suggestion, as follows, is new in that it appeals to the capacity for social learning which is suspected (Mainardi 1980; Dill 1983) for cichlids and other teleosts. We first emphasize the chief features of the "Haplochromis" flock that the hypothesis is

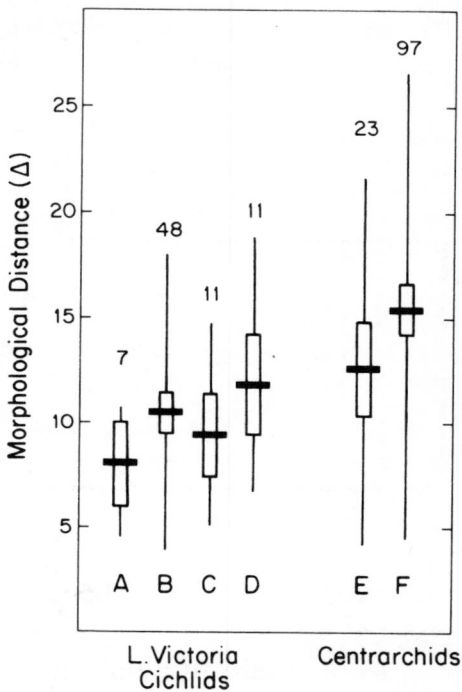

Figure 1. Morphological distances in Lake Victoria cichlid and North American centrarchid fishes. Comparisons: A, congeneric within the "Haplochromis" flock; B, intergeneric within the "Haplochromis" flock; C, *A. alluaudi* to "Haplochromis"; D, *T. nigra* to "Haplochromis"; E, congeneric within centrarchids; F, intergeneric within centrarchids. Delta values computed from data of Appendices 3, 4. Number of comparisons shown above vertical lines. Vertical lines indicate range; open bars show 95% confidence intervals for the mean.

designed to explain.

Tree and Time Scale

The tree in Figure 2 depicts the probable branching order of the lineages leading from a common ancestor to *Tilapia, Astatoreochromis,* and the "Haplochromis" flock of species. This branching order is based on the protein data and was determined by applying the Fitch and Margoliash (1967) and Farris (1972) methods to the Rogers distances (1-S) computed from the S values in Table 3. We probably can ignore alternative

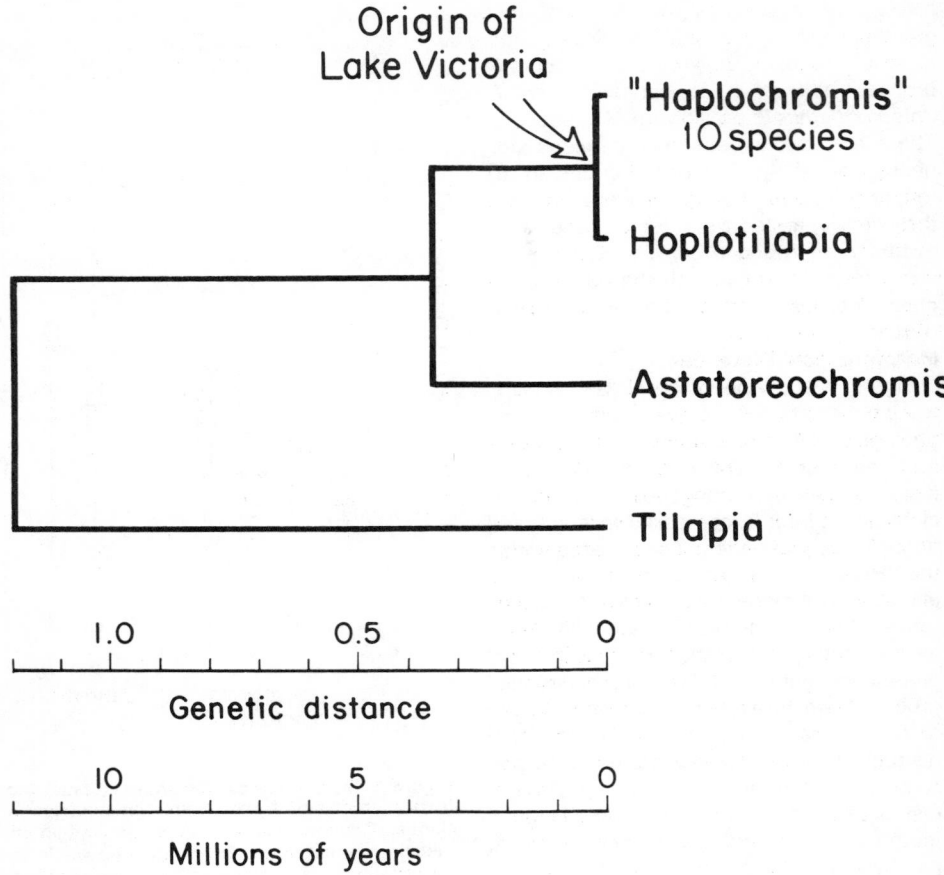

Figure 2. Tree for Lake Victoria cichlids based on electrophoretic comparisons of proteins encoded by 10 enzmyic loci. Tree length (computed by Farris' (1972) method) in Rogers' distance units is 0.868. By contrast, a branching order associating the *Hoplotilapia* lineage with *Astatoreochromis* rather than with the other "Haplochromis" species has a length of 0.997 units. Due to extremely small genetic differences among the 11 lineages of the "Haplochromis" flock, we chose to show the relationships of the most dissimilar species, namely the monotypic genus *Hoplotilapia*, and an average of the species in the remainder of the "Haplochromis" flock.

branching orders like that linking *Hoplotilapia* more closely to *Astatoreochromis* than to other members of the "Haplochromis" group, because they are much less parsimonious than the tree shown in Figure 2.

Some cladistic workers (e.g., Funk, Brooks 1981) are critical of the use of molecular distance methods for phylogenetic analysis. We agree that such methods have limited resolving power, but, nevertheless, we emphasize their phylogenetic utility, which is

evident from empirical studies, especially those on higher primates. A notable example concerns the orangutan lineage which distance methods strongly indicate to lie outside the group of lineages leading to humans and African apes (Sarich, Wilson 1967; Sarich, Cronin 1976). This placement was later established beyond statistical doubt (Templeton 1983) by cladistic analysis of restriction maps (Zimmer 1981; Ferris et al. 1981) and nucleotide sequences (Brown et al.

TABLE 3. GENETIC SIMILARITIES AMONG LAKE VICTORIA CICHLIDS

(Data from Appendix 2. Upper diagonal, Nei's genetic distance (D) corrected for sample size; Lower diagonal, Rogers' (1972) Similarity (S); Negative D values set to -.000)

Population	I	2	3	4	5	6	7	8	9	10	11	12	13
1. *H. lividus*	...	0.017	-.000	-.000	0.000	-.000	0.000	-.000	-.000	0.006	0.021	0.373	1.204
2. *H. "pectoralis"*	0.950	...	0.018	0.018	0.017	0.017	0.019	-.000	0.017	0.025	0.042	0.392	1.169
3. *L. maxillaris*	0.983	0.933	...	-.000	-.000	-.000	-.000	-.000	-.000	0.002	0.010	0.356	1.187
4. *L. "microdon"*	0.975	0.925	0.992	...	-.000	-.000	-.000	-.000	-.000	0.000	0.005	0.348	1.178
5. *L. obesus*	0.996	0.946	0.982	0.974	...	-.000	0.000	-.000	0.000	*0.005	0.020	0.369	1.200
6. *L. parvidens*	0.993	0.943	0.990	0.982	0.992	...	0.000	-.000	-.000	0.004	0.017	0.366	1.197
7. *N. nigricans*	0.974	0.924	0.963	0.955	0.976	0.973	...	-.000	0.000	0.006	0.016	0.364	1.218
8. *P. chilotes*	0.950	0.950	0.967	0.975	0.949	0.957	0.930	...	-.000	0.000	0.005	0.347	1.151
9. *P. longirostris*	1.000	0.950	0.983	0.975	0.996	0.993	0.974	0.950	...	0.006	0.021	0.373	1.204
10. *P. "macrognathus"*	0.970	0.920	0.973	0.970	0.973	0.972	0.958	0.945	0.970	...	0.017	0.348	1.179
11. *H. retrodens*	0.928	0.878	0.945	0.953	0.930	0.935	0.925	0.928	0.928	0.936	...	0.316	1.147
12. *A. alluaudi*	0.672	0.659	0.679	0.682	0.674	0.675	0.667	0.685	0.672	0.685	0.699	...	1.077
13. *T. nigra*	0.300	0.313	0.307	0.310	0.302	0.303	0.295	0.320	0.300	0.313	0.326	0.334	...

* Comparison: Genetic distance calculated from electrophoretic results in Appendix 1 is 0.006.

1982). Recent support for the use of molecular distance methods in building trees relating species also comes from Li (1981), Fitch (1982), Beverley and Wilson (1982), Tateno et al. (1981), and Nei et al. (1983).

Figure 2 also presents a tentative time scale based in part on Nei genetic distances among the taxa compared. Nei (1975) and Sarich (1977) present the advantages of using this measure of distance (D) to estimate relative times of divergence. To produce an absolute time scale, we assumed that a D value of 1 corresponds to a divergence time of 10 million years. This calibration fits with Van Couvering's (1972) recognition of *Tilapia*-like fossils from the Ngorora formation (9-12 Myr BP) in East Africa (H Schwartz pers comm). We are aware that a D value of 1 could correspond to as much as a 30-million-year divergence for enzymes like those surveyed here (Sarich 1977). The protein and fossil data therefore allow the following suggestions:

1) The *Astatoreochromis* lineage diverged from that leading to the "Haplochromis" flock more than a million years ago, before the formation of Lake Victoria.[2]

2) This flock evolved very recently from a common ancestor that lived roughly 170,000 years ago, probably after the lake formed. The simplest interpretation of this date is that the common ancestor lived in Lake Victoria, but more complex explanations are not excluded.

Contrast of Morphological and Protein Divergence.

The strikingly low degree of protein divergence within the "Haplochromis" flock is not unique. Indeed, it is a notable feature of other flocks of fish species (Kornfield 1978; Kornfield et al. 1979; McKaye et al. 1982; Echelle, Echelle 1984; Humphries 1984; Kornfield, Carpenter 1984). Moreover, since the degree of morphological divergence among "Haplochromis" species can be large (Figure 1), the apparent ratio of morphological change to protein change is far larger for cichlid flocks than for centrarchids.

Evolutionary Rates and Heterozygosity

When considered in relation to the hypothesized time scale in Figure 2, point-mutational evolution seems not to have accelerated in the proteins of East African cichlids. Yet the mean rate of anatomical evolution within the "Haplochromis" flock appears to have been high in relation to the mean rate for the Centrarchidae (Figure 3). As regards speciation, we note the existence of more than 128 "Haplochromis" species in Lake Victoria. At least 7 speciation events per lineage are required to account for the production of 128 species from one common ancestor. Considering our estimate of when this ancestor lived (Figure 2) we calculate that there have been more than 10 speciation events per lineage per million years during

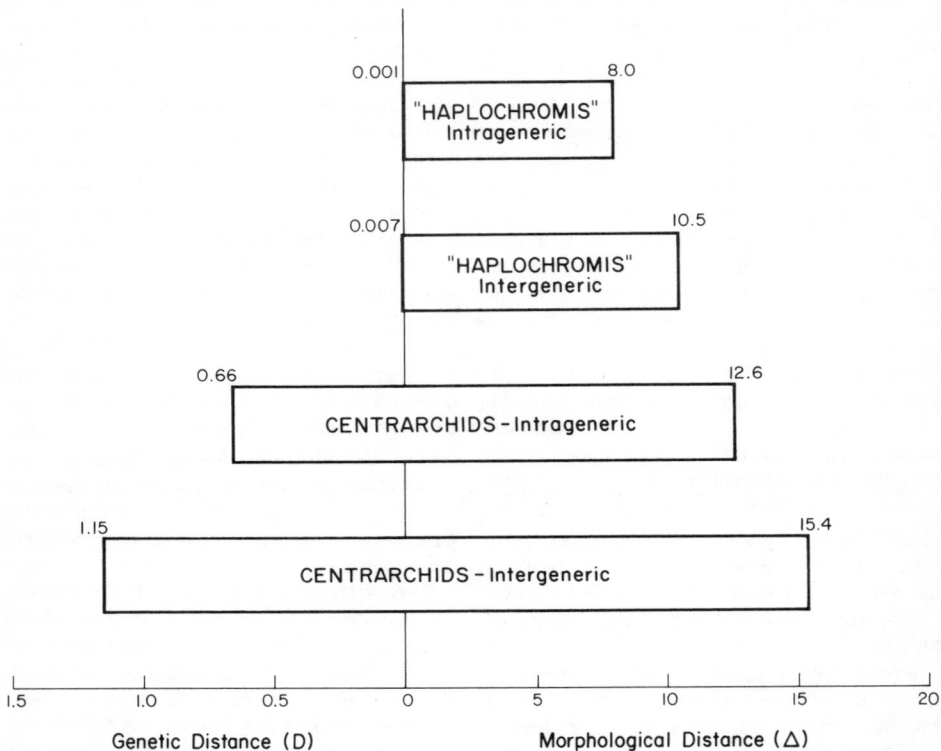

Figure 3. Contrast between morphological and protein divergence in the "Haplochromis" species flock in Lake Victoria and the Centrarchidae of North America. Bar length depicts mean intra- and intergeneric differences within each group. Mean degree of point-mutational divergence at the protein level is shown in terms of Nei's distance (D), calculated for cichlids from D values, Table 3, and for centrarchids, from published D values (Avise, Smith 1977). Mean morphological distances in Delta units calculated from Appendices 3 and 4.

the evolution of this flock, a higher value than has been estimated for mammals (Bush et al. 1977) and songbirds (Prager, Wilson 1980; Tegelstrom et al. 1983), the most rapidly speciating groups of land vertebrates.

It is possible that morphological evolution within the cichlid species flock is not concentrated in speciation events. Rather, the rapid speciation may be due mainly to sexual selection (Dominey 1984). This could be responsible for producing the many cichlid species between which the most obvious differences are in color patterns of the males (Fryer, Iles 1972; Greenwood 1981).

Other facts to remember when trying to explain the high rates of speciation and

morphological evolution in the Victoria species flock are that the levels of heterozygosity are mostly normal, and that enzyme polymorphisms are shared among species belonging to different trophic groups. Evidently, most of these species stem from populations that were never very small for extended periods (Nei et al. 1975; Motro, Thomson 1982).

Speciation and Morphological Divergence through Social Learning

We now ask whether the facts and inferences listed in the previous three sections can be explained by invoking a version of the behavioral drive hypothesis proposed by Wyles et al. (1983) to account for the rapid phenotypic evolution observed in mammals and birds. This hypothesis was extended by Larson et al. (1984) who considered also its potential for accelerating speciation and karyotypic evolution in big-brained vertebrates.[3] Cichlids are social creatures, noted for their alertness and behavioral flexibility (Lorenz 1952; Fryer, Iles 1972; Fryer 1977). Also, the effective population sizes in the Great Lakes of East Africa are probably very large for some of the species. From reported data (Fryer, Iles 1972) we estimate that the total number of cichlid individuals in the Lake Victoria flock at any one time is of the order of 10^{12} or 10^{13}. The conditions required for operation of the behavioral drive mechanism may thus exist in this flock.

Speciation by Social Selection

The rapid speciation exhibited by this flock could be due in part to the capacity for rapid, socially transmitted shifts in mate recognition or preference. Imagine two populations that are temporarily isolated as suggested by Fryer (1977) or even two populations in contact across a cline. They are expected to undergo rapid genetic divergence in traits that are the focus of intense social competition (Lande 1982; West-Eberhard 1983; Dominey 1984). Further acceleration of such divergence is expected in proportion to the capacities for behavioral innovation and social learning (Larson et al. 1984). To the extent that the cichlids in Lake Victoria have not only these capacities but also unusual geographic opportunities resulting from geologic instability of the Rift Valley and the shallowness of the lake, the potential for rapid speciation may have exceeded that in all other lakes.

Morphological Change

The morphological differences among species in the Lake Victoria flock could also result from a behavioral drive mechanism. We think it significant that the major morphological differences are related to differences in feeding behavior, some species living mainly on fish, others on snails, and others on algae, and so on (cf. Table I). The East African cichlids are known to undergo shifts in feeding behavior from one food source to another and shifts from one ecological community to another (Fryer, Iles 1972; Liem 1979, 1980; McKaye, Marsh 1983). It is possible that a novel feeding behavior, arising in a single individual, can quickly propagate by social learning to other members of the species, thereby imposing selection pressure favoring the fixation of mutations that complement the new behavior, such as mutations affecting feeding structures. Polymorphism for genetic variants that differ radically in the shape and size of mouthparts exists in some cichlid populations (Sage, Selander 1975; Kornfield, Taylor 1983; Liem, Kaufman 1984). Hence the potential for a rapid, genetically-based morphological response to an altered feeding behavior could be high in cichlids, expecially in species with a large effective population size.

Interspecific Communication

Another factor which may be a key for understanding the speed of morphological change in the species flock is that the social propagation of a new feeding behavior may extend beyond the species boundary. Several possible consequences of interspecific transmission are imaginable. 1) The probability of a morphological response to a new feeding behavior is higher if the message spreads to many species rather than being confined to the species in which it arose. 2) The spread of a new feeding behavior to several species should elicit parallel or convergent responses, producing morphologically similar species from diverse antecedent species. 3) A new habit can be tested on a variety of genetic, anatomical, and behavioral backgrounds. The probability that an advantageous modification of the habit will emerge in a species other than that of its origin may be relatively high. Hence, successful new behavioral combinations may be more likely to appear in a species flock that experiences crosstalk than in a lake with large populations of fish that do not engage in interspecific learning. Perhaps the species flock in Lake Victoria is analogous in a behavioral sense to a Wrightian population (Wright 1982).

Future Testing

Our behavioral speculations point to the need for comparative studies of relative brain size and the capacity for both intra- and interspecific learning of new feeding behaviors not only in cichlids but also in other groups of teleost fishes. There is also a need to construct, with DNA methods, a more highly-resolving tree, relating species in the Lake Victoria flock to one another and to other cichlids.

Such a molecular tree would provide a framework with which to test a prediction of our model — that the extent of morphological parallelism, reversal, and convergence has been unusually high in this flock. Properly calibrated molecular trees will also serve eventually as temporal frameworks on which to estimate and compare rates of organismal divergence among "Haplochromis" species in Lake Victoria with those of other groups of species. Such comparisons will be essential for testing hypotheses about the evolutionary mechanisms at work in this flock.

SUMMARY

We studied the proteins and body shapes of 13 species of cichlids from Lake Victoria, namely, one species each of *Tilapia* and *Astatoreochromis* which comprise the outgroup material, and 11 "Haplochromis" species. The electrophoretic comparisons show three groups: *Tilapia, Astatoreochromis*, and the "Haplochromis" species flock. The mean genetic distance within the "Haplochromis" flock is extremely small, 0.006 electrophoretically detectable substitutions per locus, despite normal amounts of intraspecific variability. The small genetic distances indicate recent times of separation, consistent with the possibility that the "Haplochromis" flock arose from a single ancestral species after Lake Victoria formed, within the past few hundred thousand years. By contrast, the *Astatoreochromis* lineage appears to have separated from that leading to the "Haplochromis" flock more than a million years ago, before the lake formed.

By using quantitative summary procedures, morphological differences among cichlids were compared with those among sunfishes (Centrarchidae). The results contrast sharply with the protein data. The apparent degree of anatomical divergence relative to protein change has been many times higher during the evolution of the Lake Victoria flock than during sunfish evolution. In seeking to explain the high rates of both speciation (by social selection) and morphological evolution in the "Haplochromis" flock, we draw attention to the potential accelerating influence of behavioral innovation and social learning on these two types of evolution, and ask whether cichlids possess these abilities.

FOOTNOTES

1. To investigate whether, in fishes, increases in morphological distance occur with increasing levels in the taxonomic hierarchy as reported for four other classes of vertebrates (Cherry et al. 1982; Wyles et al. 1983) we made 25 selected interfamilial morphological comparisons. Species 1, 9, 11-13 among cichlids were each compared to species 1, 8, 10, 12, 14 among the centrarchids. The mean morphological distance and 95% confidence interval was 17.7 $\pm$ 1.9 Delta units; the range was 8.9-27.6 Delta units. The mean value was thus larger than all the means in Figure 1, and significantly larger than all except the centrarchid intergeneric values. These results support our approach to morphological comparisons of fishes.

2. We assume with Greenwood (1981) that Lake Victoria is about 750,000 years old, but are aware that it may be much younger. According to Temple (1969), the oldest lake deposits are mid-Pleistocene about 60,000 years before present. The lake may be even younger than this. Livingston (1980) states that 14,000 years ago it decreased from its present 100 m depth to 25 m, and may even have completely dried up at this time.

3. There are three components to this mechanism: 1) behavioral innovation by one individual; 2) nongenetic propagation of new behavior by imitative learning or other kinds of social learning; 3) fixation of mutations complementary to the new behavior. The rate of evolution by this mechanism is expected to depend on the frequency with which new habits originate, the speed and accuracy of transmission, and the genetically effective size of the population (Wyles et al. 1983).

ACKNOWLEDGMENTS

We thank W Bemis, J Dawson, A Caulfield, and E M Prager for typing the manuscript, and E M Prager for checking computation, H Schwartz, A Walker, and T White for discussing fossil record in East Africa, G W Barlow, I L Kornfield, A Larson, E Mayr, E M Prager, and D B Wake for comments on the ms, M Hyder for University of Nairobi laboratory facilities, T Iwamoto for loans of specimens from the California Academy of Sciences, and M Frelow for computational assistance. Grant support: RDS, National Science Foundation

(DEB 72-02545); ACW, Guggenheim fellowship for work in Nairobi, National Science Foundation and National Institutes of Health.

REFERENCES

Avise J C, C F Aquadro 1982 A comparative study of genetic distances in the vertebrates. Patterns and correlations. *Evol Biol* 15:151-185

Avise J C, G B Kitto 1973 Phosphoglucose isomerase gene duplication in the bony fishes: an evolutionary history. *Biochem Genet* 8:113-132

Avise J C, M H Smith 1977 Gene frequency comparisons between sunfish (Centrarchidae) populations at various stages of evolutionary divergence. *Syst Zool* 26:319-335

Beverley S M, A C Wilson 1982 Molecular evolution in *Drosophila* and higher Diptera. Prt 1. Microcomplement fixation studies of a larval hemolymph protein. *J Mol Evol* 18:251-264

Brown W M, E M Prager, A Wang, A C Wilson 1982 Mitochondrial DNA sequences of primates: tempo and mode of evolution. *J Mol Evol* 18:225-239

Bush G L, S M Case, A C Wilson, J L Patton 1977 Rapid speciation and chromosomal evolution in mammals. *Proc Natl Acad Sci USA* 74:3942-3946

Cherry L M, S M Case, J G Kunkel, J S Wyles, A C Wilson 1982 Body shape metrics and organismal evolution. *Evolution* 36:914-933

Dill L M 1983 Adaptive flexibility in the foraging behavior of fishes. *Canad J Fish Aquat Sci* 40:398-408

Dominey W J 1984 Effects of sexual selection and life history phenomena on speciation in species flocks: African cichlids and Hawaiian *Drosophila*. 231-250 A A Echelle, I Kornfield eds. *Evolution of Fish Species Flocks*. Univ Maine Press. Orono

Echelle A A, A F Echelle 1984 Evolutionary genetics of a "species flock." Atherinid fishes from the Mesa Central of Mexico. 93-110 A A Echelle, I Kornfield eds. *Evolution of Fish Species Flocks*. Univ Maine Press. Orono

Farris J S 1972 Estimating phylogenetic trees from distance matrices. *Amer Nat* 106:645-668

Felsenstein J 1981 Skepticism towards Santa Rosalia, or why are there so few kinds of animals? *Evolution* 35:124-138

Ferris S D, A C Wilson, W M Brown 1981 Evolutionary tree for apes and humans based on cleavage maps of mitochondrial DNA. *Proc Natl Acad Sci USA* 78:2432-2436

Fitch W M 1981 A non-sequential method for constructing trees and hierarchical classifications. *J Mol Evol* 18:30-37

Fitch W M, E Margoliash 1967 Construction of phylogenetic trees. *Science* 155:279-284

Fryer G 1977 Evolution of species flocks of cichlid fishes in African lakes. *Z Zool Syst Evolutionsforsch* 15:141-165

Fryer G, T D Iles 1972 *The Cichlid Fishes of the Great Lakes of Africa. Their Biology and Evolution.* Oliver Boyd. Edinburgh

Funk V A, D R Brooks (eds) 1981 *Advances in Cladistics. Proceedings of the First Meeting of the Willi Hennig Society.* New York Bot Garden, Bronx

Greenwood P H 1974 The cichlid fishes of Lake Victoria, East Africa: the biology and evolution of a species flock. *Bull Br Mus Nat Hist (Zool)* Suppl 6:1-134

_____ 1979 Towards a phyletic classification of the "genus" *Haplochromis* (Pisces Cichlidae) and related taxa. Part 1. *Bull Brit Mus Nat Hist (Zool)* 35:265-322

_____ 1980 Towards a phyletic classification of the "genus" *Haplochromis* (Pisces Cichlidae) and related taxa. Part 2. The species from Lake Victoria, Nabugabo, Edward, George and Kivu. *Bull Brit Mus Nat Hist (Zool)* 39:1-101

_____ 1981 Species flocks and explosive evolution. 61-74 P L Forey ed. *The Evolving Biosphere.* Cambridge Univ Press. Cambridge

_____ 1984 African cichlids and evolutionary theories. 141-154 A A Echelle, I Kornfield eds. *Evolution of Fish Species Flocks.* Univ Maine Press. Orono

Holmes R S 1977 The genetics of α-hydroxyacid oxidase and alcohol dehydrogenase in the mouse: Evidence for multiple gene loci and linkage between *Hao-2* and *Adh-3. Genetics* 87:709-716

Humphries J M 1984 Genetics of speciation in pupfishes from Laguna Chichancanab, Mexico. 129-140 A A Echelle, I Kornfield eds. *Evolution of Fish Species Flocks.* Univ Maine Press. Orono

Karig L M, A C Wilson 1971 Genetic variation in supernatant malate dehydrogenase of birds and reptiles. *Biochem Genet* 5:211-221

King M-C, A C Wilson 1975 Evolution at two levels in humans and chimpanzees. *Science* 188:107-116

Kornfield I L 1978 Evidence for rapid speciation in African cichlid fishes. *Experientia* 34:335-336

Kornfield I, K Carpenter 1984 The cyprinids of Lake Lanao, Philippines: Taxonomic validity, evolutionary rates and speciation scenarios. 69-84 A A Echelle, I Kornfield eds. *Evolution of Fish Species Flocks.* Univ Maine Press. Orono

Kornfield I L, U Ritte, C Richler, J Wahrman 1979 Biochemical and cytological differentiation among cichlid fishes of the Sea of Galilee. *Evolution* 33:1-14

Kornfield I L, J N Taylor 1983 A new species of polymorphic fish, *Cichlasoma minckleyi*, from Cuatro Cienegas, Mexico (Teleostei: Cichlidae). *Proc Biol Soc Wash* 96:253-269

Lande R 1982 Rapid origin of sexual isolation and character divergence in a cline. *Evolution* 36:213-223

Larson A, E M Prager, A C Wilson 1984 Chromosomal evolution, speciation and morphological change in vertebrates: the role of social behavior. *Chromosomes Today* 8: In press

Li W-H 1981 Simple method for constructing phylogenetic trees from distance matrices. *Proc Natl Acad Sci USA* 78:1085-1089

Liem K F 1973 Evolutionary strategies and morphological innovations: cichlid pharyngeal jaws. *Syst Zool* 22:425-441

Liem K F 1979 Modulatory multiplicity in the feeding mechanism in cichlid fishes as exemplified by the invertebrate pickers, of Lake Tanganyika. *J Zool Lond* 189:93-125

———— 1980 Adaptive significance of intra- and and interspecific differences in the feeding repertoires of cichlid fishes. *Amer Zool* 20:295-314

Liem K F, L S Kaufman 1984 Intraspecific macroevolution: functional biology of the polymorphic cichlid species *Cichlasoma minckleyi*. 203-216 A A Echelle, I Kornfield eds. *Evolution of Fish Species Flocks* Univ Maine Press. Orono

Livingston D A 1980 Environmental changes in the Nile headwaters. 339-359 M A J Williams, H Faure eds. *The Sahara and the Nile*. Balkema. Rotterdam

Lorenz K 1952 *King Solomon's Ring*. Methuen. London

Mainardi D 1980 Tradition and social transmission of behavior in animals. 227-255 G Barlow, J Silverberg eds. *Sociobiology: Beyond Nature / Nurture?* Amer Assn Adv Sci Washington D C

Markert C L, J B Shaklee, G S Whitt 1975 Evolution of a gene. *Science* 189:102-114

Marsh A C, A J Ribbink, B A Marsh 1981 Sibling species complexes in sympatric populations of *Petrotilapia* Trewavas (Cichlidae, Lake Malawi). *Zool J Linn Soc* 71:253-264

McKaye K R, W N Gray 1984 Extrinsic barriers to gene flow in rock-dwelling cichlids of Lake Malawi: macrohabitat heterogeneity and reef colonization. 169-184 A A Echelle, I Kornfield eds. *Evolution of Fish Species Flocks*. Univ Maine Press. Orono

McKaye K R, T Kocher, P Reinthal, R Harrison, I Kornfield 1984 Genetic evidence for allopatric and symatric differentiation among color morphs of a Lake Malawi cichlid fish. *Evolution* 38:215-219

McKaye K R, T Kocher, P Reinthal, I Kornfield 1982 A sympatric sibling species complex of *Petrotilapia* Trewavas from Lake Malawi analyzed by enzyme electrophoresis (Pisces: Cichlidae). *Zool J Linn Soc* 76:91-96

McKaye K R, A Marsh 1983 Food switchings by two specialized algae-scraping cichlid fishes in Lake Malawi, Africa. *Oecologia (Berlin)* 56:245-248

Motro U, G Thomson 1982 On heterozygosity and the effective size of populations subject to size changes. *Evolution* 36:1059-1066

Nei M 1975 *Molecular Population Genetics and Evolution*. North-Holland. Amsterdam

———— 1978 Estimation of average heterozygosity and genetic distance from a small number of individuals. *Genetics* 89:583-590

Nei M, T Maruyama, R Chakraborty 1975 The bottleneck effect and genetic variability in populations. *Evolution* 29:1-10

Nei M, F Tajima, Y Tateno 1983 Accuracy of estimated phylogenetic trees from molecular data. Prt 2. Gene frequency data. *J Mol Evol* 19:153-170

Oijen M J P van 1982 Ecological differentiation among the haplochromine piscivorous species of Lake Victoria (East Africa). *Neth J Zool* 32:336-363

Oijen M J P van, F Witte, E L M Witte-Maas 1981 An introduction to ecological and taxonomic investigations on the haplochromine cichlids from the Mwanza Gulf of Lake Victoria. *Neth J Zool* 31:149-174

Prager E M, A C Wilson 1980 Phylogenetic relationships and rates of evolution in birds. *Proc Int Ornithol Cong* 17:1209-1214

Rogers J S 1972 Measures of genetic similarity and genetic distance. *Studies in Genetics 7*. Univ Texas Pub 7213:145-153

Sage R D 1978 Genetic heterogeneity of Spanish house mice (*Mus musculus* complex). 519-553 H C Morse III ed. *Origins of Inbred Mice*. Academic Press. New York

Sage R D, R K Selander 1975 Trophic radiation through polymorphism in cichlid fishes. *Proc Natl Acad Sci USA* 72:4669-4673

Sarich V M 1977 Rates, sample sizes, and the neutrality hypothesis for electrophoresis in evolutionary studies. *Nature* 265:24-28

Sarich V M, J E Cronin 1976 Molecular systematics of the primates. 141-170 M Goodman, R E Tashian eds. *Molecular Anthropology. Genes and Proteins in the Evolutionary Ascent of the Primates*. Plenum Press. New York

Sarich V M, A C Wilson 1967 Immunological time scale for hominid evolution. *Science* 158:1200-1203

Tateno Y, M Nei, F Tajima 1982 Accuracy of estimated phylogenetic trees from molecular data. Prt 1. Distantly related species. *J Mol Evol* 18:387-404

Tegelstrom H, T Ebenhard, H Ryttman 1983 Rate of karyotypic evolution and speciation in birds. *Hereditas* 98:235-239

Temple P H 1969 Some biological implications of a revised geological history for Lake Victoria. *Biol J Linn Soc* 1:363-371

Templeton A R 1983 Convergent evolution and nonparametric inferences from restriction data and DNA sequences. 151-179 B S Weir ed. *Statistical Analysis of DNA Sequence Data*. Dekker. New York

Van Couvering J A 1972 *Geology of Rusinga Island and correlation of the Kenya Mid-Tertiary fauna*. Doctoral thesis. Cambridge Univ.

Wake D B 1981 The application of allozyme evidence to problems in the evolution of morphology. 257-270 G G E Scudder, J L Reveal eds. *Evolution Today. Proceedings of the Second International Congress of Systematic and Evolutionary Biology*. Carnegie-Mellon Univ. Pittsburgh

West-Eberhard M J 1983 Sexual selection, social competition and speciation. *Quart Rev Biol* 58:155-183

Wilson A C 1975 Evolutionary importance of gene regulation. *Stadler Genetics Symp.* 7:117-134

Witte F 1984 Ecological differentiation in Lake Victoria Haplochromines: Comparison of cichlid species flocks in African Lakes. 155-168 A A Echelle, I Kornfield eds. *Evolution of Fish Species Flocks*. Univ Maine Press. Orono

Wright S 1982 Character change, speciation and the higher taxa. *Evolution* 36:427-443

Wyles J S, J G Kunkel, A C Wilson 1983 Birds, behavior and anatomical evolution. *Proc Natl Acad Sci USA* 80:4394-4397

Zimmer E A 1981 *Evolution of primate globin genes.* Doctoral thesis. Univ California. Berkeley

APPENDIX 1. ALLELE FREQUENCIES AT 13 PROTEIN LOCI IN 11 ANIMALS EACH OF *Lipochromis obesus* AND *Prognathochromis "macrognathus"* LAKE VICTORIA NEAR JINJA

Locus	Allele	Allele Frequency	
		Lipochromis obesus	*Prognathochromis "macrognathus"*
1. Ldh-C	a	1.00	1.00
2. Ldh-B	a	--	0.14
	b	1.00	0.86
3. Ldh-A	a	1.00	1.00
4. Mdh-1	a	1.00	1.00
5. Mdh-2	a	1.00	1.00
6. Iddh	a	0.27	--
	b	0.73	1.00
7. Ndh	a	1.00	1.00
8. Sod-S	a	0.23	0.14
	b	0.77	0.86
9. Sod-M	a	1.00	1.00
10. Aat-S	a	1.00	1.00
11. Aat-M	a	1.00	1.00
12. Hb-1	a	1.00	1.00
13. Hb-2	a	1.00	1.00

APPENDIX 2. ALLELE FREQUENCIES AT EIGHT ENZYME LOCI IN LAKE VICTORIA CICHLID FISHES

(Alleles alphabetically ordered by decreasing distance from gel origin)

Species	Adh			Gpd		Ldh-A		Icd-2		Pgd		Pgm			Gpi-1						Gpi-2						Sample n	
	a	b	c	a	b	a	b	a	b	a	b	a	b	c	a	b	c	d	e	f	a	b	c	d	e	f	Ldh-A	Gpi
1. *H. lividus*		1.00		1.00		1.00		1.00		1.00		1.00						1.00								1.00	1	
2. *H. "pectoralis"*		1.00		1.00		1.00		1.00		1.00		0.50	0.50					1.00								1.00	2	
3. *L. maxillaris*		1.00		1.00		1.00		1.00		1.00		1.00						1.00						0.17	0.83		3	
4. *L. "microdon"*		1.00		1.00		1.00		1.00		1.00		1.00						1.00						0.25	0.75		2	
5. *L. obesus*		1.00		1.00		1.00		1.00		1.00		1.00						0.98	0.02						0.98	0.02	39	
6. *L. parvidens*		1.00		1.00		1.00		1.00		1.00		1.00						1.00					0.07		0.93		7	
7. *N. nigricans*		1.00		1.00		1.00		0.12	0.88	1.00		1.00					0.03	0.88	0.09				0.03		0.97		17	
8. *P. chilotes*		1.00		1.00		1.00		1.00		1.00		0.75	0.25					1.00						0.25	0.75		2	
9. *P. longirostris*		1.00		1.00		1.00		1.00		1.00		1.00						1.00							1.00		11	
10. *P. "macrognathus"*		1.00		1.00		1.00		1.00		1.00		1.00						0.94	0.03	0.03					0.75	0.25	39	
11. *H. retrodens*		1.00		1.00		1.00		1.00		1.00		1.00					0.08	0.58		0.33			0.33		0.67		6	
12. *A. alluaudi*	1.00			1.00		1.00		1.00		1.00		0.31	0.69		1.00						0.94	0.06					16	
13. *T. nigra*			1.00	1.00			1.00	1.00		1.00		1.00						1.00						1.00			13	

APPENDIX 3. MEAN TRAIT LENGTHS OF LAKE VICTORIA CICHLID FISHES

Species	Body length Mean, range (mm)	Fish n	TL	HL	BH	BW	EW	PF	ES	PL	Sum: trait lengths (mm)
1 *Haplochromis lividus*	75, 68-78	5	3017	1600	1600	702	510	1430	678	454	162
2 *Harpagochromis "pectoralis"*	93, 88-98	2	3102	1705	1367	774	477	1218	802	567	194
3 *Lipochromis maxillaris*	121,117-127	3	3094	1548	1547	767	417	1280	716	639	261
4 *Lipochromis "microdon"*	98, 95-101	2	3110	1614	1422	772	396	1324	770	600	208
5 *Lipochromis obesus*	135,126-159	11	3028	1513	1627	847	359	1283	708	636	297
6 *Lipochromis parvidens*	137,125-147	5	2905	1765	1529	709	360	1145	864	714	294
7 *Neochromis nigricans*	86, 81-91	5	3235	1557	1567	739	442	1289	694	459	179
8 *Paralabidochromis chilotes*	96, 94-97	2	2943	1820	1446	698	446	1172	898	574	200
9 *Prognathochromis longirostris*	113,102-120	5	3454	1743	1189	591	429	1216	802	571	217
10 *Prognathochromis "macrognathus"*	105, 99-115	5	3014	1744	1428	687	453	1248	823	605	221
11 *Hoplotilapia retrodens*	120,108-130	5	2955	1566	1600	792	403	1421	680	580	265
12 *Astatoreochromis alluaudi*	117,112-120	5	3081	1610	1552	877	391	1263	659	563	249
13 *Tilapia nigra*	83, 76-94	5	2841	1604	1525	823	464	1571	693	482	187

APPENDIX 4. MEAN TRAIT LENGTHS OF CENTRARCHID FISHES

Species	Body length mean (mm)	range (mm)	Fish n	TL	HL	BH	BW	EW	PF	ES	PL	Sum: trait lengths, (mm)
1 *Lepomis auritus*	75	73-76	3	2588	1903	1822	861	411	1262	641	501	166
2 *Lepomis cyanellus*	121	98-132	5	2886	1754	1847	879	346	945	684	665	261
3 *Lepomis gibbosus*	113	95-128	3	2687	1560	2256	804	374	1312	589	413	266
4 *Lepomis microlophus*	195	183-203	3	2819	1548	2294	837	290	1247	552	411	446
5 *Lepomis megalotis*	74	67-90	5	2632	1793	2013	758	478	1265	620	451	168
6 *Lepomis gulosus*	116	104-126	4	2677	1795	1887	905	389	969	644	727	260
7 *Lepomis macrochirus*	118	102-131	5	2822	1527	2133	776	350	1408	570	414	270
8 *Micropterus dolomieui*	194	158-222	5	3401	1755	1475	805	333	848	665	722	376
9 *Micropterus salmoides*	148	139-156	5	3301	1774	1394	886	342	927	625	749	291
10 *Pomoxis annularis*	192	172-208	5	2982	1815	1790	656	359	1035	605	755	400
11 *Pomoxis nigromaculatus*	153	133-170	5	2995	1664	1878	659	450	1040	623	697	327
12 *Enneacanthus obesus*	60	53-69	5	2690	1633	2044	760	508	1246	552	540	139
13 *Ambloplites rupestris*	135	127-157	5	2962	1700	1822	804	455	892	649	713	290
14 *Acantharchus pomotis*	91	78-102	5	2583	1817	1841	1035	438	976	554	767	206
15 *Archoplites interruptus*	168	160-180	5	3051	1716	1869	773	394	937	596	665	352
16 *Centrarchus macropterus*	88	86-89	2	2827	1550	2148	685	492	1252	524	549	200

INTRASPECIFIC MACROEVOLUTION: FUNCTIONAL BIOLOGY OF THE POLYMORPHIC CICHLID SPECIES *CICHLASOMA MINCKLEYI*

KAREL F LIEM AND LESLIE S KAUFMAN

INTRODUCTION

Recently Kornfield et al. (1982, 1983) gave unequivocal evidence that *Cichlasoma minckleyi* is a distinctly polymorphic species, confirming an earlier hypothesis by Sage and Selander (1975). The differences between these two morphs led earlier workers to recognize them as distinct species (Taylor, Minckley 1966; Minckley 1969; La Bounty 1974; Kornfield, Koehn 1975). This striking intraspecific morphological and trophic differentiation, exceeding that found among many closely related and sympatric cichlid species has profound implications for evolutionary biology. 1) It might be associated with macroevolution and the formation of species flocks. 2) It may give us the best evidence yet as how very simple epigenetic mechanisms can produce rapid and quantum structural changes. 3) It gives us a unique opportunity to study incipient functional differentiation within a species. 4) Ecologically, such a polymorphic species behaves differently from a conventional species and may demonstrate fundamental mechanisms of the formation of species flocks by means of phenotypic tracking of the trophic resource abundance. We will review these four evolutionary implications in a detailed laboratory analysis of the deep-bodied forms of the papilliform and molariform morphs of *Cichlasoma minckleyi*.

MORPHOLOGY

Morphological differences between the two morphs can be summarized as follows: 1) The lower pharyngeal jaw in papilliform morphs is more delicate, possessing a significantly narrower outline with smaller horns (Figure 1: LPJ, PM, MM). 2) The gut in the molariform morph is significantly shorter (Smith 1982).

3) The neurocranial articulation of the upper pharyngeal jaw in the molariform morph is greatly expanded posteriorly and laterally, and is more massive (Kornfield, Taylor 1983; Figures 1, 2:AP). 4) Correlated with the hypertrophied pharyngeal jaw apparatus, head width is significantly greater in molariform morphs (Kornfield, Taylor 1983).

To these, we add the following important differences between the two morphs. The upper pharyngeal jaws in the molariform morph are more massive and significantly expanded posteriorly and laterally, with a distinctly convex posterior contour (Figure 1:UPJ). The fourth epibranchial bone in the molariform morph is not only enlarged, but has a more concave posterior outline, providing a larger tunnel for the passage of the hypertrophied fourth levator externus muscle (Figure 1:EB$_4$). The branchial musculature (Figures 2,3) exhibits the most dramatic dichotomy between the morphs. In both morphs the levator externus 4 (LE$_4$), levator posterior (LP), retractor dorsalis (RD), obliquus posterior (OP), obliquus dorsalis (OD), and transversus dorsalis anterior (TDA) are hypertrophied when compared with those of more generalized cichlids. However, in the molariform morph, the levator externus 4 (LE$_4$), levator posterior (LP), and the retractor dorsalis (RD) undergo an immense increase in cross-sectional area (Figure 2), and undergo a complex differentiation in muscle architecture with the appearance of numerous tendinous elements and pinnations (Ono, Kaufman 1983; Figure 3). The hypertrophy and increased areas of origin and insertion of the fourth levator externus, levator posterior, and the retractor dorsalis have pronounced influences on the form and architecture of the neurocranium and

papilliform   molariform

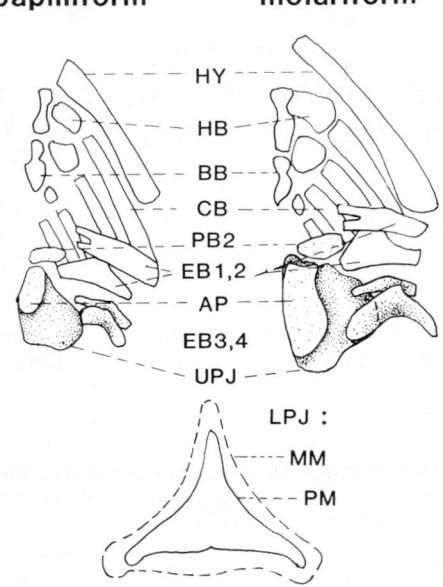

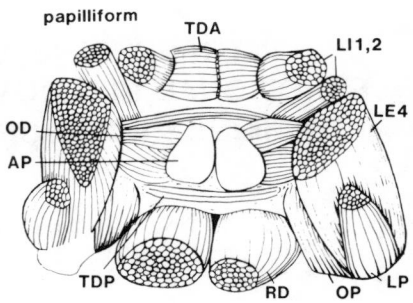

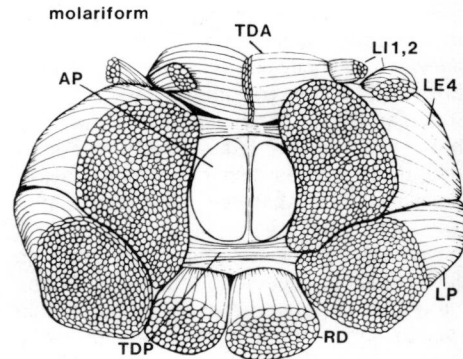

Figure 1. Dorsal view, right half of branchial arches of *Cichlasoma minckleyi*. First pharyngobranchial removed. 6.9 cm SL, each specimen. Lower pharyngeal jaw (LPJ) outlines on bottom are ventral views of molariform morph (MM) and papilliform morph (PM). Code: AP, pharyngeal apophysis of upper pharyngeal jaw; BB, basibranchial; CB, ceratobranchial; EB1-4, epibranchial 1-4; HB, hypobranchial; HY, hyoid; LPJ, lower pharyngeal jaw; MM, molariform morph, PB2, second pharyngobranchial; PM, papilliform morph.

Figure 2. Dorsal view of branchial musculature of *Cichlasoma minckleyi*. Levator posterior, levator externus 4, retractor dorsalis and levatores interni 1, 2 are sectioned. Papilliform morph, 7.2 cm SL; molariform morph, 7.0 cm SL. Code: AP, pharyngeal apophysis, LE4, fourth levator externus; LI 1, 2, levator internus 1, 2; LP, levator posterior; OD, obliquus dorsalis; OP, obliquus posterior; RD, retractor dorsalis, TDA, transversus dorsalis anterior; TDP, transversus dorsalis posterior.

lower pharyngeal jaws (Figures 1,4). The enlarged sites of origin of the fourth levator externus and levator posterior (Figure 4:LE$_4$, LP) in the molariform morph are correlated with an increase in neurocranial width behind the orbit and the expansion of the posterolateral corner of the neurocranium. Similarly, the hypertrophied retractor dorsalis (Figure 3:RD) is correlated with an enlarged fourth vertebral apophysis and changes in the form and architecture of the first four vertebrae. The insertion site of the hypertrophied fourth levator externus, levator posterior, and retractor dorsalis muscles have also undergone concomitant hypertrophy as is evident from the massive horns of the lower pharyngeal jaw in the molariform morph (Figure 1:LPJ, MM, PM).

Other dorsal branchial muscles exhibiting

distinct morphological dichotomies are the transversus dorsalis anterior, obliquus dorsalis, and obliquus posterior, all of which are hypertrophied in the molariform morph (Figure 2). However, the levatores interni 1 and 2 (Figure 2:LI$_{1,2}$) are reduced drastically in the molariform morph.

The ventral branchial musculature (pharyngocleithralis externus and internus, pharyngohyoideus) is identical in the two morphs (Figure 3:PCE, PCI, PH). The pharyngeal and branchial musculature of *Cichlasoma cyanoguttatum*, from its extreme southern and northern localities, Parque, Chapultepec, Mexico, to Chacon Creek,

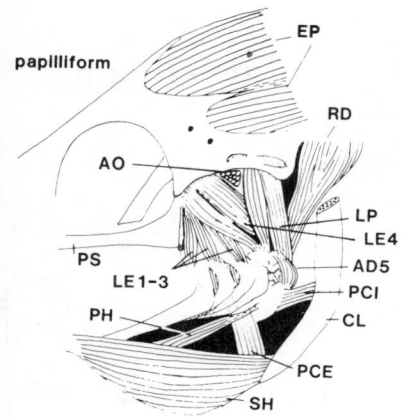

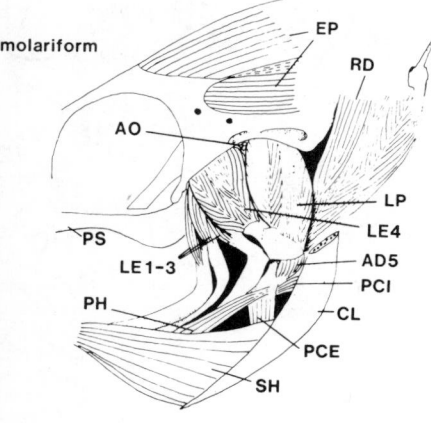

Figure 3. Lateral view of branchial musculature of *Cichlasoma minckleyi*. Code: AD5, fifth adductor branchialis; AO, adductor operculi; CL, cleithrum; EP, epaxial musculature; LE1-3, levatores externi 1-3; LE4, fourth levator externus; LP, levator posterior; PCE, pharyngocleithralis externus; PCI, pharyngocleithralis internus; PH, pharyngohyoideus; PS, parasphenoid; RD, retractor dorsalis; SH, sternohyoideus.

Texas, is virtually identical with that of the papilliform morph of *C. minckleyi*. No major intraspecific differences have been found in other myological and osteological systems. Partial molarization of the pharyngeal jaw apparatus occurs sporadically, but is not accompanied by the specializations so characteristic of the molariform morph of *C. minckleyi*.

FUNCTIONAL ANALYSIS

Prey Capture. The jaw muscles of the two *C. minckleyi* morphs and *C. cyanoguttatum* have been analyzed by high speed motion pictures and electromyography of 5 key muscles: the adductor mandibulae Part A_1, and Part A_2, sternohyoideus, levator operculi, and geniohyoideus). To determine the complete feeding repertoire, the experimental fishes were presented with a wide variety of prey in different positions in the water column (Liem 1980a,b). No major differences were found among the two morphs and *C. cyanoguttatum*. All three exhibited identical kinematic and electromyographic patterns during prey capture (Figure 5), with four feeding patterns:

1) Horizontal Suction. This most frequently employed pattern shows relatively little overlap of activity of the abductor and adductor muscles (Figure 5: sternohyoideus, SH; and levator operculi, LO; adductor mandibulae, AM_1, AM_2; geniohyoideus, GH). The upper jaw is protruded as the gape opens, and the mouth is closed with the upper jaw in a protruded condition. This pattern permits prey capture from midwater and from the surface.

2) Downwards Directed Suction: Employed during bottom feeding with extended variable and overlapping activity of the sternohyoideus and geniohyoideus muscles (Figure 5: SH, GH). Kinematically, this mode is distinguished by early upper jaw protrusion during the opening cycle. Tubifex worms, chironomid larvae, and benthic invertebrates including snails are captured by this method.

3) High Velocity Suction. This pattern is invariably employed during the pursuit and capture of agile fish prey. It is characterized by greatly overlapping actions of the abductors and adductors (Figure 5). Gape of the mouth and expansion of the buccal cavity are maximized. The feeding cycle of opening and closing the jaws is shortened to less than 50 msec.

4) Manipulating. This variable pattern can best be characterized by continuous adjustments in the neuromuscular actions and kinematic profiles. Multiple overlapping bursts of adductors and abductors which show activity of varying lengths and amplitude produce varying degrees of upper jaw protrusion during both mouth openings and closings (Figure 5). Manipulating is a complex, non-cyclical, and

papilliform   molariform

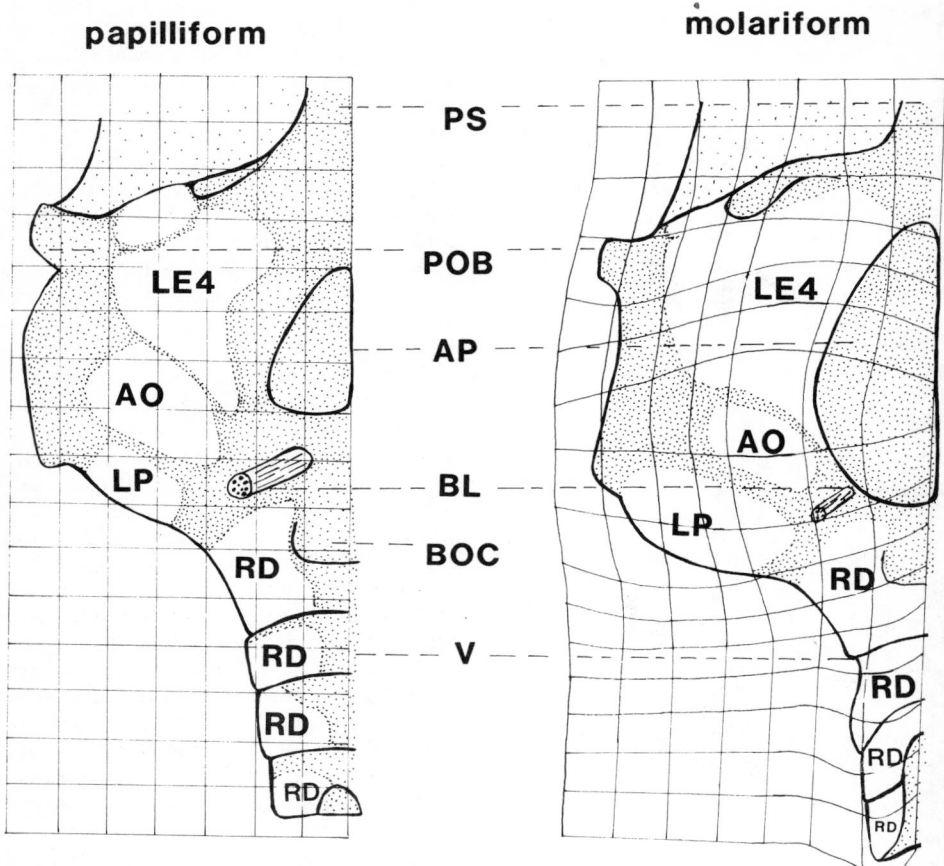

Figure 4. Ventral view, right half of otic region of neurocranium and anterior vertebrae with origins of major muscles (unshaded areas) of *Cichlasoma minckleyi* **morphs as projected on the D'Arcy Thompson transformation grid.** The papilliform morph is projected as the baseline. Distortion of the grid for the molariform morph expresses the transformation. Standard length, both specimens = 7.4 cm. Code: AO, adductor operculi; AP, pharyngeal apophysis of the parasphenoid; BL, Baudelot's ligament; BOC, basioccipital; FHM, fossae for hyomandibula; LE4, fourth levator externus; LP, levator posterior; POB, postorbital process; PS, parasphenoid; RD, retractor dorsalis; TC, trigeminofacial chamber; V, vertebra.

often extended pattern. Protrusive, retractive, and gripping movements are produced to dislodge sessile prey such as snails.

The two *C. minckleyi* morphs and *C. cyanoguttatum* exhibit an identical repertoire of the four feeding patterns. The intraspecific occurrence of multiple feeding patterns in *Cichlasoma* resembles that of the other "generalized" cichlids (Liem 1980 a,b), and it indicates that the prey capture apparatus in cichlids can meet multiple problems. Thus

profound functional shifts can be realized even if structure of the jaws and associated musculature remain relatively unaltered.

Prey Mastication and Crushing. High speed cineradiography during mastication and crushing of prey reveals a stereotypic kinematic pattern common for the two morphs and *C. cyanoguttatum* (Figure 6). The kinematic pattern of mastication starts with a forward, protrusive movement of the upper pharyngeal jaws, and a posterior, retractive

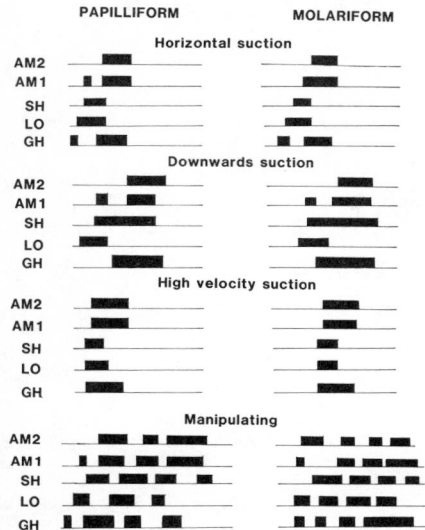

Figure 5. Block diagrams of active periods of key jaw muscles of *C. minckleyi* during capture of various prey in various positons in the water column. Code: AM1, AM2, parts A1 and A2 of adductor mandibulae muscle complex; GH, geniohyoideus; LO, levator operculi; SH, sternohyoideus.

movement of the lower pharyngeal jaw (Figure 6:30). This is followed by a retraction of the upper pharyngeals and protraction of the lower pharyngeal jaw. Both jaws are abducted at this stage (Figure 6:35). Crushing is accomplished by elevation of the lower pharyngeal jaw against the retracting upper pharyngeal jaws (Figure 6:40). After crushing, the prey is transported into the esophagus by retraction of the upper pharyngeal jaw and lowering of the lower pharyngeal jaw. This stereotypic pattern is employed regardless of whether the prey is soft, such as brine shrimp, tubifex worms, and dried fish food flakes, or tough, such as live fish, or hard, such as snails. It appears that the pharyngeal jaw movements are governed by a pattern generator in all three forms studied. Further evidence for the concept of a pharyngeal jaw pattern generator can be found in the electromyographic profiles of some key muscles associated with the pharyngeal jaw apparatus of the two morphs and *C. cyanoguttatum* (Figure 7). The electromyographic patterns are

virtually identical, and are independent of the nature of the prey. In all three forms the levator posterior and levator externus muscles are dominant in activity patterns (Figure 7:LP, LE_4). Crushing occurs by synchronous activity of the levator posterior, levator externus 4 (elevation of the lower pharyngeal jaw), and retractor dorsalis (retraction of the upper pharyngeal jaw) muscles in an identical pattern in all three forms, regardless of whether the prey is soft, tough, or hard. The electromyographic pattern resembles that of other cichlids, except that the levator posterior plays a more prominent role (Liem 1974, 1978).

LABORATORY FEEDING BEHAVIOR

Patterns of food utilization in relation to food abundance, food type, and predator morphology were analyzed by offering the predators, *C. cyanoguttatum* and the two morphs of *C. minckleyi*, varied combinations of prey items. Abundant food supply and satiation of the fish were indicated when the different food types were no longer taken by the fish after 10 minutes. Low food abundance was indicated when all food types were ingested before satiation within 3 minutes. Intermediate abundance was indicated when fish were satiated within 10 minutes with all food depleted. Fish were conditioned to the particular abundance level of food seven days prior to the recording of feeding cycles. The number of feeding cycles or "bites" on each type of food were counted during a 10-minute observation period; the cycle was defined as the opening and closing of the oral jaws. Stable estimates of each individual's preference in a given experimental situation were obtained by averaging frequency data from 20 trials. Chi square tests for heterogeneity were applied to test the null hypothesis that patterns of food utilization were random with respect to food type.

Three Individuals of Both Morphs: Two Food Types. Paired food types were presented sequentially to three each of the papilliform and molariform morphs of *C. minckleyi*, 6.9 - 7.5 cm SL, in a single 76 l aquarium. When food is abundant, no significant preferences occurred between the morphs for any of the paired food combinations. Behavioral variations sometimes occurred within morphs. Significant within-morph

CINERADIOGRAPHY OF PJA OF MOLARIFORM

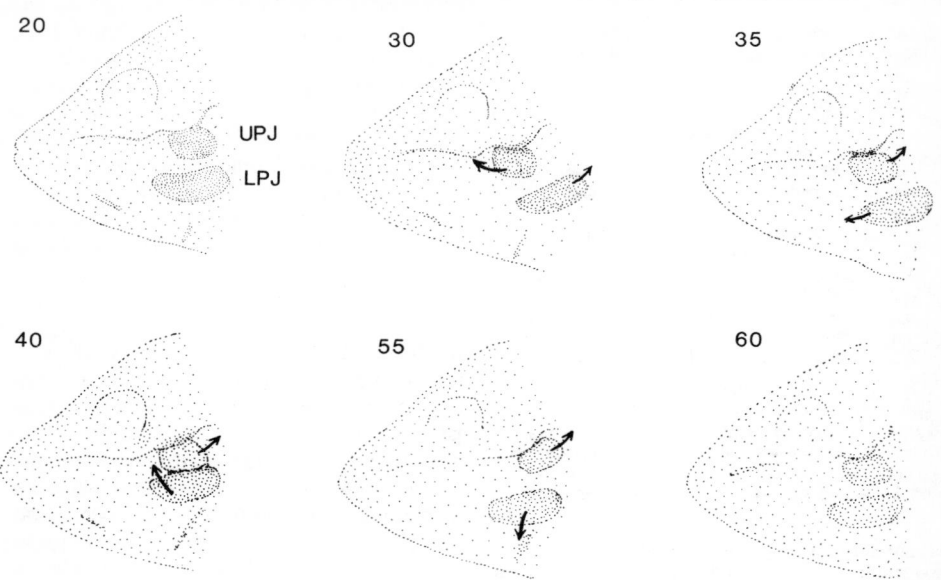

Figure 6. Tracings of images of individual frames of cineradiographic film taken at 150 frames/sec (300 MA, 50 KV) of *Cichlasoma minckleyi* molariform morph during mastication of a snail. Frame numbers indicated. Arrows indicate principal direction of movements on basis of positions of elements prior and just after the frame shown here. Code: LPJ, lower pharyngeal jaw; PJA, pharyngeal jaw apparatus; UPJ, upper pharyngeal jaw.

deviation from random feeding by Chi square test was: snails vs. shrimp, 25.16, p ◄ .005; shrimp vs. fish, 11.76, p ◄ .05; and fish vs. flakes, 14.78, p ◄ .025.

At low food abundance, a different pattern emerged. When offered a choice between fish or snails in low abundance, the papilliform morph fed proportionately on very few snails, while the molariform morph fed heavily on snails as well as fish (Chi square = 11.28; p ◄ .05). When the choice was between snails and shrimp in low abundances, the papilliform morph fed preferentially on shrimp, while the molariform morph switched to snails (Chi square = 23.56; p ◄ .005). Thus, the effect of lowered food availability is to polarize the feeding behavior of the fishes in accordance with their pharyngeal jaw phenotypes.

Three Individuals, One Morph: Five Food Types. To examine food preference when only one morph is present, three specimens of a single phenotype were offered a range of five food types in abundant supply (Table 1.1). The results indicate a high degree of behavioral uniformity. Interestingly, snails were the least desirable food types for both the papilliform and the molariform morph.

Three Individuals, Both Morphs: Five Food Types. When three individuals of both morphs in a single 76 l aquarium were offered five food types in high abundance, the feeding behavior did not differ significantly from random expectations either within or between morphs (Table 1.2). However, with low abundance of food the between-morph heterogeneity became highly significant. Papilliform morphs fed more than expected by chance on tubifex worms and dried fish food flakes, while the molariform morphs concentrated their feeding efforts on snails (Table 1.3).

The effects of varied food concentrations on the feeding behavior of the papilliform and molariform morphs under laboratory conditions are shown in Figure 8. The sharp

Table 1. Food preference of papilliform and molariform *Cichlasoma minckleyi* when fed separately, and together. (Six specimens)

Table 1.1 Fed separately; Food abundant

Morph		Shrimp	Flake	Fish	Snail	Tubifex	
Papilli:	1	26	34	39	11	40	
	2	40	36	41	15	36	
	3	28	33	33	11	34	
	Total	94	103	113	37	110	475
	Ratio	.20	.23	.25	.08	.24	1.00
		Chi square =		3.15; p = .92			
Molari:	1	31	37	32	12	33	
	2	32	32	36	12	33	
	3	23	37	31	17	31	
	Total	95	106	99	41	97	438
	Ratio	.22	.24	.23	.09	.22	1.00
		Chi square =		2.17; p = .98			

Table 1.2 Fed together; Food abundant

Morph		Shrimp	Flake	Fish	Snail	Tubifex	
Papilli:	1	36	29	41	13	24	
	2	22	29	29	14	26	
	3	21	34	29	20	21	
	Total	79	92	99	47	71	388
	Ratio	.20	.24	.26	.12	.18	1.00
		Chi square =		8.65; p = .37			
Molari:	1	27	19	33	17	28	
	2	11	26	26	17	27	
	3	17	21	31	11	21	
	Total	55	66	90	45	76	332
	Ratio	.16	.20	.27	.14	.23	1.00
		Chi square =		9.19; p = .33			
Between morph: Chi square = 22.23 p = .33							

Table 1.3 Fed together; Food low

Morph		Shrimp	Flake	Fish	Snail	Tubifex	
Papilli:	1	9	7	11	8	11	
	2	7	15	11	1	12	
	3	8	15	10	8	10	
	Total	24	37	32	17	33	143
	Ratio	.17	.26	.22	.12	.23	1.00
		Chi square =		9.56; p = .30			
Molari:	1	7	7	11	17	2	
	2	3	3	11	16	2	
	3	4	1	6	13	3	
	Total	14	11	28	46	7	106
	Ratio	.13	.10	.26	.44	.07	1.00
		Chi square =		5.55; p = .70			
Between morph: Chi square = 54.83 p ◄ .001							

dichotomy in feeding behavior appeared only at a low food concentration. The degree to which papilliform or molariform morphs prey on snails is clearly dependent on the concentration of various food resources.

TROPHIC ECOLOGY IN THE FIELD

Smith (1982) made extensive field observations of the behavioral ecology of the papilliform and molariform morphs in Cuatro Cienegas, Coahuila, Mexico. His major conclusions are: 1) There is a lack of dietary specialization in the season when food is presumably most abundant. 2) The ecological differences are not as great as expected on the basis of distinct trophic morphologies. 3) Snails may constitute a competitive refugium for molariform morphs as food becomes scarce. These conclusions are congruent with our data on feeding behavior in the laboratory.

DISCUSSION

Our data and the unequivocal evidence that *Cichlasoma minckleyi* is a distinctly polymorphic species shed light on the pattern and processes underlying morphological, functional, and ecological diversification that could be involved in the process of speciation. Because the different morphs belong to the same species, they represent unquestionably a monophyletic clade, providing us the best possible scale for a well-founded evaluation of the processes by which such diversifications evolve. Of the two, the molariform morph is hypothesized to be the more specialized. Morphologically, the papilliform morph resembles most forms of *Cichlasoma cyanoguttatum* very closely. In our analysis we have considered *C. cyanoguttatum* the primitive sister-group of *C. minckleyi*, even though Kornfield and Taylor (1983) proposed that *C. labridens, C. bartoni,* and *C. steindachneri* may be the species most closely aligned with *C. minckleyi*. However, because *C. cyanoguttatum* exhibits a sufficiently generalized, primitive morphology, we think that using it as a primitive outgroup is justifiable. Polymorphism in *C. minckleyi* has been considered a differentiating process with the molariform morph as a specialized daughter form, possibly under the influence of a recessive allele for the molariform character (Kornfield et al. 1982). The alternative hypothesis is that the morphs represent disappearance of two or more species by introgression.

Because no other *Cichlasoma* species studied has the complete set of fully developed specializations of the molariform morph,

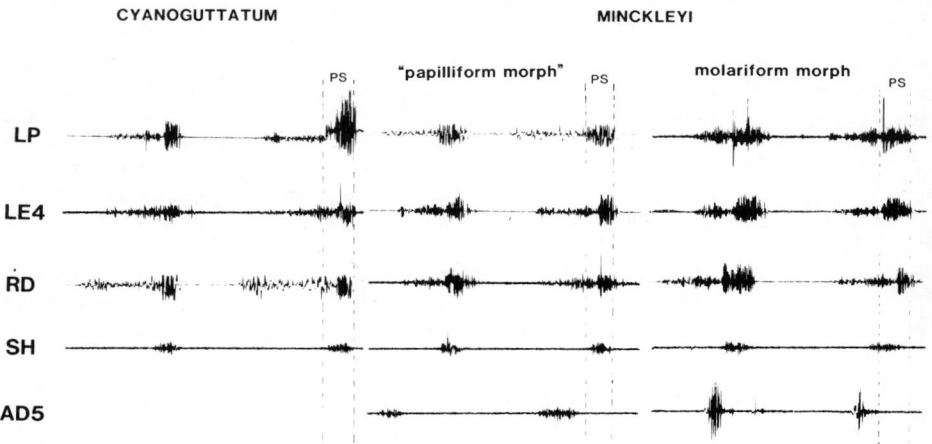

Figure 7. Electromyograms of key pharyngeal jaw muscles during mastication in *Cichlasoma cyanoguttatum* and the two morphs of *Cichlasoma minckleyi*. During PS, the upper pharyngeal jaw occludes against the retracting upper pharyngeal jaw, resulting in crushing action. Time scale, 200 msec. Code: AD5, fifth adductor; LE4, fourth levator externus; LP, levator posterior; PS, power stroke resulting in crushing action; RD, retractor dorsalis; SH, sternohyoideus.

introgression seems less likely than diversification. However, *C. regani* and *C. labridens* do show a high degree of molarization of the pharyngeal jaw apparatus (Miller 1974; Taylor, Miller in prep), but information on the degree of other specializations is lacking. Thus, a hybridization scenario for the origin of *C. minckleyi* cannot be dismissed.

Morphological Transformations: Pattern and Process. It is clear that the specializations in the molariform morph transcend the morphological limits of related species and even of genera. This pattern of extreme intraspecific morphological differentiation strongly supports Greenwood's (1974) notion that simple epigenetic mechanisms influencing ontogeny underlie both the rapidity and diversity of morphological radiations in the cichlid species flock of Lake Victoria. Evolutionary biologists have hypothesized that small changes in the timing of ontogenetic events can lead to distinctly discontinuous changes in phenotype, not just to graded phenotypic changes (Waddington 1969; Alberch 1980, 1982a,b; Gould 1982). The intraspecific morphological shift from papilliform to molariform morph is one of the best examples among vertebrates of Gould's concept that "small inputs give rise to big outputs." Morphological

discontinuity between the morphs involves multiple features. Head structural elements in fishes are coupled in specific patterns so that a perturbation in one element will elicit accommodating structural changes in many other component parts (Dullemeijer 1974, 1980; Hoogerhoud, Barel 1978; Liem 1980; Lauder 1981,1983b). We postulate that a series of regulatory responses are triggered by the genetically induced hypertrophy of the pharyngeal jaws and associated principle muscles, including reduction of the adductor operculi and levatores interni 1-2 muscles, hypertrophy of the fourth epibranchial and basipharyngeal apophyses, and increased cranial width in the postorbital region. This hypothesis is based on the well-documented evidence of regulatory interactions between muscle and bone during ontogeny (Scott 1957), the ability of bones to react to changes in stress regimes (Currey 1968), and Greenwood's (1965) studies on *Astatoreochromis alluaudi*. In the last study, "reduced atypical populations" and normal hypertrophied populations show, respectively, the full extent of genetic control over the development of molariform specializations and the environmentally influenced hypertrophy of the genetically controlled condition.

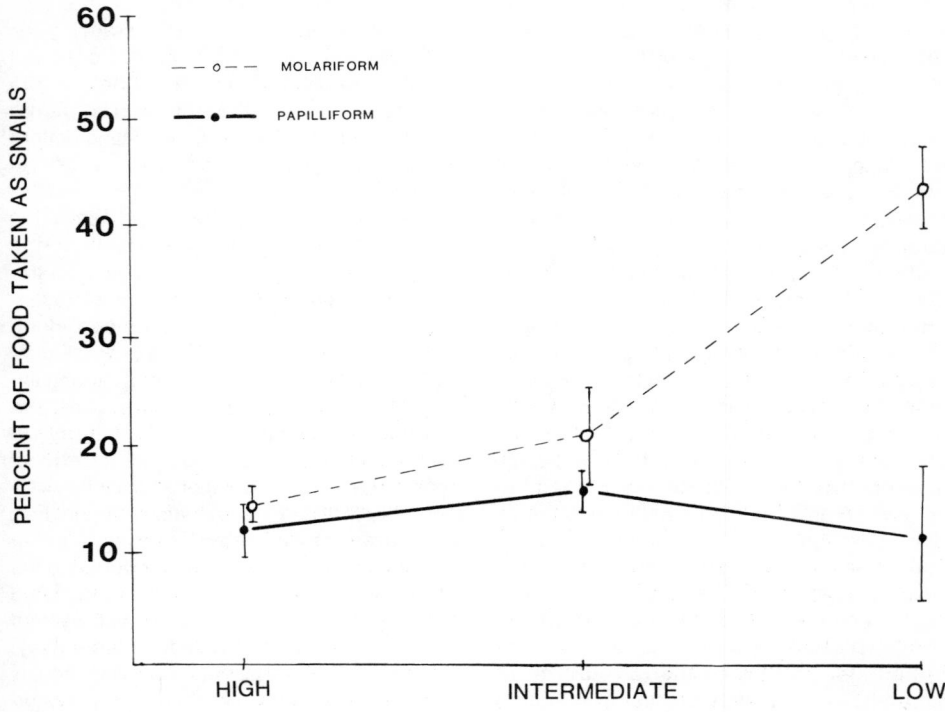

Figure 8. Graphic presentation of feeding behavior of the papilliform and molariform morphs of *Cichlasoma minckleyi* with varying food abundance. Graph expresses the percent of food taken as snails, from the choice of five food types: brine shrimp, flakes, fish, tubifex worms, and snails. Bars denote one standard deviation for three individuals of each morph.

Ontogenetically, the pharyngeal jaw apparatus becomes specialized first, and is followed secondarily by changes in surrounding elements (Greenwood 1965). As growth proceeds, all specializations in the molariform morph become more accentuated. Thus, the morphological dichotomy between the papilliform and molariform morphs is the result of interplay among intrinsic and extrinsic factors (Alberch 1982b) in development. It shows how an intrinsic change in a parameter such as the specialized pharyngeal jaw apparatus in the molariform morph can produce a concomitant series of regulatory changes (head width increase; drastic reduction of spatially related muscles). Phenotypically, the resulting molariform is so different from the papilliform

morph that this could be interpreted as a saltatory or macroevolutionary event.

Strong supporting evidence that small inputs generate big outputs can be derived from strikingly convergent specializations in other molariform cichlid species of Africa (Hoogerhoud, Barel 1978). All specialized characters present in the molariform morph of *C. minckleyi* can be found in a nearly identical condition in the following cichlids that habitually feed on snails by crushing their shells: *Lamprologus tretocephalus* from Lake Tanganyika (Liem 1974); *Tylochromis polylepis* from Lake Tanganyika (Greenwood pers comm); *Astatoreochromis vanderhorsti* from the Luiche River, Tanzania, East Africa (Liem 1974); *Astatoreochromis alluaudi* from

Lake Victoria (Greenwood 1965; Liem, Osse 1975); *Cyrtocara placodon* from Lake Malawi; *Labrochromis mylodon, L. mylergates, L. pharyngomylus,* and *L. teegelaari* from Lake Victoria (Hoogerhoud, Barel 1978).

The morphological pattern of the morphs of *C. minckleyi* has important systematic imiplications. In respect to specialized pharyngeal characters, the molariform morph of *C. minckleyi* resembles the mollusc-crushing cichlids of East Africa more closely than its conspecific papilliform morph. This demonstrates the kind of problems that multiple detailed convergences can create in phylogenetic analyses. The occurrence in *C. minckleyi* and in African cichlids, of the same coadaptive character complex of the molariform morph demonstrates how a suite of specialized characters might be inappropriately used as a set of separate synapomorphies. These striking convergences also furnish indirect evidence on how developmental interactions (Alberch 1982b) and the network of interacting constraints (Liem 1980; Lauder 1981) impose limits on the direction and nature of phenotypic transformations during evolution. **Functional Patterns and Evolution of Trophic Specialization.** Electromyography of prey capture in the papilliform and molariform morphs of *C. minckleyi* (Figure 5), and in *C. cyanoguttatum* reveals no significant differences among the three forms. All three conform to the cichlid pattern of having a repertoire of varying numbers of programs in their neuronal circuits, producing a wide range of finely tuned movements of the jaws. Each of the three forms have four different programs (Figure 5), and conform to the notion that non-piscivorous cichlids specializing on prey which cannot be collected by simple suction, tend to expand the number of options in their feeding repertoire (Liem 1980 a,b 1984). This increases their versatility as they specialize. Intraspecific repertoires of programs affecting the motor output to the musculature involved in prey capture may be very important in the evolution of specialized behavioral patterns (Lauder 1983a). Nonpiscivorous cichlids often become specialists but remain generalists. This characteristic functional flexibility may have constrained the extent of polymorphism in the prey capture apparatus of *C. minckleyi.*

The acquisition of specialized prey capture mechanisms appears to be functional and related to controlling movements of the prey capture apparatus, rather than morphological. Using an oral jaw apparatus that does not deviate morphologically from that of the papilliform morph, the molariform morph would simply employ the manipulating pattern to capture snails attached to the substrate (Figure 5).

A radically different pattern is present in the masticatory mechanism composed of the pharyngeal jaw apparatus. The electromyographic patterns are constant and identical in *C. cyanoguttatum* and the papilliform and molariform morphs of *C. minckleyi* (Figure 7). There is no flexibility of the neuromuscular pattern mediating motor output to the pharyngeal muscles. This similarity in neuromuscular output so characteristic of cichlids (Liem 1974, 1978) provides a basis for morphological change. In the pharyngeal mill, specializations evolve by extensive morphological changes, while the motor output to the pharyngeal muscles remains unmodified (Figures 1,2,3,7). In a functional system characterized by neuromuscular stereotypy, the pathway to functional specialization is through morphological variation, i.e. polymorphism. In the molariform morph of *C. minckleyi* and the molariform *Astatoreochromis alluaudi,* the kinematics of the pharyngeal jaws exhibit a characteristically strong dominance of the crushing power stroke during which the lower pharyngeal jaw is pulled dorsally and anteriorly to occlude against the posteriorly moving upper pharyngeal jaws (Figure 6). Thus, the shearing phase characterized by occlusion of the lower pharyngeal jaw against forward moving upper pharyngeal jaws so characteristic of other cichlids is de-emphasized or absent in molariform cichlids. A shearing phase is still discernible in the papilliform morph when the lower pharyngeal jaw approaches the forward tilting upper pharyngeal jaws. Correlated with the absence of the shearing phase during pharyngeal mastication in molariform cichlids is the reduction of the levatores interior 1, 2, which play an important role in generating the shearing forces. Thus, the functional shift to a purely crushing pharyngeal jaw apparatus in cichlids is accomplished by morphological alterations rather than by changes in the patterns or sequences of muscle actions.

Implications of Polymorphism on Trophic Ecology. Our data provide a link between functional morphology and ecological tracking of food resources by a polymorphic species. Kornfield et al. (1982) have proposed that simple communities composed of trophically polymorphic species may be more responsive to environmental and biotic perturbations than standard communities. Our laboratory results on feeding preferences and behavior indicate that a different set of selection pressures operates when food is abundant than when food is scarce (Table 1). Differences in feeding behavior between papilliform and molariform morphs occur only at low levels of food availability (Figure 8). These findings are consistent with the notion that the evolutionary trajectory of vertebrate species is largely determined by occasional bottlenecks of intense selection during a small portion of their history (Wiens 1977; Boag, Grant 1981; McKaye, Marsh 1983). Our observations demonstrate that carefully controlled laboratory experiments can reproduce the proximate effects of bottlenecks. Only during low food abundance does the molariform morph switch to feeding selectively on hard-shelled snails for which its specialized morphology is well adapted (Figure 8). During high food abundance in laboratory conditions, both morphs switch opportunistically to a diversity of energy-rich food types. Again the findings from field studies on *C. minckleyi* (Smith 1982) are congruent with the laboratory data. The capacity of the polymorphic *C. minckleyi* to specialize on available foods when rich ones are scarce insures survival of the polymorphic species under varying conditions. Thus the extremely specialized pharyngeal jaw apparatus of the molariform morph does not necessarily increase efficiency of feeding on preferred food, but enhances the use of a secondary, less preferred food, such as snails in the laboratory and *Mexipyrgus* snails in the field (Smith 1982). McKaye and Marsh (1983) came to the same conclusion in their analysis of nonpolymorphic algae-scraping cichlids in Lake Malawi. Our studies strongly support the suggestion that a community of polymorphic species is more responsive to environmental and biotic perturbations than is a community of conventional species (Kornfield et al. 1982).

CONCLUSION AND SUMMARY

Polymorphism as an Incipient Intraspecific Macroevolutionary Transition. Polymorphism in *Cichlasoma minckleyi* could be interpreted as a stage of divergence in speciation with the molariform morph as a specialized daughter form that shows a radically different morphology, function, ecology, and behavior when compared with the ancestral papilliform morph. This polymorphism does not exhibit assortative mating (Kornfield et al. 1982). However we suggest that some degree of trophically based differences in habitat selection may occur during "bottlenecks" or "ecological crunches" (Maynard Smith 1966; Wiens 1977). McKaye et al. (1982) have suggested that this incipient stage of stable polymorphism without assortative mating may be followed first by incomplete and then by complete assortative mating stages under influence of intrinsic behavioral mechanisms in an essentially sympatric setting. Quantum morphological specialization has arisen intraspecifically in *C. minckleyi* by quantitative changes in some basic parameters of development which have caused an interrelated series of alterations in functionally correlated elements. The striking convergences in several different mollusc-crushing cichlids from both Africa and the Americas must be interpreted as results of developmentally canalized variation. Apparently the realm of possible morphologies is determined by the internal network of interacting constraints. The intraspecific functional shift to a purely crushing pharyngeal jaw apparatus is accomplished by a phenotypically discontinuous suite of structural specializations rather than by changes in the motor output of the pharyngeal jaw muscles. In centrarchids, mollusc-crushing is accomplished by a change in the motor output of the pharyngeal jaw muscles (Lauder 1983a,b). The extremely specialized morphology of the molariform morph does not increase efficiency of feeding on preferred food, but enhances exploitation of the secondary, less preferred food during ecological bottlenecks. Our studies of *Cichlasoma minckleyi* have shown how stable polymorphism provides major variants that lead to a saltatory alteration of the phenotype for certain key features of great importance for macroevolutionary transitions.

ACKNOWLEDGMENTS

We thank K Hartel, C Fox, and I Kornfield for help in many ways, R Sage, R K Selander, K W Thompson, B Chernoff, W Fink, R R Miller J Taylor, and W Courtenay for specimens, P H Greenwood G V Lauder, P Anton, P Wald, F Irish, M Stiassny, A Launer and others mentioned here for helpful comments. Support: National Science Foundation Grant DEB 82-06888 to KFL and the John Simon Guggenheim Foundation.

REFERENCES

Alberch P 1980 Ontogenesis and morphological diversification. *Amer Zool* 20:653-667

_____ 1982a Developmental constraints in evolutionary porcesses. 313-332 J Bonner ed *Evolution and Development*. Springer Verlag. Berlin

_____ 1982b The generative and regulatory roles of development in evolution. 19-36 D Mossakowski, G Roth eds. *Environmental Adaptation and Evolution*. Fischer. Jena

Boag P T, P R Grant 1981 Intense natural selection in a population of Darwin's finches (Geospizinae) in the Galapagos. *Science* 214:82-85

Currey J D 1968 Adaptation of bones to stress. *J Theor Biol*20:91-106

Dullemeijer P 1974 *Concepts and Approaches in Animal Morphology*. Assen Van Gorcum. The Netherlands

_____ 1980 Functional morphololgy and evolutionary biology. *Acta Biotheor* 29:151-250

Greenwood P H 1965 Environmental effects on the pharyngeal mill of a cichlid fish, *Astatoreochromis alluaudi* and their taxonomic implications. *Proc Linn Soc Lond* 176:1-10

_____ 1974 The cichlid fishes of Lake Victoria, East Africa: The biology and evolution of a species flock. *Bull Br Mus Nat Hist (Zool)* Suppl 6:1-134

Gould S J 1982 Change in developmental timing as a mechanism of macroevolution. 333-346 J Bonner ed. *Evolution and Development*. Springer Verlag. Berlin

Hoogerhoud R J C, C D N Barel 1978 Integrated morphological adaptations in piscivorous and mollusc-crushing *Haplochromis* species. 52-56 *Proc Zodiac Symp on Adaptation* Wageningen, The Netherlands

Kornfield I L, R K Koehn 1975 Genetic variation and evolution in New World cichlids. *Evolution* 29:427-437

Kornfield I L, D C Smith, P S Gagnon, J N Taylor 1982 The cichlid fish of Cuatro Cienegas, Mexico: Direct evidence of conspecificity among distinct morphs. *Evolution* 36:658-664

Kornfield I L, J N Taylor 1983 A new species of polymorphic fish, *Cichlasoma minckleyi* from Cuatro Cienegas, Mexico (Teleostei: Cichlidae). *Proc Biol Soc Wash* 96:253-269

La Bounty J F 1974 Materials for the revision of cichlids from northern Mexico and Southern Texas, U.S.A. PhD thesis. Arizona State Univ

Lauder G V 1981 Form and function: Structural analysis in evolutionary morphology. *Paleobiol* 7:430-442

_____ 1983a Neuromuscular patterns and the origin of trophic specialization in fishes. *Science* 219:1235-1237

_____ 1983b Functional design and evolution of the pharyngeal jaw apparatus in euteleostean fishes. *Zool J Linn Soc*77:1-38

Liem K F 1974 Evolutionary strategies and morphological innovations: Cichlid pharyngeal jaws. *Syst Zool* 20:425-441

_____ 1978 Modulatory multiplicity in the functional repertoire of the feeding mechanism in cichlid fishes. Part I: Piscivores. *J Morph* 158:323-360

_____ 1980a Adaptive significance of intra- and interspecific differences in the feeding repertoires of cichlid fishes. *Amer Zool* 20:25-314

_____ 1980b Acquisition of energy by teleosts: Adaptive mechanisms and evolutionary patterns. 299-334 M A Ali ed. *Environmental Physiology of Fishes*. Plenum. New York

_____ 1984 Functional versatility, speciation and niche overlap: Are fishes different? *Trophic Dynamics in Aquatic Ecosystems*. D Myers, R Strickler eds. *Proc AAAS Symposium* In press

Liem K F, J W M Osse 1975 Biological versatility, evolution and food resource exploitation in African cichlid fishes. *Amer Zool* 15:427-454

Maynard Smith J 1966 Sympatric speciation. *Amer Nat* 100:637-650

McKaye K R, A Marsh 1983 Food switching by two specialized algae-scraping cichlid fishes in Lake Malawi, Africa. *Oecologia* (Berlin) 56:245-248

McKaye K R, T Kocher, P Reinthal, I Kornfield 1982 A sympatric sibling species complex of *Petrotilapia* Trewavas from Lake Malawi analyzed by enzyme electrophoresis (Pisces: Cichlidae). *Zool J Linn* 76:91-96

Miller R R 1974 *Cichlasoma regani*, a new species of cichlid fish from the Rio Coatzacoalcos basin, Mexico. *Proc Bio Soc Wash* 87:465-472

Minckley W L 1969 Environments of the bolson of Cuatro Cienegas, Coahuila, Mexico, with special reference to the aquatic biota. Texas Western Press, Univ Texas El Paso, Sci Ser 2:1-63

Ono R D, L S Kaufman 1983 Muscle fiber types and functional demands in feeding mechanisms of fishes. *J Morph* 177:69-87

Sage R D, R K Selander 1975 Trophic radiation through polymorphism in cichlid fishes. *Proc Nat Acad Sci* 72:4669-4673

Scott J H 1957 Muscle growth and function in relation to skeletal morphology. *Amer J Phys Anthrop* 15:197-234

Smith D C 1982 Trophic ecology of the cichlid morphs of Cuatro Cienegas, Mexico. MS thesis. Univ Maine at Orono

Taylor D W, W L Minckley 1966 New world for biologists. *Pacific Disc* 19:18-22

Waddington C J 1969 Paradigm for an evolutionary process. 2:106-124 C Waddington ed. *Towards a Theoretical Biology.* Edinburgh Univ Press. Edinburgh

Wiens J A 1977 On competition and variable environments. *Amer Sci* 65:590-597

ALLOMETRY AND FUNCTIONAL FEEDING MORPHOLOGY IN HAPLOCHROMINE CICHLIDS

RICHARD E STRAUSS

INTRODUCTION

The superficial jaw musclature and associated bony elements in vertebrates have together been assumed to form a complex adapted for specific kinds of food and foraging (Bowman 1961; Alexander 1967). In teleosts the size and configuration of the various skeletal and muscular elements of the head vary greatly (Gregory 1933; Gosline 1971). Comparisons of superficial features of fish skulls led Dullemeijer (1974) to postulate that of the structural components of the head, the eye and jaws are evolutionarily dominant and conservative elements while the mandibular adductor and other cephalic muscles are evolutionarily subordinate, having configurations which are dependent on the space available between the eye and upper jaw (see also Otten 1981). Later, Dullemeijer and Barel (1977) attempted to quantify the degree of evolutionary plasticity of these components among the closely related haplochromine cichlids and to show that the relative sizes of the components are highly correlated with the type of prey normally consumed.

This paper presents a reevaluation of these data and inferences, with a further analysis of cephalic morphometry in haplochromine cichlids. I will attempt to show the following: 1) Rather than being specific structural reorganizations in response to trophic adaptations, the relative sizes and shapes of cephalic components lie in simple allometric relationships with one another and represent simple scalings of shape based on size. 2) The allometric relationships among species in many cases are simple extensions of those within species, indicating that no inferences can be made about the evolutionary plasticity of various components. 3) Functional and evolutionary inferences derived from morphological analyses must be derived within

the context of a null hypothesis of functional neutrality or functional equivalence.

TROPHICALLY CORRELATED VARIATION IN HAPLOCHROMINE MORPHOLOGY

With an estimated 200-350 species, the species flocks of haplochromine cichlids dominate the fish fauna of the East African Lakes Victoria and George. Together these fishes exploit a wide variety of food sources and habitats. Each species, however, tends to have specialized food and habitat preferences, at least under modal conditions, with remarkably little diversity in body form (Greenwood 1973, 1974, 1981a; Witte 1981; Oijen et al. 1981). Greenwood (1974) has found that head morphology seems to be most diverse among species, presumably in response to trophic specialization.

The lateral aspect of the head in cichlids is comprised superficially of several major components (Figure 1): the orbit; the adductor mandibulae muscle complex, which extends from a broad insertion on the lateral face of the suspensorium to the lower and upper jaws and which serves primarily to close the mouth; and the levator arcus palantini and the dilatator operculi muscles, which connect the postorbital region of the cranium with the posterodorsal edge of the opercle, and aid in expanding the buccal and opercular cavities and opening the mouth (Winterbottom 1974; Anker 1978). Anterior to the orbit and adductor mandibulae complex is a relatively large lachrymal bone which partially overlies the maxilla.

To investigate the evolutionary plasticity of these components and their relationship to trophic specialization, Barel and Voogt-Kokx (in Dullemeijer, Barel 1977:101) measured the surface area (in lateral projection) of the orbit and the two muscle complexes in 39 species

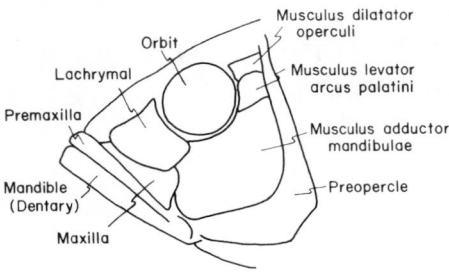

Figure 1. Superficial lateral cephalic components of a generalized haplochromine cichlid.

of haplochromines (Figures 2, 3). These measures describe only the gross morphology, but on a level equivalent to that of many taxonomic and ecological studies. From their data they drew several major conclusions. First, they found that the ratio of largest to smallest absolute size of each component, a figure which Dullemeijer and Barel (1977) use as a measure of the component's relative evolutionary plasticity, is approximately 12 for the orbit, 35 for the levator-dilatator complex, and 68 for the adductor complex. Thus, they conclude that the eye is stable in relative size among species (a "dominant" feature in the terminology of Dullemeijer 1974) while the adductor complex is plastic in expression (a "subordinate" feature), its size being dependent on the available space between the orbit and upper jaw. The levator complex is intermediate in plasticity, dependent to a more moderate extent on the size and position of the orbit.

Secondly, Barel and Voogt-Kokx observed a significant association between the relative component size and the preferred prey type of each species (Figures 3, 4). They found that planktivores, insectivores, and feeders on benthic crustaceans are characterized by relatively large eyes and hence, small adductors. In contrast, piscivores have relatively small eyes and large adductors, while algal grazers, paedophages, and molluscivores have intermediate morphologies. These results presumably can be accounted for by any number of adaptive explanations invoking selective responses to competitors and to varying size of prey.

However, further examination of Barel's

and Voogt-Kokx's data reveals a confounding factor in these explanations — varying body size. Size is manifested in two ways, in the obviously continuous variation of the cephalic components among species (Figure 2), and in the tendency for the size of the predator to be correlated with the size of its prey (Figure 3). To the extent that variation in body size alone accounts for variation in either head morphology or prey preference, it is a more parsimonious explanation, one that invokes neither notions of evolutionary plasticity nor post hoc adaptive arguments about trophic specialization.

FUNCTIONAL EXPLANATIONS IN THE CONTEXT OF ALLOMETRY

Before examining the importance of allometry in accounting for variation among haplochromines, it is worthwhile to review the manner in which variation in an underlying factor can be said to "account for" variation in morphololgy. If we were to sample two populations of fishes and then discover a significant morphological difference between them, we might be tempted to explain the difference in terms of some adaptive response to different ecological conditions, such as prey availability, competition, or predator pressure. However, if we later sampled a number of added populations at intermediate localities and determined that the observed variation was clinal, we would have no need to explain independently the difference between any two populations. Particular differences could be incorporated with a more general explanation of the clinal pattern, which could result from variation in temperature, historical patterns of gene flow, and other factors. If a population were encountered which did not fit the clinal pattern, then the difference between its observed morphology and that predicted on the basis of the cline would demand a special explanation, perhaps the adaptive one.

This example is simplistic, but biologists have often been led astray by situations not much more complex. Grant (1972), for example, in response to a proliferation of examples of character displacements of apparent ecological importance in resource partitioning, noted that a character difference between two taxa may simply be the result of independent clinal, allopatric variation trends

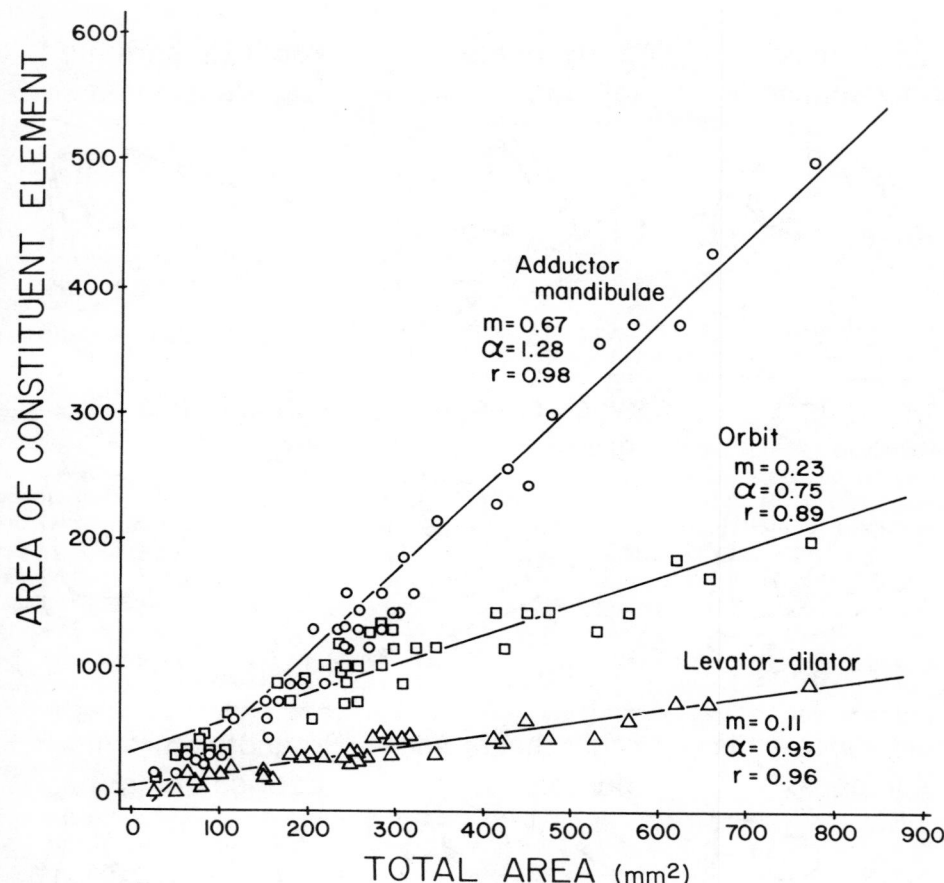

Figure 2. Scatter of relationship between surface area (lateral aspect) of each of three cephalic components and their combined surface area for 39 species of haplochromine cichlids. (Dullemeijer, Barel 1976) Allometric coefficients (α) computed from log-log regressions of same data: slope, m; correlation, r.

extrapolated into an area of sympatry. In such cases, morphological variation in allopatry predicts or accounts for morphological differences in sympatry, and there is no need to resort to less parsimonious explanations of selectively advantageous character shifts.

Similarly, ever since paleontologists recognized a tendency for body size in mammals to increase during phylogeny, comparative morphologists have searched for regularities that would characterize the change in form that accompanies change in size (Huxley 1932; Gould 1971). McMahon (1975), for example, found that the dimensions of the ulna and other structural supports in

adult ungulates scale allometrically across species over a size range that encompasses the deer mouse and the moose, with the single exception of the giraffe. Thus, the differences in shape of the ulna between most pairs of ungulates can be accounted for by the general explanation of static allometric scaling necessary to maintain functional equivalence at different sizes. Only the giraffe requires an additional, perhaps adaptive explanation to account for the shape of its ulna.

In this context we must consider hypotheses of adaptation in a comparative sense only (Clutton-Brock, Harvey 1979). We might formulate a number of evolutionary scenarios

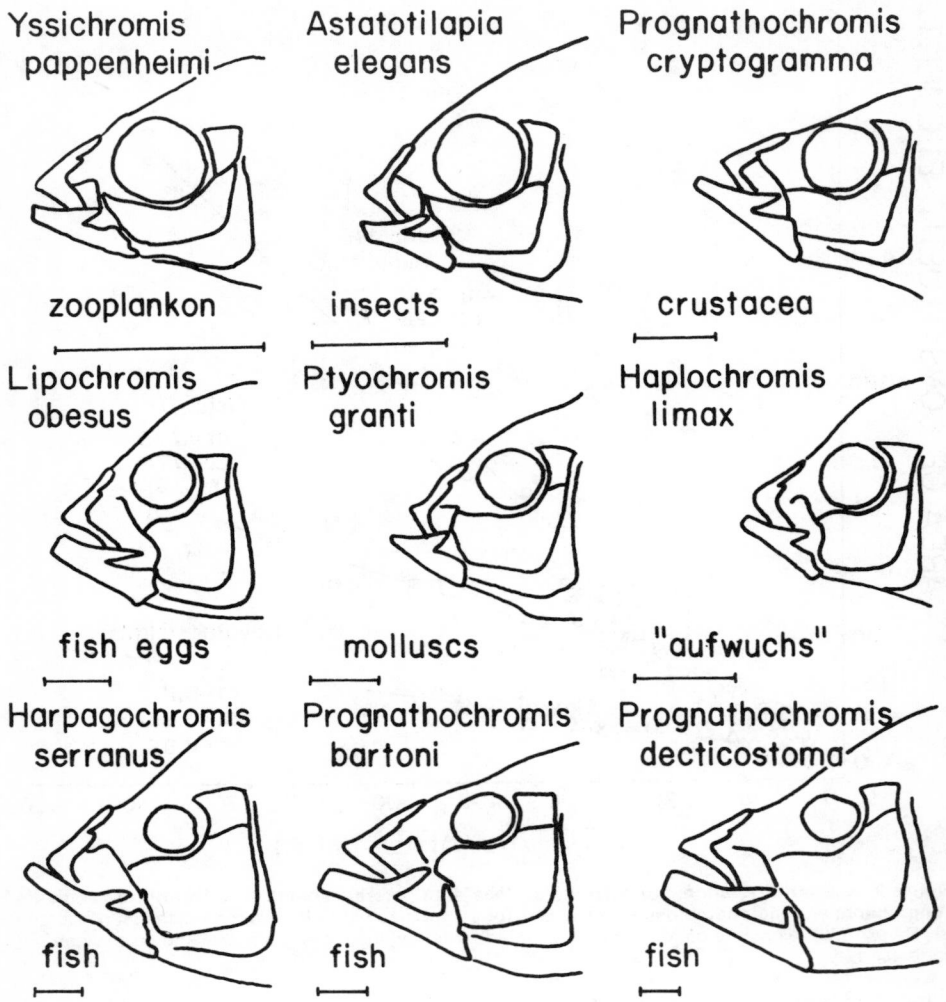

Figure 3. Schematics showing relative size of orbit, muscle complexes, and mouth parts for adult specimens of 9 haplochromines, with their preferred prey. (Dullemeijer, Barel 1977 Fig 8, modified) Generic designations follow Greenwood (1981b). Scale bar = 10 mm.

accounting for the adaptive value of an ulna per se. They would be only scenarios. But if we restrict ourselves to closely related species, we may approach the subject in a more objective manner. We can compare the morphology of two species, observe, and preferably quantify, the differences between them, and then try to decide whether the observed differences are the result of divergent functional adaptations or of some other "neutral" effect. For this decision we need some context in which to decide whether an observed difference is adaptive, or whether it may simply be accounted for by some underlying factor such as temperature or body size. This is not to say that a difference in body size might not itself be adaptive, but only that we do not have to search for other explanations to account for a difference in shape of correlated structures. An alternation of shape

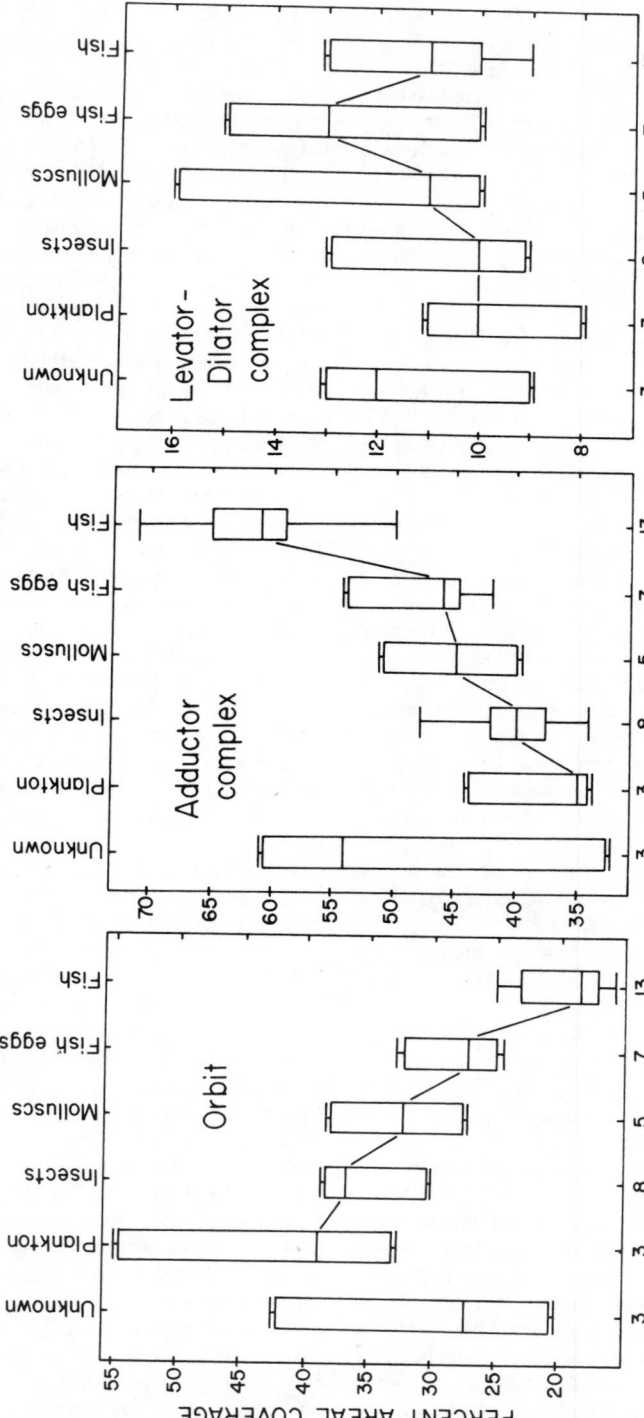

Figure 4. Variability in relative size of orbit, adductor complex and levator-dilatator complex as a function of preferred prey. Symbols represent the median (central bar), quartiles (box), and range (outer bars) of relative size of components for specimens in each preferred-prey group. Summarized from Dullemeijer and Barel (1977, Table 1). Values at bottom of graphs are numbers of species in each group.

is not necessarily an adaptation conferring special advantages independent of other cor-related trends. If it can be shown that the relative shape or size of some structure con-forms to an expected dependence on body size, then the dependence can be subsumed in the more general explanation (Sweet 1980). Conversely, notable departures from other-wise consistent allometric trends may be of particular value in identifying and understand-ing specific functional adaptations.

ALLOMETRY IN HAPLOCHROMINES

The data of Barel and Voogt-Kokx show very strong patterns of static allometry among the 39 species they examined. The allometric coefficient of 1.28 for the adductor complex is rather high, indicating strong positive allometry among species. In contrast, the size of the orbit is strongly negatively allometric (0.75), while the levator-dilatator complex is nearly isometric. Because these allometric coefficients measure the relative exponential change in size of each component in relation to the combined surface area of all three, they must necessarily average to unity. They will be true measures of allometry in relation to total head size only if the combined area is highly correlated with head or body size, however measured. Because 1) we are con-cerned primarily with head morphology, 2) Barel and Voogt-Kokx did not provide other measures of size for their specimens, and 3) the combined area of the components is highly correlated with standard length in other haplochromines, I will assume that com-bined surface area is an adequate descriptor of body size.

To understand these patterns it is first necessary to determine whether they hold for other cephalic structures, and then to com-pare them with corresponding patterns of intraspecific allometry. For this purpose I repeated and extended the mensural pro-cedure of Voogt-Kokx on 63 specimens representing seven haplochromine species. (Specific epithets given are catalog names which, due to the constant flux of taxonomic work on lacustrine cichlids and difficulty of identification, could be incorrect. Misidentifica-tions would be unfortunate but would not alter the analyses or conclusions discussed here.) Figure 5 shows resulting values for the

surface area of the orbit, adductor complex, levator complex, lachrymal bone, and oper-cle. Also given are the lengths of the mandi-ble, premaxilla, and maxilla in relation to head length, measured from snout to posterior margin of the opercle. All of these structures are highly allometric across specimens, but it is particularly noteworthy that the allometric trends among species are consistent with those within (Figure 6), and in fact seem to be simple extrapolations of intraspecific patterns. It is not necessarily appropriate to equate within-group allometry with longitudinal growth patterns when the mensural data have been derived from a series of individuals varying in size and age (Cock 1966). Nevertheless, we can speculate that the quantitative patterns of growth among these species may be almost entirely plesiomorphic, with only minor shifts in allometry or timing in some derived forms. This could be investigated experimentally.

Such developmental shifts, if present, might not be resolved in simple bivariate and trivariate analyses. A principal components analysis (Figure 7) of the eight logarithmical-ly transformed characters of Figure 5 reveals such a case. The first component accounts for the joint allometric correlation of the eight traits with increasing body size (Table 1) and accounts for 96 percent of the total variation within and among species. (The correlation of the first component with standard length is 0.97.) Excluding a single species, *"Haplochromis" strigigena*, the first compo-nent is also correlated with increasing prey size, from phytoplankton to fish. The second component accounts only for an additional 2 percent of the variance, but is strictly cor-related with prey size. The third and subse-quent components seem to express only ran-dom noise in the data. Thus for six of the seven species, a single trend in body size ac-counts simultaneously for variation in 1) the relative sizes of the major superifical struc-tures of the head, and 2) trophic specializa-tion as it correlates with the size of preferred prey. *"Haplochromis" strigigena* is an exception to this generalized trend in that it possesses the same morphological configura-tion as two other piscivores, *Harpagochromis squamulatus* and *H. spekei*, but at a smaller body size. If the allometric trend among the six species becomes a general one when we

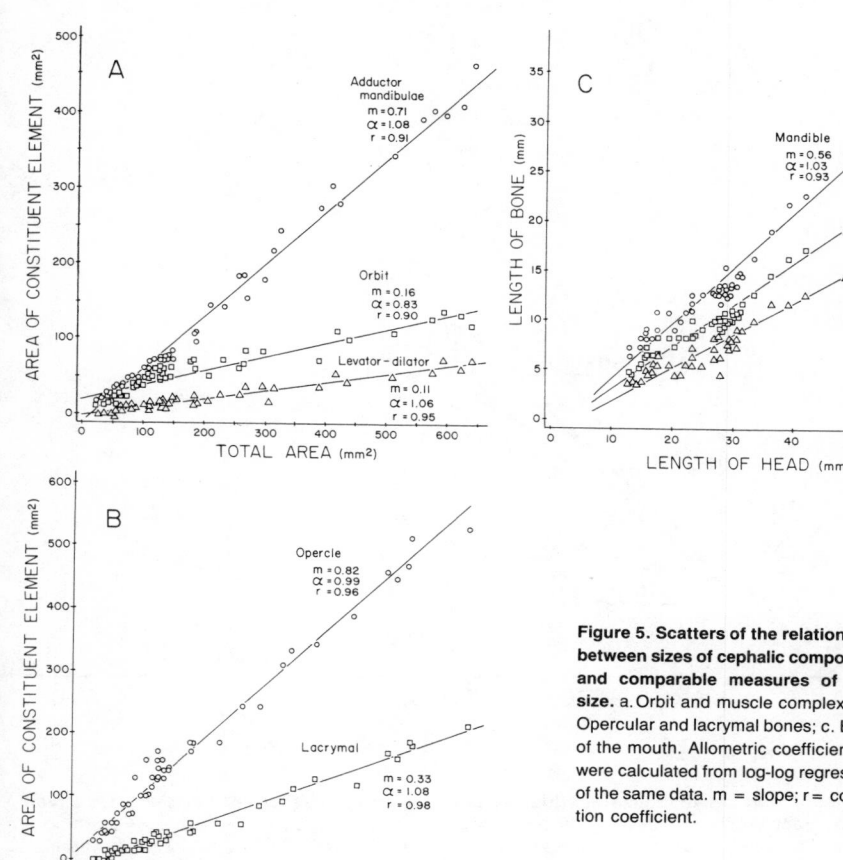

Figure 5. Scatters of the relationships between sizes of cephalic components and comparable measures of head size. a. Orbit and muscle complexes; b. Opercular and lacrymal bones; c. Bones of the mouth. Allometric coefficients (α) were calculated from log-log regressions of the same data. m = slope; r = correlation coefficient.

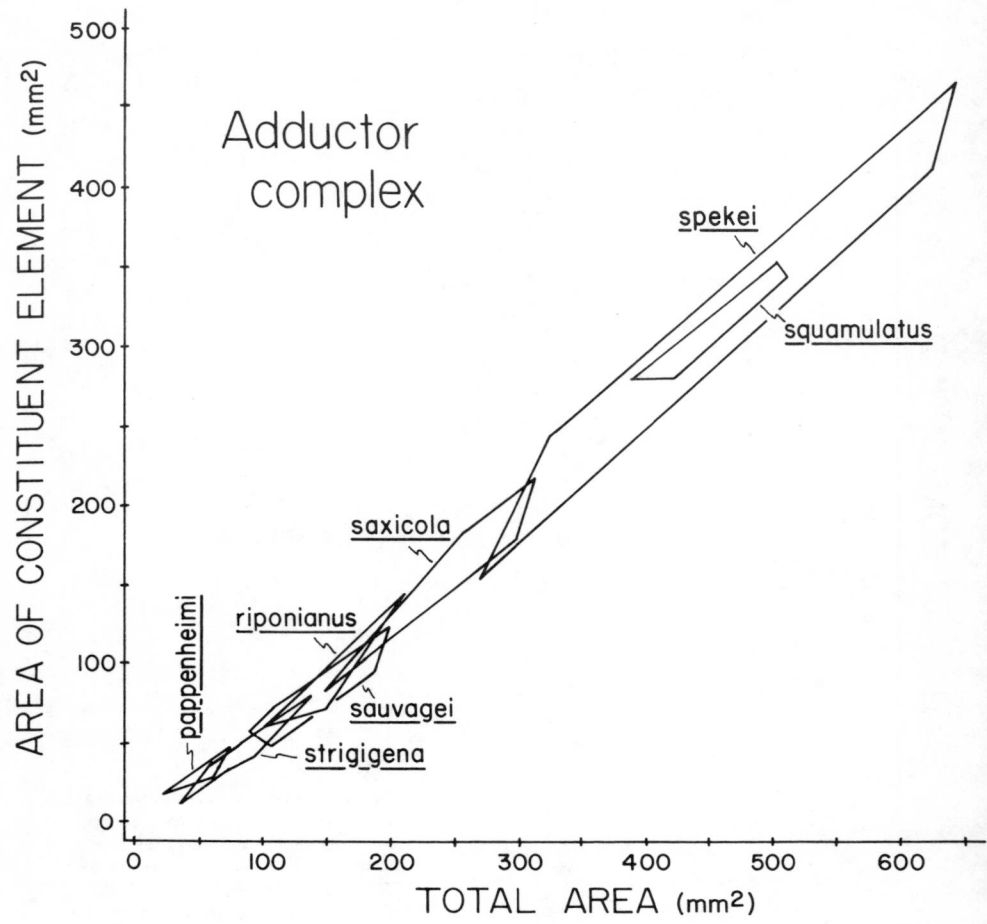

Figure 6. Relationship between size of adductor complex and a measure of head size for seven species of haplochromines. *Yssichromis pappenheimi, Psammochromis riponianus, Pytochromis sauvagei, Psammochromis saxicola, Harpagochromis spekei, H. squamulatus,* and *"Haplochromis" strigigena.* Generic designations follow Greenwood (1981b). Size ranges were limited by available collections. Plots for the other seven traits of Figure 5 show similar consistency of intraspecific trends.

look at many more haplochromines, then *"H." strigigena* could very well represent an adaptive derived morphology. It might also be a member of a sister group characterized by a somewhat different pattern of ontogenetic allometry.

Reconsidering the data of Barel and Voogt-Kokx, we have seen that the relative sizes of the three cephalic components display strong trends of static allometry. Dullemeijer's and Barel's measure of the relative evolutionary plasticity of these components is then just a statistically weak measure of this allometry, taken in areal rather than log-areal dimensions, and dependent on both the slope and the intercept of the trend and on the sizes of the smallest and largest individuals. If the interspecific trend of Figure 2 is as much an extrapolation of composite intraspecific trends as that of Figure 6, then an appeal to evolutionary plasticity or dominance and subordination of morphological structures is unnecessary and misleading. We can only say that there is a marked tendency for the

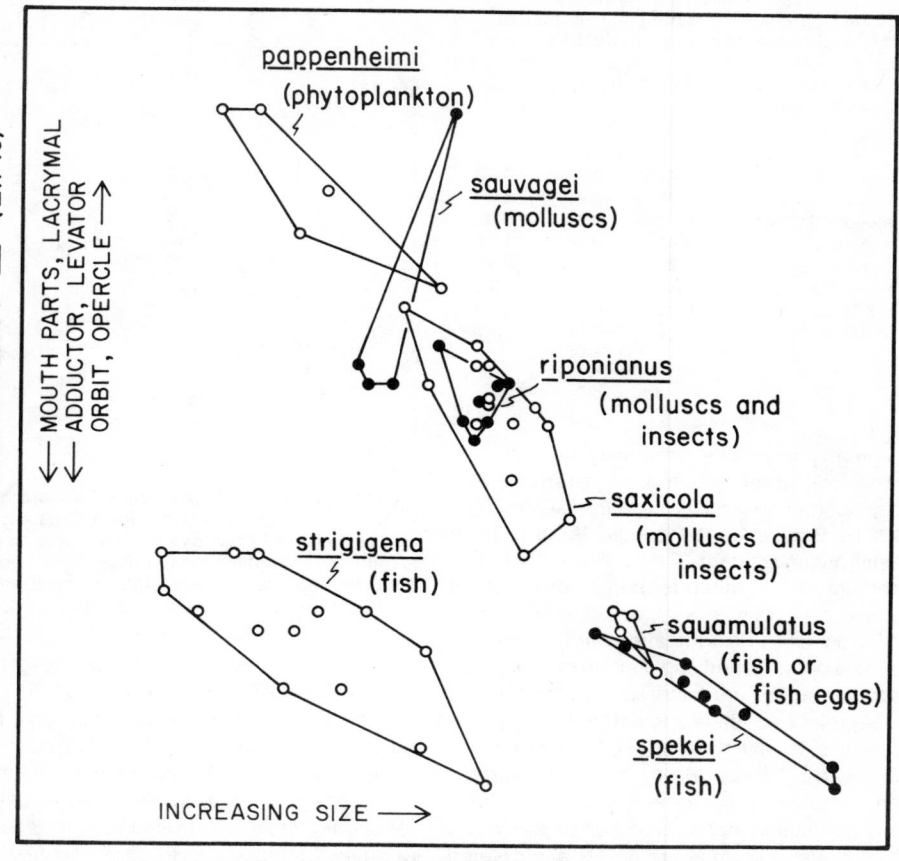

Figure 7. Scatter of specimens of Figure 5 on first two principal components; species of Figure 6.

growth rates of different parts to bear a constant relationship to one another, both within and among species.

Although body sizes are not available for all of the specimens measured by Voogt-Kokx, the correlation of preferred prey with body length for the 39 species may be indirectly estimated by 1) regressing the log-transformed data of Figure 5a against the known log-transformed standard length of each specimen; 2) adjusting the data of Figure 2 so that the slopes and intercepts are equal to those of Figure 5a, i.e., adjusting the data to compensate for differences in scale and mensural technique; and 3) using the resultant log-linear equation from step 1 to predict a standard length for each of the species of Figure 2. The results, in Figure 8, can be only approximate, but do show the high expected correlation between body size and preferred prey. Paedophages (egg-eating predators) fall into an adult-size group approaching piscivores, perhaps because their usual mode of feeding involves engulfing the snout of mouth-brooding cichlids (Wilhelm 1980).

DISCUSSION
Analysis of the effects of body size and

Table 1. Loadings on Principal Components of Eight Cephalic Characters.
(Loadings are scaled so that their squares sum to unity. Multivariate allometric coefficients, α, are the loadings on component I rescaled so that their squares sum to the number of characters, Strauss, Bookstein 1982.)

Character	I	II	α
Orbit	0.25	0.43	0.72
Adductor	0.41	-0.25	1.17
Levator-dilatator	0.32	-0.14	0.90
Lachrymal	0.42	-0.26	1.20
Opercle	0.31	0.19	0.87
Premaxilla	0.35	-0.33	0.98
Maxilla	0.37	-0.60	1.04
Mandible	0.36	-0.39	1.03

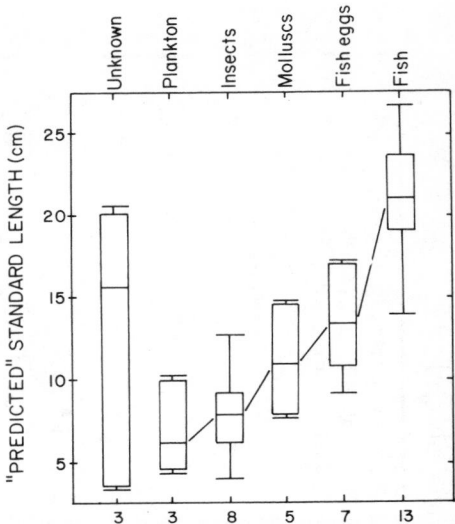

Figure 8. Variability in predicted standard body length of the specimens of Figure 2 as a function of preferred prey. Symbols represent median (central bar), quartiles (box), range (outer bars) of predicted lengths for each preferred-prey group. Sample sizes shown below graph.

allometry have been seriously neglected in morphological investigations of haplochromine cichlids (see Greenwood 1956, 1981a, Barel et al. 1977, Galis, Barel 1980 for limited discussions). To the extent that morphological variation is continuous across space, time, or size, it is unproductive to treat differences as discrete and adaptive until the effects of the underlying continuity have been accounted for and factored out. This appeal to simpler hypotheses is certainly not a new one. It is, for example, one of five alternatives to the "adaptationist programme" enumerated by Gould and Lewontin (1979): "no adaptation and no selection on the [body] part at issue; form of the part is a correlated consequence of selection directed elsewhere," in this case at the level of body size. Almost all anatomical, physiological, and behavioral attributes are size-related in some way (Clutton-Brock, Harvey 1979). The size and form of many individual structures may constrain other structures in related ways, introducing complexities into comparative arguments. If a change in body size takes place, for neutral or selective reasons, a large number of morphological, histological, physiological, behavioral and developmental relations will change concomitantly. While it is important that we ask functional questions about size itself, if we wish to examine relations between traits that are inevitably affected by body size, we must first account for its effects. The importance of taking size differences into account is sufficiently obvious

to be a truism were it not so frequently ignored (e.g., Alexander 1967; Bock 1976).

The allometry of functional scaling may thus be used as an objective criterion, specifically a null hypothesis, for determining whether evolutionary changes in morphology may be required correlates of trends in size variation, or special adaptations to particular features of the organisms' modes of life. It must be emphasized that the null hypothesis of the allometric model is not *constancy* of shape, but rather, size-correlated *change* in shape (Sweet 1980).

The primary conclusion of this study is that, rather than being specific structural reorganizations in response to trophic adaptation, the relative sizes and shapes of many cephalic elements in these cichlids lie in simple allometric relationships with one another. Further, the differences between species are extrapolations of those within species. Thus, both spatial-component relationships and relatively specialized food habits are expected correlates of intra- and interspecific differences in absolute body size, i.e., functional equivalence. I do not wish to imply that all morphological differences in haplochromines are of this sort, nor especially that all variation in cichlids can be accounted for by size differences. The cichlids of the older African Rift Lakes are much more diverse morphologically (Fryer, Iles 1972), and even among Victoria haplochromines there seems to be much size-independent variation in oral and pharyngeal dentition (see Greenwood 1974, figs 4-5). Nevertheless, we should concentrate on the exceptions to observed patterns, because they stand the greatest chance of being special adaptations. The burden of evidence is clearly with the functional morphologist.

These findings also support Greenwood's (1974, 1981a) contention that differences among haplochromines are the result of subtle changes in relative growth rates with few morphological gaps. Ontogeny at the organismic level has two aspects: 1) relative size, or allometry, and 2) relative timing, or growth rate. Small heterochronic changes in development may lead to notable differences in form among species, even when the allometries remain constant or nearly so (Frazetta 1975). A single growth program can thus be the source of considerable diversity in form, but can also limit the range of adaptations to certain directions and amounts (Riedl 1978).

Consideration of size-dependent trophic changes, both within and among species (Greenwood 1974; Greenwood, Barel 1978), might also contribute to a better understanding of the size-selective constraints allowing (or driving) diversification in these and other fishes. The specimens examined for this study were limited to available museum collections but were chosen to represent "typical" adult size distributions (Figures 6,7). But fishes grow continuously and indeterminately, albeit more slowly with age, and the ontogenetic development of a large-bodied individual must encompass the entire size range of any smaller species. We might speculate, then, that within a given habitat it is likely that the expected derived condition for a newly evolving species would be an increase in body size, with a concomitant increase in the size of consumable prey, because the adults of the new species may then avoid direct trophic competition with extant species and with their own young. Of course, the size of an individual affects many other aspects of its life, including the predators that can feed upon it, the holes in which it can hide, the size of its home range and foraging area, and diverse features of its reproductive and social behavior, so that this generalization is likely to be simplistic and misleading. However, regulation of body size is obviously important in maintaining community structure (Keast 1978; Smith 1978) and must be considered in process-oriented questions about speciation and diversification.

Allometry is a greatly simplified abstraction of an intricate developmental process. It does not reveal the mechanisms of development, and can only provide superficial clues about the nature of the process. Nevertheless, examining functional adaptation in the context of allometry leads us away from having to invoke explanations for every particular difference in shape that we observe. It instead channels our questions into two categories. The first concerns the results of allometric scaling that yield functional equivalence at varying sizes. What underlying conservative developmental patterns and processes are responsible for maintaining the allometric regularities that we observe within and among species? This problem has been attacked both with reductionist approaches of physiology and developmental biology (Goss 1964), and with holistic approaches such as biomechanical solutions to optimal design criteria (Oxnard 1973; Weijs 1980). An elegant study spanning these extremes is McMahon's (1975) demonstration that the allometric scaling of metabolic rate and breathing rates are derivable from criteria of elastic similarity in locomotor systems of terrestrial vertebrates.

The second category of questions concerns those structures that do not conform to general allometric trends. What is the nature of the particular adaptations involved? For this question we have available the methods of classical functional morphology developed over the past hundred years (Dullemeijer 1974; Gans 1974) as well as, for example, the electromyographic and cineradiographic methods applied by Liem (1973, 1978a,b, 1979; Liem, Greenwood 1981). An adequate assessment of ontogenetic and static allometry can only improve our understanding of haplochromines in particular and of functional morphology in general.

SUMMARY

Various investigators have observed important correlations between the relative sizes and shapes of cephalic and muscular components and trophic specializations in Lake Victoria haplochromines. Quantitative analyses of these components, in the context of a null hypothesis of functional neutrality or functional equivalence, reveal the following: 1) Rather than being specific structural reorganizations in response to trophic adaptation, the relative sizes of most components occur in simple allometric relationships with one another, thus representing scalings of shape due to differences in size. 2) The allometric patterns among species are simple extrapolations of those within species, indicating that no inferences can be made about the relative evolutionary "plasticity" of the components. 3) Except for differences in dentition and in the shape of the pharyngeal jaw elements, spatial-component relationships and accompanying trophic specializations in haplochromines may parsimoniously be attributed to differences in adult body size. Consideration of size-related effects is necessary for an adequate understanding of speciation, diversification, and adaptation in cichlids and other organisms.

REFERENCES

Alexander R McN 1967 *Functional Design of Fishes.* Hutchinson. London

Anker G Ch 1978 The morphology of the head muscles of a generalized *Haplochromis* species: *H. elegans* Trewavas 1933 (Pisces, Cichlidae). *Neth J Zool* 28:234-271

Barel C D N, M J P van Oijen, F Witte, E L M Witte-Maas 1977 An introduction to the taxonomy and morphology of the haplochromine Cichlidae from Lake Victoria. *Neth J Zool* 27:333-389

Bock W J 1976 Adaptation and the comparative method. M Hecht, P Goody, B Hecht eds. *Major Patterns in Vertebrate Evolution.* NATO Adv Study Inst Ser A 14:57-82

Bowman R I 1961 Morphological differentiation and adaptation in the Galapagos finches. *Univ Calif Publ Zool* 58:1-302

Clutton-Brock T H, P H Harvey 1979 Comparison and adaptation. *Proc R Soc London B* 205:547-565

Cock A G 1966 Genetical aspects of metrical growth and form in animals. *Quart Rev Biol* 41:131-190

Dullemeijer P 1974 *Concepts and Approaches in Animal Morphology.* Van Gorcum. Assem Netherlands

Dullemeijer P, C D N Barel 1977 Functional morphololgy and evolution. M Hecht, P Goody, B Hecht eds. *Major Patterns in Vertebrate Evolution. NATO Adv Study Inst Ser A,* 14:83-117

Frazetta T H 1975 *Complex Adaptations in Evolving Populations.* Sinauer. Sunderland MA

Fryer G, T D Iles 1972 *The Cichlid Fishes of the Great Lakes of Africa.* TFH. Neptune City NJ

Galis F, C D N Barel 1980 Comparative functional morphology of the gills of African lacustrine Cichlidae (Pisces, Teleostei): an ecomorphological approach. *Neth J Zool* 30:392-430

Gans C G 1974 *Biomechanics: An Approach to Vertebrate Biology.* Sinauer. Sunderland MA

Gosline W A 1971 *Functional Morphology and Classification of Teleostean Fishes.* Univ Hawaii Press. Honolulu

Goss R J 1964 *Adaptive Growth.* Academic Press. New York

Gould S J 1971 Geometric scaling in allometric growth: a contribution to the problem of scaling in the evolution of size. *Amer Natur* 105:113-136

Gould S J, R C Lewontin 1979 The spandrels of San Marco and the Panglossian paradigm: a critique of the adaptionist programme. *Proc R Soc London B* 205:581-598

Grant P R 1972 Convergent and divergent character displacement. *Biol J Linn Soc* 4:39-68

Greenwood P H 1956 The monotypic genera of cichlid fishes in Lake Victoria. *Bull Brit Mus Nat Hist (Zool)* 3:295-333

_____ 1973 A revision of the *Haplochromis* and related species (Pisces, Cichlidae) from Lake George, Uganda. *Bull Brit Mus Nat Hist (Zool)* 25:139-242

_____ 1974 The cichlid fishes of Lake Victoria, East Africa: the biology and evolution of a species flock. *Bull Brit Mus Nat Hist (Zool)* Suppl 6 1-134

_____ 1981a Species flocks and explosive evolution. 61-74 P H Greenwood, P L Forey eds. *Chance, Change, and Challenge — The Evolving Biosphere.* Cambridge Univ Press

_____ 1981b *The Haplochromine Fishes of the East African Lakes.* Cornell Univ Press. Ithaca NY

Greenwood P H, C D N Barel 1978 A revision of the Lake Victoria *Haplochromis* species (Pisces, Cichlidae), Part VIII. *Bull Brit Mus Nat Hist (Zool)* 33:141-192

Gregory W K 1933 Fish skulls: a study of the evolution of natural mechanisms. *Trans Amer Phil Soc* 23 Pt 2:75-474

Huxley J S 1932 *Problems of Relative Growth.* Methuen. London

Keast A 1978 Trophic and spatial interrelationships in the fish species of an Ontario temperate lake. *Env Biol Fish* 3:7-31

Liem K F 1973 Evolutionary strategies and morphological innovations: cichlid pharyngeal jaws. *Syst Zool* 22:425-441

_____ 1978a Modulatory multiplicity in the functional repertoire of the feeding mechanism in cichlid fishes, I. Piscivores. *J Morph* 158:323-360

_____ 1978b Adaptive significance of the intra- and interspecific differences in the feeding repertoires of cichlid fishes. *Amer Zool* 20:295-314

Liem K F 1979 Modulatory multiplicity in the feeding mechanism in cichlid fishes, as exemplified by the invertebrate pickers of Lake Tanganyika. *J Zool Soc London* 189:93-125

Liem K F, P H Greenwood 1981 A functional approach to the phylogeny of the pharyngognath teleosts. *Amer Zool* 21:83-101

McMahon T A 1975 Allometry and biomechanics: limb bones in adult ungulates. *Amer Nature* 109:547-563

Oijen M J P van, F Witte, E L M Witte-Maas 1981 An introduction to ecological and taxonomic investigations on the haplochromine cichlids from the Mwanza Gulf of Lake Victoria. *Neth J Zool* 31:149-174

Otten E 1981 Vision during growth of a generalized *Haplochromis* species: *H. elegans* Trewavas 1933 (Pisces, Cichlidae). *Neth J Zool* 31:650-700

Oxnard C E 1973 *Form and Pattern in Human Evolution: Some Mathematical, Physical, and Engineering Approaches.* Univ Chicago Press

Riedl R 1978 *Order in Living Systems.* Wiley. New York

Smith C L 1978 Coral reef fish communities: a compromise view. *Env Biol Fish* 3:109-128

Strauss R E, F L Bookstein 1982 The truss: body form reconstructions in morphometrics. *Syst Zool* 31: 113-135

Sweet S S 1980 Allometric inference in morphology. *Amer Zool* 20:643-652

Weijs W A 1980 Biomechanical models and the analysis of form: a study of the mammalian masticatory apparatus. *Amer Zool* 20:707-719

Wilhelm W 1980 The disputed feeding behaviour of a paedophagous haplochromine cichlid (Pisces) observed and discussed. *Behaviour* 74:310-323

Winterbottom R 1974 A descriptive synonomy of the striated muscles of the Teleostei. *Proc Acad Nat Sci Phila* 125:225-317

Witte F 1981 Initial results of the ecological survey of the haplochromine cichlid fishes from the Mwanza Gulf of Lake Victoria (Tanzania): breeding patterns, trophic and species distribution. *Neth J Zool* 31:175-202

EFFECTS OF SEXUAL SELECTION AND LIFE HISTORY ON SPECIATION: SPECIES FLOCKS IN AFRICAN CICHLIDS AND HAWAIIAN *DROSOPHILA*

WALLACE J DOMINEY

INTRODUCTION

Interest in mating systems and sexual selection has increased dramatically in the last decade. One of the most important recent developments is a growing appreciation of the role of sexual selection in speciation (Lande 1981; West-Eberhard 1983a; Thornhill, Alcock 1983). Combined with population genetics (Templeton 1980,1981), sexual selection theory shows particular promise in explaining the origin of "species flocks": clusters of closely related species endemic to narrow geographic areas. Such arguments have already been applied to the Hawaiian *Drosophila* (Ringo 1977; Carson 1978; Lande 1981), but have not yet been extended to the cichlid fishes of the Great Lakes of Africa. This is not to diminish the importance of repeated suggestions of the potential role of parental care and mating systems to speciation in these fishes (Kosswig 1947, 1963; Fryer 1959, 1977; Lowe-McConnell 1959, 1969; Fryer and Iles 1972), nor explicit comparisons of the African cichlids to the Hawaiian *Drosophila* (Fryer and Iles 1972; Fryer 1977; Greenwood 1981a).

First, I will review several aspects of population genetics, mating systems, and sexual selection. I then discuss parallels in the life histories, mating systems, and ecological circumstances of the Hawaiian *Drosophila* and the African Great Lakes cichlids which might account for the enormous proliferation of species in these groups. Finally, I suggest the amphipods of Lake Baikal as a suitable test case for the ideas presented here. These three distantly related groups were chosen because they are the most impressive of the known examples of species flocks.

POPULATION GENETICS

Speciation is intimately related to the genetic differentiation of populations (Templeton 1981). First, species must represent different gene pools with the corollary that most of the organic diversity we observe is based on genetic differences between species. Regardless of their degree of morphological differentiation, ecophenotypes and developmental stages are not recognized as separate species because they are part of the same gene pool.

Second, speciation occurs by the splitting and genetic differentiation of lineages. Selection, mutation, and drift are potentially differentiating forces, while interbreeding (gene flow) acts to genetically homogenize diverging populations. A balance between these forces exists. For instance, strong selection can differentiate populations even when considerable gene flow occurs, while in the absence of selection even a few interbreeding migrants can prevent genetic differentiation (Endler 1973). In general, reduced intrinsic tendency to disperse, as well as environments that isolate local populations should favor speciation by reducing interbreeding of genetically dissimilar individuals (migrants).

Third, the most important step in the divergence of populations into recognizable species is reproductive isolation. This has long been recognized and was formalized with the biological species concept (Mayr 1963). Sympatric populations which fail to interbreed freely are more or less assured of separate future evolutionary fates because they are unlikely to develop traits which promote interbreeding in the future. It is this separateness or independence of evolutionary lineage which characterizes a species, and which permits genetic divergence.

Allopatry is a form of forced reproductive isolation. Populations in allopatry may, but will not necessarily differentiate genetically (Ehrlich, Raven 1969). It is important to separate the events producing reproductive isolation (historical or genetic) from the genetic, behavioral, and morphological differentiation which may follow such isolation. In the case of allopatry, and perhaps in most cases, reproductive isolation permits, but does not cause divergence.

The amount of genetic differentiation necessary to ensure reproductive isolation between groups cannot be determined *a priori*. Failure to interbreed freely may result from fixation at a single locus or at hundreds of loci. Obviously, genetic changes affecting mating behavior will be crucial, and minor changes in the mating system could effectively isolate sibling species, whereas populations from widespread species might show numerous genetic differences but still interbreed freely.

There is some controversy regarding the amount and type of genetic change which normally occurs before, during, and after speciation. Classically, the accumulation in allopatry of changes to the genome in response to changed ecological conditions was considered to be the first step in speciation (see West-Eberhard 1983a). Such a process can be called *adaptive* speciation (Ringo 1977) because adaptation to differing ecological circumstances is the causal agent. Premating isolating mechanisms evolve either as by-products of this adaptive change (Mayr 1963), or may result from the ecological inadequacy of the hybrids following recontact (Dobzhansky 1970). The alternative view, *non-adaptive* speciation, suggests that premating isolation evolves in allopatry, but as a direct consequence of sexual or social selection rather than as a by-product of adaptation to changed ecological conditions (Paterson 1978; Thornhill, Alcock 1983; West-Eberhard 1983a).

THE MATING SYSTEM

When, where, and how mating takes place comprise the mating system of a species. As shown below, each of these factors can initiate speciation. The spring and fall spawning races of many fishes (Svardson 1953; Smith, Todd 1984) demonstrate the reproductive isolating capacity of timing of reproduction. New genetic variants breeding at different times might be instantaneously reproductively isolated. Indeed, the genetic differentiation necessary to produce temporal isolation can be small. Single allele differences at two loci provide an effective temporal reproductive barrier between sibling species of lacewing, *Chrysopa* (Tauber, Tauber, Nechols 1977). Colonization of tropical lacustrine or Pacific island environments by seasonally breeding organisms might provide opportunities for allochronic speciation.

The location of breeding activity can also promote reproductive isolation. In a polymorphic South American cichlid, *Cichlasoma citrinellum*, the gold morph breeds in deeper water than the dark morph, resulting in considerable assortative mating. McKaye (1980) proposed this as a potential example of ongoing sympatric speciation due to spatial segregation of breeding sites. Cichlids breeding at different depths and partially reproductively isolated might evolve different courtship signals, perhaps in response to different visual environments.

In addition to the timing and location of breeding, purely ethological differences can result in reproductive isolation. Paterson (1978) discussed the importance of changes in the specific mate recognition system (SMRS) to speciation. The SMRS can be defined as the coadaptation of male and female behavior by which mates are recognized and mating is accomplished. Any change in the SMRS in an isolated population could enforce reproductive isolation upon recontact.

Examples of changes in SMRS which could lead to reproductive isolation are well known in *Drosophila*. Mutations altering courtship behavior have been induced (Schilcher 1977), and artificial selection on courtship characteristics has been successful (Wood, Ringo 1982). In addition, selection in other contexts can pleiotropically alter courtship behavior. Fourteen generations of selection for mating in darkness in *Drosophila subobscura* produced males which no longer gave the typical courtship dance. Rather, they located females by "tapping" and then attempted forced copulation (Pinsker, Doschek 1980). Females no longer required the courtship dance to be receptive. Another case involved selection for small body size in *Drosophila melanogaster* (Ewing 1961). In addition to small body size, males showed a quantitative increase in the number of wing vibrations during courtship similar to the differences that exist between related but sexually isolated species. Both cases indicate that selection in a new socioecological context can rapidly alter the SMRS.

SEXUAL SELECTION

Darwin (1871) proposed both natural and sexual selection as mechanisms of evolutionary change. Sexual selection is defined as differential mating success due to competition within one sex for mates (usually male-male competition), or choice by members of one sex for members of the other (usually female choice). Genetic variants which differ in mating characteristics, such as when, where, or how mating takes place may be selectively favored or disadvantaged in mate competition or mate selection.

Sexual selection is responsible for the elaborate courtship displays, bright colorations, and otherwise useless structures found in many species. Such characteristics are often assumed to be deleterious under strictly natural selection, but to confer a counterbalancing advantage due to sexual selection. Sexual selection has special significance with respect to speciation because sexually selected traits are likely to diverge rapidly, and when they diverge they should be particularly effective at facilitating reproductive isolation.

West-Eberhard (1983a) argues convincingly that sexually selected traits, and other socially competitive traits, are more likely to undergo rapid divergence in allopatry than are nonsexually selected traits. She makes the following points: 1) Sexual and social competition exert relentless and potentially strong selective pressures. Thus, their capacity to produce evolutionary change is great. 2) Changes in sexually selected traits are often relatively arbitrary with respect to the physical environment. Thus, there are many possibilities for change, often unpredictable change. A corollary is that differences in the allopatric environment are not necessary to trigger

changes in sexually selected traits. This is not to suggest that environmental change, e.g., in the acoustic or visual environment, cannot trigger sexual selection. 3) Many aspects of sexual selection are coevolutionary in nature. In a coevolutionary contest, the goal is to surpass an opponent rather than to "perfect" an adaptation. Thus, rapid escalation without natural limit, except by natural selection in other contexts, can occur. Also, because success may depend on the traits of others, new initial group compositions may alter selective pressures, and hence lead to new bouts of evolution. 4) There may be selection for novelty *per se*, such as the rare male effect in *Drosophila* (Spiess 1982) in which females prefer atypical males as mates. Rare-male advantage may lead to the establishment of different rare mutants in different allopatric populations, thus altering the socioecological environment which may in turn cause sexual selection. Also, the high variance in male mating success typical of sexually selected species will increase the impact of successful variants on the evolution of sexually related characters (Thornhill, Alcock 1983; Lowe-McConnell 1959). Finally, the spread of successful variants within a population may create a new group composition and produce new sexual selection pressures; thus change may be self-sustaining (see also West-Eberhard 1979). Because changes in sexually selected traits will often directly affect assortative mating, such changes should have a high potential for ensuring reproductive isolation upon recontact with parental populations.

In addition to the general tendency of sexual selection to promote rapid and critically important divergence, three specific interactions between sexual selection and speciation can be identified. 1) Sexual selection can maintain intraspecific variation in mating behavior and morphology which can later contribute to rapid speciation in allopatry (West-Eberhard 1983b). 2) Following founder events, sexually selected traits can undergo genetic reorganizations which can rapidly produce reproductive isolation effective upon recontact (Carson 1978). 3) Female choice of male secondary sexual characteristics and the male characteristics themselves can undergo "runaway" accelerating selection (Fisher 1958) or drift rapidly in isolated populations which can result in reproductive isolation upon recontact (Lande 1981).

Sexual selection can maintain within-population variability including obvious examples such as alternative mating tactics (Dominey 1984), color polymorphisms (Endler 1980), and "rare male" genotypes (Spiess 1982). Alternative mating tactics in which males "sneak" spawnings rather than compete with territorial males occur in a lekking Malawi cichlid, *Cyrtocara eucinostomus* (McKaye 1983a). Such tactics indicate the intensity of sexual selection (Dominey 1980). Color morphs occur in many fishes, usually resulting from a balance between sexual selection and natural selection (Endler 1980). Color varieties in two African Great Lakes cichlids, *Pseudotropheus zebra* and *Petrotilapia tridentiger*, show significant differences in allelic frequencies at polymorphic loci (McKaye et al. 1982). Assortative mating of color forms may indicate that they currently have reached sibling species status (Holzberg 1978; Schroder 1980; Marsh et al. 1981). Lewis (1982) suggests that color varieties of *Labidochromis, Aulonocara, Lethrinops,* and *Pseudotropheus* are also sibling species.

West-Eberhard (1983b) hypothesized that sexually selected polymorphisms and other polymorphisms, such as the feeding morphs of some Mexican cichlids (Sage, Selander 1975), may represent the variation necessary to promote rapid speciation. West-Eberhard suggests that once a population is isolated in changed ecological circumstances, one or the other morph might be eliminated. This would lead to ecological or physiological release from the absent morph, and consequently to rapid evolution for perfection of the existing form. The divergence might be sufficient to insure reproductive isolation upon recontact. An alternative scheme utilizing the variability maintained by sexual selection would be disruptive selection on morphs resulting in sympatric speciation (Maynard Smith 1966).

Sexual selection can maintain the variability necessary for polymorphic speciation, but may also produce traits prone to genetic revolutions (Mayr 1963) following founder events. Founder events occur when small populations become geographically isolated from parental populations. Such populations are likely to have atypical frequencies for many genes (genetic drift). The complex equilibrium maintained by sexual selection and natural selection may make sexually selected traits particularly subject to genetic reorganization following this type of perturbation to the genetic system.

Initial close inbreeding followed by the fixation of some alleles and the selection for alleles with positive effects in homozygous rather than heterozygous condition may also contribute. Carson (1978) proposed genetic reorganization to explain the proliferation of species in Hawaiian *Drosophila*. Templeton (1980) outlined the basic population genetics models with predictions on when genetic reorganizations are expected. Even if sexually selected traits were no more subject to genetic reorganizations than other traits, when sexually selected traits do undergo reorganization the likelihood of reproductive isolation would be high because such traits directly affect differential mating.

Experimental evidence on reproductive isolation by genetic reorganization is available. *Drosophila pseudoobscura* populations achieved some degree of reproductive isolation following founder-flush cycles in the laboratory (Powell 1978). Controls indicated that both founder (inbreeding) and flush (rapid proliferation) periods were necessary. Similarly a laboratory population of the Hawaiian *Drosophila silvestris* which underwent several severe population reductions became partially ethologically isolated from stock derived from a natural population (Ahearn 1980).

In addition to polymorphic speciation and genetic reorganization, a third specific interaction between sexual selection and speciation can be identified. Male mating characteristics and female preferences for such characteristics are subject to accelerating "runaway" sexual selection (Fisher 1958). At equilibrium these traits are prone to rapid genetic drift and destabilizing selection following shifts away from equilibrium (Lande 1981). There is an equilibrium phenotype for any male trait (advantageous only in sexual selection and disadvantageous in natural selection) and any set of female choice characteristics (Lande 1981). Lande's models suggest that a "line of equilibria" exists in the absence of direct selection upon female choice (e.g., males must not provide parental care). The male trait can then drift rapidly in reproductively isolated populations along this line of equilibria. Also, following any perturbation of the equilbrium, as by the isolation of a small population with its unique genetic configuration, the return to equilibrium would likely take the isolated population to a position away from the parental population. At the new equilibrium the particular configuration of male and female traits might be sufficiently distinct to result in reproductive isolation upon recontact.

Perhaps even more interesting are cases of unstable equilibria in which perturbations from the equilibria result in "runaway" selection with the isolated population evolving away from genetic equilibrium, and cases of speciation along clines (Lande 1981, 1982). In all models, Lande assumes continuous heritability (due to mutations) in both the male trait and female choice characteristics. The model applies only to cases in which males provide no parental investment except for gametes and females do not benefit directly from superior male (parental) genotypes. Otherwise, there is direct selection on female preference, violating the necessary condition that female preferences evolve only as a correlated response to selection on males. Arnold (1983) provided a clear discussion of Lande's models.

Predictions. From the above arguments one expects proliferation of species in groups with 1) complex and labile specific mate recognition systems; 2) sexual selection; and 3) low powers of dispersal coupled with environments which provide opportunities for founder populations to develop in isolation. Examination of the Hawaiian drosophilid flies and the African Great Lakes cichlids supports these predictions.

THE HAWAIIAN *DROSOPHILA*

The endemic Hawaiian drosophilid flies are the largest group of animals in Hawaii, and represent the most remarkable species flock known among invertebrates. The number of endemic species may reach 800 (Carson, Kaneshiro 1976). Compared with continental drosophilids, there is extreme morphological variety in the Hawaiian flies, including stalk eyes, spoon-shaped tarsi, long and prominent bristles, bright colorations and elaborate patterns on wings and bodies (Ringo 1977). Much of this morphological variation is present only in males, and male dimorphisms are invariably related to male courtship behavior (Spieth 1974a).

The primitive mating pattern in dipterans is copulation in flight, while the most common derived pattern is copulation on substrates without preliminary courtship (Spieth 1974b). Some dipteran families, however, including Drosophilidae, have independently evolved more complex courtship, including male display prior to copulation. The courtship behavior of *Drosophila* includes species-specific movements of various body parts, and tactile, auditory, and olfactory cues (Spieth 1974b). The endemic Hawaiian *Drosophila* carry this trend towards courtship complexity to an extreme beyond that of continental species (Spieth 1974a). The Hawaiian endemics are also characterized by territoriality, stereotyped agonistic displays, fighting, social dominance, and intricate dance-like communal displays (Ringo 1977).

Many of the Hawaiian *Drosophila* have elaborate lek mating systems in which males aggregate and hold small territories. Males then display visually and chemically to attract females (Spieth 1974b). Leks promote sexual selection, exaggerated male characteristics, and female choice. Because of the intensity of sexual selection, and the resultant complexity of morphological and behavioral variation, microgeographic isolation on mountaintops apparently has permitted the evolution of new species of Hawaiian *Drosophila* (Spieth 1974a).

In general, *Drosophila* are weak fliers and thus weak dispersers. Also, many local populations occur in "kipukas", islands of mature forest surrounded by new lava which reduces dispersal (Carson 1978). The Hawaiian islands themselves are powerful isolating factors.

Female Hawaiian *Drosophila* oviposit away from mating sites, and neither sex provides parental care. About 75 percent of the Hawaiian drosophilids are monophagous, using a single host family as the larval substrate (Carson, Kaneshiro 1976). Patchy distribution of host plants may lead to the isolation of local populations as may selection of novel lekking sites (Ringo 1977).

Of great interest given the remarkable diversity and species richr s of the endemic Hawaiian *Drosophila* is their apparent monophyly. The entire group may be derived fom a single gravid female (Carson 1978). The rapidity of speciation in these flies can be deduced because their evolution is temporally constrained by the age of the islands. The largest and youngest island, Hawaii, has a diverse endemic *Drosophila* fauna and probably is not over 700,000 years old, while the oldest island is about 6,000,000 years old (Carson 1978).

AFRICAN GREAT LAKES CICHLIDS

The hundreds of species of endemic cichlid fishes of the African Great Lakes are the most remarkable vertebrate species flocks known (Fryer, Iles 1972). More than half of the world's cichlids occur in these lakes. The three largest Great Lakes, Victoria, Tanganyika and Malawi, have more species of fishes than are found in any other lakes. Most are endemic and most are cichlids. By far the largest radiation has occurred in *Haplochromis* and *Haplochromis*-derived cichlids.

Cichlids are well known for complex courtship behavior (Baerends, Baerends van Roon 1950). Visual signals including color pattern are emphasized in cichlid courtship and aggression in ways not remotely approached by other fishes of the African Great Lakes (Fryer 1977). Divergence in male breeding coloration undoubtedly has played a special role in cichlid speciation (Fryer, Iles 1972). One repeatedly finds distinctive male breeding colors characterizing morphologically similar species, as in the large flock of Lake Victoria *Haplochromis* (Greenwood 1965). Male breeding coloration in some species such as *Tropheus moorii (Fryer 1977), Lebeotropheus, Lethrinops, Pseudotropheus,* and *Petrotilapia* (Lewis 1982) varies geographically, indicating the potential for local isolation to lead to race formation. Many color varieties actually may be sibling species (Lewis 1982).

The apparent absence of naturally occurring hybrids in African Great Lakes cichlids (Fryer, Iles 1969, 1972; Greenwood 1974) is striking. Investigators repeatedly suggest near complete mating fidelity even among color varieties. This is surprising given the large number of closely related, sympatric species,

and contrasts with the situation in many other freshwater fishes where hybridization is common, albeit often due to introductions. If lack of hybridization among members of cichlid species flocks is upheld it might suggest a unique mechanism of speciation. Alternatively, the potentially long time these cichlids have coexisted in stable tropical environments might explain the apparent completeness of reproductive isolation.

The rapidity with which breeding coloration can change in allopatry is indicated by the divergence of Lake Nabugabo *Haplochromis* from their recent (4000 years) and morphologically similar ancestors in Lake Victoria (Greenwood 1965). Such divergence in male coloration could not be due to selection for reproductive isolating mechanisms against closest relatives, because recontact with the putative parental species in Lake Victoria has not occurred. Greenwood (1965) suggests that upon such contact, the Lake Nabugabo *Haplochromis* would behave as biological species. If so, this would be an impressive example of rapid, apparently nonadaptive speciation.

Many African cichlids are lekking species. As in lekking Hawaiian *Drosophila*, male courtship is intense, ritualized and exaggerated (Fryer, Iles 1972) but brief (Lowe-McConnell 1959). Males compete vigorously for display sites (nests) and alternative mating tactics occur (McKaye 1983a). Behaviors resulting in complex changes in nest appearance can differentiate sibling species, as in the related forms of *Tilapia macrochir* (Fryer, Iles 1972). Nest characteristics may also be subject to "runaway" sexual selection, as in *Cyrtocara eucinostomus* in which females select males guarding the largest nests (McKaye 1983b).

In lekking species, males do not provide parental care. All endemic cichlids from Lakes Malawi and Victoria are female mouthbrooders (Fryer, Iles 1972) without male parental care. In some endemic species of Lake Tanganyika, most notably a large species flock of *Lamprologous,* and in certain nonendemic species, the eggs are not collected in the mouth by the female, but are left on the substrate and guarded by both parents, typically in holes or crevices (Fryer, Iles 1972). In these species, sexual selection may be of reduced intensity, but may still be a potent evolutionary force. Recent field studies show that in some species such as *Lamprologous brichardi*, males are not strictly monogamous, but rather guard territories in which one or more females raise broods (Taborsky, Limberger 1981). African hole-brooders are more similar to mouthbrooders than to open substrate guarders in having reduced egg number, more sexual dichromatism and a loosening of the pair bond (Fryer, Iles 1972). These trends favor the mechanisms of speciation proposed here.

Even in truly monogamous species mate selection may play a role in speciation. Kosswig's (1974) original conception of the importance of mate choice to speciation in African cichlids dealt with biparental monogamous species. He suggested that individual mating preferences might lead to the isolation of family groups. Perhaps more likely, microgeographic divergence in the complex and protracted courtship of these species (Fryer and Iles 1972) may have led to speciation.

Because in most monogamous African cichlids both sexes are brightly colored, an alternative hypothesis is suggested (West-Eberhard 1983a). Social rather than sexual competition may be responsible for both the rate of speciation and for the maintenance and divergence of monomorphically brilliant species-typical colorations. That both members of pairs defend spawning sites which, as discussed below, may be limiting resources is consistent with the possible social significance of bright monomorphic coloration. A similar argument may explain the relatively rapid rate of obviously allopatric speciation among Central American cichlids (*Cichlasoma*), which are typically biparental monomorphic species. These fishes invaded the depauperate fish faunas of Central America from South America during the late Pliocene and have produced nearly a hundred species with high endemism and a low degree of sympatry (Barlow 1974).

The different history of Lake Victoria from the other two Great Rift Lakes must be considered in discussing cichlid speciation (Fryer, Iles 1972). Lake Victoria probably formed and re-formed as a series of shallow basins, permitting true allopatric speciation. Formation of an endemic cichlid fauna in Lake Nabugabo

following its recent separation from Lake Victoria is an example of this process. In contrast, species diversity in Lakes Tanganyika and Malawi may require an intralacustrine model (Fryer, Iles 1972), but even here temporary shore pools occur. In one year, five generations of *Haplochromis burtoni* bred near Lake Tanganyika in an isolated pool (Fernald, Hirata 1977). The high frequency of "barless" male color morphs indicates that biologically meaningful genetic changes can occur in these pools. Such shallow pools may be too transient to account for the evolution of new species, however, and are not likely to account for speciation of deepwater forms.

The isolation of local populations in Lakes Tanganyika and Malawi probably has played an important role in intralacustrine speciation (Fryer, Iles 1972). A weak intrinsic tendency to disperse and intralacustrine barriers to dispersal favor microgeographic isolation. Cichlids generally are weakly dispersing fishes, due to their small size, fully lacustrine breeding habits, low fecundity, and mouthbrooding or hole-nesting habits. In contrast, most noncichlids in the lakes are large, highly fecund fishes which spawn in rivers and streams and move widely even during the non-breeding season (Greenwood 1974). None are mouthbrooding.

The highly developed parental guarding of young is a key feature which reduces dispersal in cichlids compared with noncichlids. In mouthbrooding species, few young are produced and they are released when fully formed, eliminating the larval dispersal so characteristic of most fishes. Similarly, the speciose hole-nesting, non-mouthbrooding cichlids of Lake Tanganyika have relatively small clutches and apparently weak dispersal of young. The young of some species even remain with their parents and aid in the rearing of a second generation (Taborsky, Limberger 1981). Such "helping" behavior may indicate severe shortage of new breeding sites and extreme philopatry with offspring eventually replacing parents at natal sites (Dominey, Blumer 1984).

Even within mouthbrooding cichlids the importance of weak dispersal in promoting speciation is indicated. The impressive difference in species richness of mouthbrooding *Tilapia* and the speciose *Haplochromis*-

derived mouthbrooding cichlids may be due to the greater dispersal of *Tilapia* (Lowe-McConnell 1959; Fryer, Iles 1969). Of the African Great Lakes, only Malawi has a flock of *Tilapia*, and this flock is small (Lowe-McConnell 1959). Other small flocks exist elsewhere, e.g. in the small crater lake in Cameroon, Barombi Mbo (Trewavas, Green, Corbet 1972), but *Tilapia* nowhere have produced flocks on the scale of the haplochromines.

Tilapia are generalist colonizing species which often undergo seasonal intralacustrine breeding migrations. Female *Tilapia* bearing young have been observed to disperse many kilometers after spawning (Lowe-McConnell 1959; Fryer, Iles 1972). *Tilapia* are generally larger species, and produce larger numbers of less well developed offspring which in turn are more likely to disperse. As *Tilapia* develop they tend to move from one ecological zone to another while haplochromines remain in their natal ecological zone (Lowe-McConnell 1959). The greater tendency of young and adult *Tilapia* to disperse may have reduced their tendency to form new species, although much better information is needed on dispersal.

In addition to weak dispersal, an environment which favors isolation of local populations should be conducive to speciation. In Lakes Malawi and Tanganyika long stretches of rocky and sandy substrates occur in alternating fashion along the shoreline. Many species, even those in deeper water, are closely associated with one or the other substrate, and thus adult dispersal across substrates may be severely restricted (Fryer 1959, 1977). Models of isolation-by-distance and speciation along clines (Endler 1977) may apply to speciation along the hundreds of kilometers of shoreline in each Great Lake. Rocky islands also isolate distinct local populations (Fryer, Iles 1972). Indeed, the distributions of some African cichlids are restricted geographically within a lake, while other species show geographic variation, in some cases apparently reaching sibling species status (Lewis 1982). As hypothesized (Fryer, Iles 1972), it has been shown that changes in the water level of Lake Malawi can break up continuous habitats and create new habitats as rocky and sandy areas are disrupted (McKaye, Gray 1984).

Barriers to dispersal are less easily imagined for some deepwater forms, although these species may be spatially isolated by temperature, light, or depth preferences. Changes in lake water depth provide very steep ecological gradients (Smith, Todd 1984), although the deepest waters of Lakes Malawi and Tanganyika are uninhabitable (Fryer, Iles 1972). Even ecologically sympatric deepwater species, however, may be effectively reproductively isolated by their breeding habits. For example, among planktivorous Malawian Utaka, which school together to feed, some species breed on sandy shores, others on rocky shores (Fryer, Iles 1972). Similarly, temporal segregation of breeding occurs, as between the closely related *Tilapia saka* and *T. squamipinnis* (Lowe-McConnell 1959), and *Haplochromis virginalis* and *H. mloto* (Fryer, Iles 1972). Homing to restricted areas to breed (Lowe-McConnell 1959; Fryer, Iles 1972) may also isolate local populations even when these populations are ecologically sympatric. Unfortunately, the spawning habits of most deepwater species are not known.

As in the Hawaiian *Drosophila*, speciation and adaptive radiation in African Great Lakes cichlids has apparently been rapid, as evidenced by the low degree of genetic differentiation among species (Kornfield 1978). For example, the adaptive radiation of the 170 haplochromines in Lake Victoria (Greenwood 1981a) has occurred in the past 750,000 years. In contrast, although many families of noncichlid fishes were undoubtedly present in the developing lake, none produced species flocks. Likewise, the cichlids of Lake Victoria have undergone considerable trophic specialization (Greenwood 1974), whereas the noncichlids remain undifferentiated from their riverine counterparts (Corbet 1961).

Though striking when compared to the lack of change in noncichlids, the morphological diversity in Lake Victoria cichlids is actually not very great, given the potentially long history of the lake. Three quarters of a million years is a long period for evolution, even for fishes. Significant genetic changes can occur in hundreds of generations, let alone hundreds of thousands of generations. The entire history of modern teleosts, the largest and most diverse group of modern vertebrates, has occurred in the past 75-100 million years (Lagler et al. 1977). The possibility that Lake Tanganyika may be 6 million years old (Brooks 1950) means that the Tanganyikan cichlids potentially have had almost a tenth as long as the entire history of the teleosts to acquire their adaptive specializations. Lake Victoria cichlids probably have had an order of magnitude less time, but compared to the overall diversity of teleosts, the diversity of Lake Victoria cichlids is meager. The uniqueness of African Great Lakes cichlids may not rest so much on the rate of adaptation, or even the rate of speciation. What is truly unique is that the extensive speciation and adaptive radiation occurred in relatively small geographic areas (single lake basins), and resulted in the coexistence of extremely diverse and speciose monophyletic assemblages. That one family could so dominate the process on three separate occasions makes the situation even more extraordinary.

TEST CASE: LAKE BAIKAL AMPHIPODS

Fryer and Iles (1972) considered only the amphipods of Lake Baikal to represent greater intralacustrine speciation and diversity than the cichlids of the African Great Lakes. These amphipods comprise more than a third of the world's total species of amphipods (Kozhov 1963). Although hundreds of species in 34 genera are now recognized, the closeness of relationship among these endemic amphipods is suggested by the original placement of 97 of 98 species in a single genus, *Gammarus*. The Lake Baikal amphipods are extremely diverse trophically, with herbivores, including macrophyte eaters, carnivores, detritivores, and scavengers (Kozhov 1963). By all criteria the amphipods represent an extraordinary example of a species flock.

The importance of sexual selection to species flock formation indicated by comparing the Hawaiian *Drosophila* with the African cichlids would be supported if the amphipods also fit the patterns proposed. Unfortunately, little is known about the reproductive biology of the Baikal amphipods, but there are many suggestive similarities. In the genus having the greatest species diversity, *Elimnogammarus*, there is extreme interspecific color

variation including blood red, green, violet, pink, and other colors which often appear in complex patterns (Kozhov 1963). These color variations perhaps indicate complex mating systems and sexual selection, although other forms of social competition (West-Eberhard 1983a), interspecific competition or aposematism are untested alternative explanations.

As in the African cichlids and the Hawaiian *Drosophila*, breeding is year around with some species showing peaks during certain months. Also reminiscent of the African cichlids, females carry their young in brood pouches and give birth to developed offspring. It is extremely unlikely that males provide parental care. Adults of many species are closely associated with substrate, temperature, light, and pressure conditions and thus vertical (depth) habitat partitioning is a potentially important isolating mechanism (Brooks 1950). Though dispersal patterns are not directly known, their geographic distribution is very well known (Brooks 1950). The restriction of many species to confined geographic areas and narrow adaptive zones indicates weak dispersal. As in African cichlids, the bearing of young will limit clutch size and probably also limit the tendency of young to disperse upon release. The presence of only a single pelagic species (Brooks 1950; Kozhov 1963) provides added evidence of the potential importance of local population isolation to speciation. Study of mate choice and mate competition in these organisms, as well as in the Lake Baikal turbellarians, which are similar in many respects, is suggested. Similarities between the Baikal amphipods, and the African cichlids and Hawaiian *Drosophila*, in mating system and sexual selection would provide strong added support for the importance of sexual selection in species flock formation.

COMPETITION, DEPAUPERATE FAUNAS, AND ADAPTIVE RADIATION

Two contrasting features of both the Hawaiian drosophilid flies and the African cichlids deserve special mention. 1) The most recently diverged species share similar ecological requirements. 2) Within the species flocks as a whole, there is trophic diversity and specialization beyond the normal limits shown by both groups elsewhere. The first observation is predicted by rapid nonadaptive speciation, while the second observation is in part due to ecological/competitive release resulting from the presence of both groups in initially depauperate faunas.

The conventional "adaptive" hypothesis of speciation requires that closely related species differ ecologically (Ringo 1977). This is not the case with the Hawaiian drosophilid flies (Heed 1971) or the African cichlids (Fryer, Iles 1972) in which closely related forms usually have similar ecological adaptations. The pattern of speciation observed in African cichlids has been described as *cladistic gradualism* (Greenwood 1981a) in which closely related species show varying degrees of specialization, but belong to the same guild. This observation is consistent with rapid nonadaptive speciation, but not with an adaptive mechanism of speciation.

In contrast to the ecological similarity of closely related forms, the total trophic diversity found in both groups is unusual, including supralimital (Meyers 1960) specializations. The Hawaiian *Drosophila* (Carson, Kaneshiro 1976) and the African cichlids (Greenwood 1981a) occupy niches which they occupy nowhere else in the world. The cichlids have radiated trophically into zooplankton feeders, benthic pickers, algae scrapers, piscivores, mollusc crushers, insectivores, and specializations such as scale eating and paedophagy (Greenwood 1981a). Some Lake Tanganyika cichlids have converged on characteristics usually found only in other families of fishes as diverse as Blenniidae, Girellidae, and Percidae (Meyers 1960). Likewise, predatory, parasitic, nectivorous, detritivorous, and herbivorous Hawaiian *Drosophila* occur. Some Hawaiian *Drosophila* are so unusual that at first sight they scarcely resemble *Drosophila* to someone having experience only with non-Hawaiian Diptera (Hardy 1974).

Adaptive radiation has perhaps been promoted by the presence of both groups in initially depauperate faunas: newly formed lakes or islands. Two questions are particularly relevant. 1) Were the invading faunas actually depauperate? 2) What is the effect of a depauperate fauna on the production of a species flock? The insect fauna of the Hawaiian Islands is clearly an impoverished, disharmonic fauna, perhaps resulting from as

few as 250 original immigrants with 63 percent of the insect orders unrepresented (Zimmerman 1948). Of the Hawaiian insect species, excluding introduced species, 99 percent are endemic, indicating the high degree of isolation of this fauna (Zimmerman 1948). Carson and Kaneshiro (1976) argued for the importance of a depauperate fauna to the trophic radiation of the Hawaiian *Drosophila*. Barlow (1974) made a similar argument for the trophic diversity of Central American cichlids.

In contrast, Greenwood (1981a) and Fryer and Iles (1972) conclude that the noncichlid fauna of the African Great Lakes is not now depauperate, and that all the trophic and ecological niches provided by the developing lake could have been occupied by species from other families. I consider the original invading fauna to have been depauperate, however, because there were no species specialized for lacustrine existence, especially deepwater existence. Arguments as to the potential for members of other families to fill lacustrine niches are relevant to competition with cichlids in these niches if, and only if, species in other families successfully colonized the newly available lacustrine niches. If the youngest, most recently invaded large Great Lake, Lake Victoria, is any indication, the noncichlids were extremely slow to specialize in lacustrine niches (Greenwood 1974). The radiation of cichlid fishes into progressively deeper habitats, free from potential competitors, as suggested for ciscoes (Smith, Todd 1984) and the Lake Baikal fauna (Brooks 1950) has possibly played an important role in speciation and adaptive radiation. Thus, for the cichlids, competition was limited not by the scarcity of potential competitors, but by the failure of potential competitors to adapt to lacustrine conditions.

Given that both invading faunas were depauperate, what was the effect on species flock formation? There are four potential effects: 1) the effect on speciation (splitting of lineages); 2) the effect on continued reproductive isolation following recontact; 3) the effect on the survival of newly formed lineages; 4) the effect on adaptive radiation.

Presence in a depauperate fauna may or may not alter rates of speciation, depending on the particular mechanism of speciation. For instance, microallopatric speciation without a shift in adaptation may be little affected by presence in a depauperate fauna. In contrast, rates of speciation due to some mechanisms, such as competitive speciation (Rosenzweig 1978) may be greatly increased.

Competitive speciation requires high heritable phenotypic variance in trophic-related traits. In the absence of competitiors, and in new adaptive zones to which species are not primarily adapted, unusual phenotypes may not be selectively disadvantaged. As the colonizing species expands its niche space, intermediate phenotypes representing the parental form may be less able to utilize the available environment than the developing extreme phenotypes. Selection for homogamy of extreme phenotypes may follow. Sexual selection and the complexity of the mating system should play a role in the ease with which homogamy evolves. The evolution of homogamy cannot be taken for granted as the failure of the populations of noncichlids inhabiting the lakes to mate assortively and specialize for lacustrine conditions indicates. Although the cichlid mating system undoubtedly was essential to the evolution of homogamy, the cichlid mating system alone is insufficient to produce species flocks. This is shown by the failure of the ancestral cichlids to produce flocks in rivers. Thus, something like competitive speciation or some other basic ecological difference between lakes and rivers (Smith and Todd 1984), perhaps providing unusual chances for microallopatry, must also have been essential.

The second potential effect of a depauperate fauna is on the likelihood that the nascent species will remain reproductively isolated following recontact with its parental population. This effect is only relevant to cases of geographic or microgeographic speciation since a species evolving sympatrically is always in contact with its parental population. Populations initially isolated in depauperate faunas may tend to remain separate upon recontact more often than populations isolated with the usual complement of competitors. Ecological divergence may occur more readily in depauperate faunas. Then, upon recontact with the parental population, the ecological inadequacy of hybrids, according to the Dobzhansky model, would force the evolution of premating isolating mechanisms. Isolation in

a new adaptive zone to which the species is poorly adapted, regardless of the number of competitors present, would have a similar effect.

Third, presence in a depauperate fauna may increase the survival of species once formed. However, competition from species in other families should have less impact on species survival than competition with closely related species. The observed pattern of speciation, with clusters of closely related, ecologically similar species demonstrates that species formation and survival can outrun competitive exclusion.

Finally, depauperate faunas may promote adaptive radiation. Had the cichlids or the *Drosophila* invaded mature communities, I would expect much less trophic diversity. This requires only that a full-blown specialized competitor from another family can prevent divergence into its niche by a newly differentiating form. Imagine a newly formed species derived from an algal-scraping cichlid which has a tendency to inhabit deeper water, and to eat considerable numbers of insects. The presence of a deepwater insectivorous noncichlid might depress the fitness of individuals tending to eat insects below that of individuals tending toward algal scraping. Thus, further development of insectivory away from the primary adaptation would be prevented.

In contrast, in an insectivore-free environment a newly-formed species might continue to develop insectivory. Competition with newly recontacted and ecologically similar parental species should favor such divergence, although competition has apparently been insufficient to force divergence in many cases, as the coexistence of many closely related, ecologically similar species shows. Finally, by a purely stochastic process, a large number of independently evolving lineages will favor trophic diversity.

DISCUSSION

Lake Nabugabo provides a truly unique situation to examine the intrinsic tendency of African freshwater fishes to speciate following the cessation of gene flow. Here the sexual selection hypothesis simultaneously receives strong support and yet encounters a difficulty. The support is in the rapidity (4000 years) with which five of six Lake Nabugabo

Haplochromis diverged from putative parental species in Lake Victoria, primarily by changes in male breeding coloration (Greenwood 1965). This rapid divergence contrasts with the failure of the Lake Victoria noncichlid fauna, 75 percent of which occurs in Lake Nabugabo, to produce a single endemic species. This fauna includes members of Mormyridae, Lepidosirenidae, Characidae, Clariidae, Schilbeidae, Mochokidae and Mastacembelidae (Greenwood 1966).

The difficulty for the sexual selection hypothesis is that other cichlids found in Lake Nabugabo which are female mouthbrooding, lekking species with bright male colorations did not change: *Tilapia esculenta, T. variabilis, Astatoreochromis alluadi,* and *Hemihaplochromis multicolor.* The conventional argument concerning the failure of *Tilapia* to speciate in the African Great Lakes appeals to the greater intrinsic tendency of *Tilapia* to disperse when compared to haplochromines. However, this argument only holds for intralacustrine speciation, not speciation between lake basins. In a small shallow lake like Nabugabo (30 km²), intralacustrine speciation is unlikely, although one *Haplochromis* species pair is suspiciously similar.

Fryer (1959, 1977) has repeatedly suggested that the generalized nature of *Tilapia,* compared to *Haplochromis,* permits it to exist unchanged in many varied environments. Thus, both *Tilapia esculenta* and *T. variabilis* are found in the Victoria Nile, and are widespread in the Victoria drainage. Changed environmental conditions may produce genetic restructuring in specialized species more readily than in generalist species which are adapted to inhabiting a diversity of habitats. Lake Nabugabo, with its shallow depth (avg 1.5-4 m), alkaline water, and low salt concentration, provides a very different environment from Lake Victoria (Greenwood 1965). The Lake Nabugabo environment may have exerted stronger selective pressures on the Lake Victoria-specialized *Haplochromis,* even those specialized for life in shallow bays, than on *Tilapia.* As suggested by experimental evidence using *Drosophila* (Ewing 1961; Pinsker and Doschek 1980), components of the mating system may change pleiotropically during selection in other contexts.

That generalized cichlids are less likely to undergo genetic reorganizations in new environments than specialized ones gains further support upon examination of the three *Haplochromis*-like cichlids which did not speciate in Lake Nabugabo: *Haplochromis nubilus, Astatoreochromis alluaudi,* and *Hemihaplochromis multicolor.* All three have wide distributions in the Lake Victoria drainage including the Victoria Nile (Greenwood 1966). All three habitually enter rivers and streams, in contrast to all other haplochromines present in Lake Victoria (Greenwood 1974). Each species is generalized (Fryer, Iles 1972) with *Astatoreochromis alluaudi* sharing with *Tilapia* the colonizing-adapted capacity for stunting (Fryer, Iles 1972).

In contrast to the generalized haplochromines which did not speciate in Lake Nabugabo, the ancestors of Lake Nabugabo endemics were almost certainly lacustrine specialists. Of the 9 Lake Victoria *Haplochromis* suggested by Greenwood (1965) as ancestral to the Lake Nabugabo endemics, only one, *Haplochromis obliquidens,* has definite records outside Lake Victoria (Greenwood 1981b). In this case, the endemic Victoria *Haplochromis lividus* is given equal status as a potential ancestor (Greenwood 1965).

The absence or weakness of ecological selection in the new environment is one explanation for the failure of the generalist cichlids to speciate in Lake Nabugabo. Other potential explanations include gene exchange between populations in Lake Nabugabo and its inflow and associated swamps, or introductions by man (Greenwood 1965). A different type of explanation is also consistent with the observed pattern. Carson (1982) suggested that the mature "old" species may be locked into obligatory genetic balances which prevent them from producing new species. Generalist, fluvial-adapted gene complexes in cichlids should be much older than the relatively recent gene complexes likely to be responsible for lacustrine specializtion. Both the age of the gene complexes as well as their suitability for the Lake Nabugabo environment may have contributed to the generalist species' failure to produce endemics.

I have argued that complex courtship increases the probability of speciation following geographic or microgeographic isolation. Why should complexity *per se* have this effect? 1) Having complex courtship may put more of the genetic system in the "closed system" subject to genetic reorganizations (Carson 1978). 2) Complex courtship usually indicates the importance of courtship to successful reproduction. Thus, in species with complex courtship, changes in courtship should have a greater impact on reproductive isolation than changes of equal magnitude in species with simple courtship. For example, changes in coloration or wing movements in a dipteran which does not court, but mates forcibly, will be much less critical to reproductive isolation than those same changes in a typical *Drosophila*. The possibility that the courtship elements of complex courtships (important to reproductive success) will be less subject to change is a separate, but important issue. It may be that courtship complexity which results from the primitive species-recognition or mate synchrony functions will tend to be under strong stabilizing selection, and hence will be conserved in isolated populations, whereas courtship complexity due to sexual selection may be subject to rapid divergence. 3) The more complex the courtship the greater the number of potential changes which can occur, and which may produce reproductive isolation. The above arguments should apply to complexity in the SMRS and all other aspects of the mating system, not just courtship.

Courtship complexity as a cause of rapid divergence is somewhat unsatisfying. If sexual selection is the key to rapid change in mating characteristics, then the expense of courtship probably is even more relevant than courtship complexity. Expensive courtship implies some advantage in either mate competition or mate choice. Otherwise, individuals with less expensive displays should predominate. As argued throughout this paper, once sexual selection is established there are many possibilities for rapid divergence in sexual characteristics. In measuring expense all types of mating effort should be considered (sexual dichromatism, dimorphism, weaponry, fighting, nest construction, etc.), not just courtship behavior.

Probably not coincidentally, both the African cichlids and the Hawaiian *Drosophila* have mating systems conducive to courtship complexity and expense. Expensive male mating efforts predominate in the many lekking species of both groups. Any similarities between these groups purported to explain rapid speciation, however, must be judged according to the uniqueness of the cichlids and *Drosophila* when compared to the other groups also present which did not undergo explosive speciation. It is admittedly difficult to make a quantitative measure of either courtship complexity or mating effort. That ethologists have shown a disproportionate interest in cichlid and *Drosophila* courtship behavior, as opposed to other types of behavior, may be an indication of the obvious nature and importance of courtship and mating effort in these organisms.

Most of the members of the other families of fishes present in the African Great Lakes are dull in color, and not particularly sexually dimorphic or dichromatic. Some families have members which are more brightly colored, perhaps indicating social or sexual competition, as in the small flock of Mastacembelidae in Lake Tanganyika (Travers pers. comm.). Other species, like the mormyrids, may have considerable complexity in cryptic courtship signals. The electrocommunication of mormyrids is known to be used in courtship, and is sexually dimorphic and species-specific (Hopkins, Bass 1981). In these species, unrecognized cryptic species may exist based on differences in electric discharge. Alternatively, mormyrid courtship may be relatively simple and inexpensive. Taken as a whole, little is known about courtship and mating in African Great Lakes noncichlids although guesses can be made from the behavior of related forms elsewhere.

In addition to having relatively complex courtship or expensive male mating efforts, a further similarity between the endemic African Great Lakes cichlids and the Hawaiian *Drosophila* is the absence of male parental care. The only exceptions are in some Tanganyikan biparental species. Male parental care reduces the intensity of sexual selection by decreasing the variance in male mating success and equalizing the investment of the sexes in offspring. Also, in species in which males provide only gametic investment, the primary basis for mate choice may be the genetic traits of males. Lande's (1980, 1981) models of rapid divergence in sexually selected characters due to female preference require the absence of male parental care, and the potential importance of any model based on sexual selection is increased when male care is absent. *Drosophila* are typical insects in that neither sex provides parental care, but the pattern of female-only parental care in African cichlids contrasts sharply with the African noncichlids, many of which have male-only parental care, and with New World cichlids, most of which are biparental (Barlow 1974). Although sexual selection occurs in fishes without care or with male-only care, female-only care should increase the intensity of sexual selection.

No African Great Lakes noncichlid has female-only parental care. Eight families, including Cyprinidae, Lepidosirenidae, Gymnotidae, Clariidae, Citherinidae, Cyprinodontidae, Pantodontidae, and Tetrodontidae, have members without parental care or with male-only care. Anabantidae has members with male-only care or biparental care. Seven families, Clupeidae, Mastacembelidae, Centropomidae, Sheilbeidae, Mochokidae, Amphiliidae, and Kneriidae apparently do not give care. Four families, Malapteruridae, Polypteridae, Gymnarchidae, and Mormyridae, have some members which give care, but the sex of the care giver is not known (Blumer 1982). Judging from the typical teleost pattern, males in these latter four families probably provide care when care is given.

Excepting the Cichlidae, only three families present in the African Great Lakes are known to have any members (worldwide) with female-only care: Osteoglossidae, Characidae, and Bagridae (Blumer 1982). The only African osteoglossid, *Heterotis*, has male-only care (Breder, Rosen 1966). Likewise, female-only care in Bagridae is known only in non-African bagrids (Breder, Rosen 1966) which have structures for care-giving not found in African species. Care is apparently male-only in some African bagrids (Fryer, Iles 1972), whereas *Bagris meridionalis* is biparental (McKaye, Oliver 1980). In the two characid genera present, *Hydrocynus* and *Alestes*, parental care fecund *Hydrocynus*, unlikely. Thus the only known African Great Lakes fishes having female-only care are cichlids.

I have argued that sexual selection has accelerated speciation in both African cichlids and Hawaiian *Drosophila*. However, one might reasonably argue the reverse. Thus, rapid speciation, occurring for other reasons, might lead to the coexistence of many closely related species which in turn would select for exaggerated species isolating mechanisms.

I can offer two predictions from such a model which are not substantiated. 1) Cichlids or *Drosophila* found only outside species flocks should have less exaggerated male characteristics than those found inside flocks. Thus, the East African riverine and lake cichlids which do not occur in the Great Lakes, and thus coexist with many fewer congeners, should show less evidence of sexual selection. Although I have not looked at this systematically, it is not obvious (as it should be) that this occurs. In *Drosophila* there is evidence that mating characteristics are much more exaggerated in the Hawaiian flocks than in flies occurring elsewhere, but this may result from predation pressure (Spieth 1974a) or other ecological factors, rather than selection for strong species isolating mechanisms. Certainly the evidence in cichlids tends to support the notion that sexual selection is an accelerator of speciation and not the reverse. 2) Perhaps a stronger objection to viewing sexual selection as a consequence of speciation and subsequent species packing is that in situations like Lake Nabugabo one predicts stability (or atrophy) in sexually selected characters, rather than change. The flock members from Lake Victoria which became isolated in Lake Nabugabo should have already achieved adequate exaggeration of sexually selected characters to ensure coexistence, thus further change would be unnecessary. In contrast, rapid change occurred in sexually selected characters in those species isolated in Lake Nabugabo. Again, a model viewing sexual selection as a result of rapid speciation and species packing fails to predict the observed pattern.

In addition to sexual selection, I have argued that a weak intrinsic tendency to disperse contributed to species flock formation in both the Hawaiian *Drosophila* and African cichlids. The importance of weak dispersal results from the tendency of weakly dispersing organisms to form isolated populations. However, depending on the mechanism of speciation, weak dispersal may be irrelevant or even inhibitory. For instance, dispersal within a lake basin would not influence the rate of speciation between populations isolated in separate basins. Also, Templeton (1980) argues that founders from large panmictic populations are more likely to undergo genetic reorganizations than are founders from subdivided, highly structured populations. Small inbred populations have already undergone the reduction in genetic variability and selection for alleles in the homozygous condition which gives rise to genetic reorganizations. in founder populations. Even so, opportunities for speciation in allopatry or microallopatry must usually be fewer in a widely dispersing organism. The stronger tendency for isolation of local populations and of founder populations in a weakly dispersing organism may more than compensate for the decreased likelihood of genetic reorganization following each founder event. In the African cichlids, Hawaiian *Drosophila*, and Baikal amphipods, founder events themselves are likely because single females, brooding or gravid, can establish new isolated populations. The species richness of the weakly dispersing rock-dwelling cichlids (Mbuna) of Malawi versus the relatively species-poor, more widely dispersing *Tilapia* (Lowe-McConnell 1959; Fryer, Iles 1969), as well as the general tendency to consider the African cichlids and Hawaiian *Drosophila* as poor dispersers indicates that weak dispersal may promote, not retard explosive speciation. High vagility and panmixis may often prevent the production of daughter species, even in groups predisposed for rapid speciation due to the complexity of their mating system or sexual selection.

The trophic diversity achieved by the cichlids in the African Great Lakes is in part due to the absence of competition from lacustrine specialized competitors in other families. In feeding habits, ecology and behavior the Lake Victoria noncichlids have changed little from their fluviatile counterparts. The slowness with which noncichlids have adapted to lacustrine conditions is clearly shown by the paucity of deepwater noncichlids in the youngest Great Lake, Victoria. Nearly all the noncichlids are restricted to the shallow

littoral zone (Greenwood 1974). Only one noncichlid, *Xenoclarius*, is confined to deeper water, compared to 40 or more haplochromines (Greenwood 1974). The situation in the other two Great Lakes is consistent with a rate-limited model of noncichlid specialization. Lake Tanganyika has the oldest and best differentiated noncichlid fauna, including some endemic genera (Meyers 1960), while Malawi is intermediate both in age and in the development of its noncichlid fauna.

Why have the cichlids acquired lacustrine specializations more rapidly than noncichlids? Again differences in breeding habits seem critically important. The cichlid habit of breeding within lakes contrasts with most lake noncichlids which spawn in streams, rivers, or flooded grasslands (Corbet 1961). These fishes include most of the highly successful catfishes, Bagridae, Shilbeiidae, Clariidae, and Mochokidae, which typically enter lakes only at large sizes, following development of piscivorous habits (Corbet 1961). Also included are the cyprinids, *Barbus, Barilius,* and *Labeo,* and one of the two genera of characins, *Alestes.*

Mouthbrooding was undoubtedly an important preadaptation promoting lacustrine breeding in cichlids (Lowe-McConnell 1959; Fryer, Iles 1972; Fryer 1977). Although many non-mouthbrooding fishes elsewhere breed in lakes, as do some African noncichlids and non-mouthbrooding cichlids, mouthbrooding species would have many fewer adjustments to make for fully lacustrine breeding (Fryer 1977). Mouthbrooding frees the offspring from many environmental stresses and makes the choice of spawning substrate less critical to their survival. It is ironic that many mouthbrooding cichlids are still very selective about spawning sites, and that elaborate "nest" construction sometimes occurs.

The importance of lacustrine spawning to lacustrine specialization is that it promotes isolation of lacustrine populations from their ancestral fluvial-adapted counterparts. Spawning alongside river-dwelling conspecifics ensures gene flow between lacustrine and fluvial-adapted individuals which prevents lacustrine specialization. Very different adaptations are needed for success in lakes versus rivers. Rivers are generally unstable, seasonally fluctuating environments which require many generalist adaptations (Corbet 1961; Lowe-McConnell 1969; Fryer 1959).

Examples of adaptive change in noncichlids are invariably linked to lacustrine reproduction. *Bagrus docmac* and *Mormyrus kannume* in Lake Victoria do not undertake spawning migrations, and thus are believed to spawn in the lake (Corbet 1961). Both species were preadapted for lacustrine spawning because they spawn in wave-washed shores similar to those used in riverine habitats (Corbet 1961). Probably as a result of lacustrine spawning, they are the only members of their families present which are adapted for lacustrine conditions. As one measure of their success, these two fishes are the most important noncichlid food fishes in Lake Victoria (Corbet 1961). Consistent with the possibility of gene exchange between lacustrine and fluviatile populations in this species (Corbet 1961),*Bagrus docmac* is the less specialized of the two. Of all the mormyrids in Lake Victoria, only *Mormyrus kannume* is a lake specialist, feeding on mud bottoms (Corbet 1961). All other mormyrids utilize the lacustrine equivalents of riverine habitats. Ontogenetic changes in the shape of the head and snout are associated with the lacustrine specialization of *Mormyrus kannume.*

Another interesting example of lacustrine specialization occurs in the small, deepwater clariid catfish species flock, *Dinotopterus* of Lake Malawi. The closest relatives of *Dinotopterus* are in *Clarius*, and are restricted to shallow waters of lakes, rivers, and swamps where they are partially dependent on aerial respiration (Greenwood 1961). This dependence would prevent invasion of the deeper waters to which the *Dinotopterus* species flock is now confined. Specialization for deepwater habits has involved compensatory changes in the suprabranchial cavity and gills (Greenwood 1961), and probably was coincident with fully lacustrine spawning.

Fully lacustrine breeding is perhaps best regarded as a necessary condition for lacustrine specialization, but not a sufficient condition for species flock formation. Several noncichlids which have not produced species flocks have lacustrine breeding habits. In some cases, the same spawning habitat occurs in both lakes and rivers. Examples

include the Anabantidae, Lepidosirenidae, Polypteridae and some Mormyridae which breed in the marginal swamps of both lakes and rivers. This situation would promote gene exchange with fluviatile counterparts. Others, like *Engraulicypris* (Cyprinidae) and perhaps *Lates* (Centropomidae), have pelagic eggs or larvae, and thus their dispersal may be lakewide, reducing the tendency for intralacustrine speciation. The distinction between lacustrine adaptation and lacustrine speciation is clearly seen in these highly lacustrine-adapted forms which are species poor. Mastacembelidae provides perhaps the most interesting comparison with Cichlidae. A small flock of mastacembelids occurs in Lake Tanganyika, and because of their secretive habits, there may be many uncollected species. These highly colorful, relatively small, and probably sedentary fishes apparently scatter eggs in the rocks and crevices where they live. More should be learned of their mating system and dispersal pattern.

To sum, fully lacustrine breeding, weak dispersal, environments conducive to the isolation of local populations, and presence in a depauperate fauna all promoted adaptive radiation and the formation of species flocks in cichlids. With the exception of lacustrine breeding, the same factors were important in *Drosophila* radiation. However, all the invading organisms, not just cichlids and *Drosophila* entered depauperate faunas. Also, many insects are at least as sedentary as *Drosophila*, and several noncichlids spawned in lakes along with the cichlids. I must again emphasize that the African cichlids and the Hawaiian *Drosophila* are most distinct from their rivals in other groups in the complexity of their mating system, and most particularly in the intensity of sexual selection. In the case of the Lake Tanganyikan biparental cichlids, social (West-Eberhard 1983a) rather than sexual competition may have taken a leading role.

The African Great Lakes provide three relatively independent natural experiments on rates of speciation among African freshwater fishes. Many alternatives to the sexual selection hypothesis have been proposed as to why only the cichlids formed large species flocks. These include hypotheses as diverse as small body size, high mutation rates, and short generation time (reviewed in Fryer 1977).

The major competing alternative explanation not consistent with the sexual selection hypothesis is that the cichlid "bauplan" is a unique substrate for anatomical and dental specializations (Liem 1973; Greenwood 1981a) . However, Characidae and Cyprinidae show diversity equal to the African cichlids elsewhere, as do African mormyrids. Greenwood (1981a) attributes the failure of Characidae to produce such diversity to the specialized nature of the African species present. Likewise, the failure of Cyprinidae is attributed to their lack of jaw teeth and the fact that only the lower pharyngeal bones are toothed. Greenwood (1981a) mentions that cyprinids elsewhere compensate for these deficiencies by modifying mouth, lips, and pharyngeal teeth, although "in this family too it would seem reasonable to hypothesize a greater level of genetic reorganization than was required by cichlids." Speciation in the cichlids, however, apparently has taken place without adaptive shifts involving morphology. Nearest relatives often are morphologically and ecologically similar even though, when taken as a whole, a species flock represents considerable diversity. If Characidae or Cyprinidae had had an equal tendency to produce diverging lineages, I would expect to see a species flock of 20 closely related , endemic cyprinids similarly equipped and searching the same lake bottom together for the same foods, or a species flock of 15 piscivorous *Hydrocynon*. This we do not find. What we do find are dozens of closely related, presumably monophyletic cichlids, in one case quietly scraping the rocks for algae, in another case searching for fish prey. Although the suitability of the "bauplan" may explain in some way the total diversity achieved by cichlids, differences in disperal pattern, sexual selection, and mating system are seemingly much more important to explain why the potential diversity was tapped by the production of many independently evolving lineages at all. The strength of this explanation rests not only on logical arguments, but also on the surprising concordance with the situation observed in the Hawaiian *Drosophila*. Tests of this concordance should be made using the Lake Baikal amphipods.

SUMMARY

To explain explosive speciation and the production of species flocks, I have focused on how changes in the mating system can produce rapid reproductive isolation between diverging populations. By preventing interbreeding, reproductive isolation permits genetic divergence and, thereby, morphological, ecological and behavioral divergence. The formation of recognizable species follows.

Organisms with complex mating systems are predisposed to become reproductively isolated during allopatry or micro-allopatry. Sexual selection is a particularly potent force producing reproductive isolation. The coevolutionary nature of sexual selection leading to escalated contests, the potentially strong impact of successful variants, selection for novelty *per se*, and other features make divergence in mating characteristics likely to occur and rapid in isolated populations of sexually selected species. Social (West-Eberhard 1983a), rather than sexual competition, may be critically important in some cases, as in the monomorphically brilliant, biparental Lake Tanganyikan cichlids.

Three additional interactions between sexual selection and speciation are suggested. 1) Sexual selection can maintain variation which can later become the basis for "polymorphic speciation" following the fixation of one morph in a new environment. 2) Founder events may trigger genetic reorganizations in sexually selected traits. 3) Unstable equilibria or rapid drift in sexually selected traits may occur.

An examination of the two most impressive species flocks, the Hawaiian drosophilid flies and the African Great Lakes cichlids, supports the importance of sexual selection to species flock formation. The amphipods of Lake Baikal are proposed as a test case.

Other similarities between the cichlids and *Drosophila* which have contributed to species flock formation are discussed. These include a weak intrinsic tendency to disperse and presence in environments which promote isolation of local populations. The considerable trophic diversity represented by each species flock is attributed to the relative absence of potential competitors in the case of the Hawaiian *Drosophila*, and to the failure of potential competitors to adapt to lacustrine conditions in the case of the African cichlids.

The failure of noncichlids to adapt to lacustrine conditions is attributed to continued gene flow with fluvial-adapted populations.

ACKNOWLEDGMENTS

My interest in species flocks, and in the interaction between sexual selection and speciation, was stimulated by seminars run by G R Smith and by M J West-Eberhard and W G Eberhard, respectively, at the University of Michigan. I thank the organizers of these seminars, and the seminar participants, and R D Alexander, G W Barlow, P V Loiselle and the participants of the 1983 species flocks symposium at Tallahassee, Florida for useful discussion.

REFERENCES

Ahearn J N 1980 Evolution of behavioral reproductive isolation in a laboratory stock of *Drosophila silvestris*. *Experientia* 36:63-64

Arnold S J 1983 Sexual selection: interface of theory and empiricism. P P G Bateson ed. *Mate Choice* Cambridge Univ Press

Baerends G P, J M Baerends van Roon 1950 An introduction to the study of the ethology of the cichlid fishes. *Beh Suppl* 1

Barlow G W 1974 Contrasts in social behavior between Central American cichlid fishes and coral-reef surgeon fishes. *Amer Zool* 14:9-34

Blumer L S 1982 A Bibliography and categorizaton of bony fishes exhibitng parental care. *Zool J Linn Soc* 76:1-22

Breder C M, D E Rosen 1966 *Modes of Reproduction in Fishes*. Natural History Press. Garden City NY

Brooks J L 1950 Speciation in ancient lakes. *Q Rev Biol* 25:30-60, 131-176

Carson H L 1978 Speciation and sexual selection in Hawaiian *Drosophila*. 93-107 P F Brussard ed. *Ecological Genetics: The Interface*. Springer Verlag. New York

_____ 1982 Speciation as a major reorganization of polygenic balances. 411-433 *Mechanisms of Speciation*. Liss. New York

Carson H L, K Y Kaneshiro 1976 *Drosophila* of Hawaii: systematics and ecological genetics. *Ann Rev Ecol Syst* 7:311-345

Corbet P S 1961 The food of non-cichlid fishes in the Lake Victoria basin, with remarks on their evolution and adaptation to lacustrine conditions. *Proc Zool Soc Lond* 136:1-101

Darwin C 1871 *The Descent of Man, and Selection in Relation to Sex*. Murray. London

Dobzhansky T 1970 *Genetics of the Evolutionary Process*. Columbia Univ Press. NY

Dominey W J 1980 Female mimicry in male bluegill sunfish -- genetic polymorphism? *Nature* 280:546-548

_____ 1984 Alternative mating strategies and evolutionarily stable strategies. *Amer Zool* 24 (2)

Dominey W J, L S Blumer 1984 Cannibalism of early life stages in fishes. G Hausfater, S Blaffer Hrdy eds. *Infanticide: Comparative and Evolutionary Perspectives*. Aldine. Hawthorne NY

Ehrlich P R, P H Raven 1969 Differentiations of populations. *Science* 165:1228-1232

Endler J A 1977 *Geographic Variation, Speciation and Clines*. Princeton Univ Press

_____ 1980 Natural selection on color patterns in *Poecilia reticulata*. *Evolution* 34:76-91

Ewing A W 1961 Body size and courtship behaviour in *Drosophila melanogaster*. *Anim Behav* 9:93-99

Fernald R D, N R Hirata 1977 Field study of *Haplochromis burtoni*: habits and co-habitant. *Env Biol Fish* 2:299-308

Fisher R A 1958 *The Genetical Theory of Natural Selection*. Dover. New York

Fryer G 1959 Some aspects of evolution in Lake Nyasa. *Evolution* 13:440-451

_____ 1977 Evolution of species flocks of cichlid fishes in African lakes. *Z Zool Syst Evolutforsch* 15:141-165

Fryer G, T D Iles 1969 Alternative routes to evolutionary success as exhibited by African cichlid fishes of the genus *Tilapia* and the species flocks of the Great Lakes. *Evolution* 23:359-369

_____ 1972 *The Cichlid Fishes of the Great Lakes of Africa: Their Biology and Evolution*. Oliver Boyd. Edinburgh

Greenwood P H 1961 A revision of the genus *Dinotopterus* Blgr. (Pisces, Clariidae) with notes on the comparative anatomy of the suprabranchial organs in the Clariidae. *Bull Brit Mus Nat Hist (Zool)* 7:215-241

_____ 1965 The cichlid fishes of Lake Nabugabo, Uganda. *Bull Brit Mus Nat Hist (Zool)* 12:315-357

_____ 1966 *The Fishes of Uganda*. Uganda Society. Kampala

_____ 1974 Cichlid fishes of Lake Victoria, East Africa: the biology and evolution of a species flock. *Bull Brit Mus Nat Hist (Zool)* Suppl 6:1-134

_____ 1981a Species-flocks and explosive evolution. P H Greenwood, P L Forey eds. *Chance, Change, and Challenge: The Evolving Biosphere*. Cambridge Univ Press; Brit Mus Nat Hist. London

_____ 1981b *The Haplochromine Fishes of the East African Lakes*. Cornell Univ Press. Ithaca NY

Hardy D E 1974 Introduction and background information. M J D White ed. *Genetic Mechanisms of Speciation in Insects*. Reidel. Boston

Heed W B 1971 Hosts plant specificity and speciation in Hawaiian *Drosophila*. *Taxon* 20:115-121

Holzberg S 1978 A field and laboratory study of the behaviour and ecology of *Pseudotropheus zebra* (Boulenger), an endemic cichlid of Lake Malawi (Pisces; Cichlidae). *Z Zool Syst Evolut.-forsch* 16:171-187

Hopkins C D, A H Bass 1981 Temporal coding of species recognition signals in an electric fish. *Science* 212:85-87

Kornfield I L 1978 Evidence for rapid speciation in African cichlid fishes. *Experientia* 74:335-336

Kosswig C 1947 Selective mating as a factor for speciation in cichlid fish of East African Lakes. *Nature* 159:604-605

_____ 1963 Ways of speciation in fishes. *Copeia* 1963:238-244

Kozhov M 1963 Lake Baikal and its Life. *Monographiae Biologicae* 11:1-344

Lagler K F, J E Bardach, R R Miller, D R May-Passino 1977 *Ichthyology*. Wiley. New York

Lande R 1981 Models of speciation by sexual selection on polygenic traits. *Proc Natl Acad Sci USA* 78:3721-3725

_____ 1982 Rapid origin of sexual isolation and character divergence in a cline. *Evolution* 36:213-223

Lewis D S C 1982 Problems of species definition in Lake Malawi cichlid fishes (Pisces: Cichlidae). *J L B Smith Inst Ichthyol Spec Publ* 23:1-5

Liem K F 1973 Evolutionary strategies and morphological innovations: cichlid pharyngeal jaws. *Syst Zool* 22:425-441

Lowe (McConnell) R H 1959 Breeding behaviour patterns and ecological differences between *Tilapia* species and their significance for evolution within the genus *Tilapia* (Pisces: Cichlidae). *Proc Zool Soc Lond* 132:1-30

Lowe-McConnell R H 1969 Speciation in tropical freshwater fishes. *Biol J Linn Soc* 1:51-75

Marsh A C, A J Ribbink, B A Marsh 1981 Sibling species complexes in sympatric populations of *Petrotilapia* Trewavas (Cichlidae, Lake Malawi). *Zool J Linn Soc* 71:253-264

Maynard Smith J 1966 Sympatric speciation. *Am Nat* 100:637-650

Mayr E 1963 *Animal Species and Evolution.* Belknap. Cambridge Mass.

McKaye K R 1980 Seasonality in habitat selection by the gold color morph of *Cichlasoma citrinellum* and its relevance to sympatric speciation in the family Cichlidae. *Env Biol Fish* 5:75-78

_____ 1983a Ecology and breeding behavior of a cichlid fish, *Cyrtocara eucinostomus* on a large lek in Lake Malawi, Africa. *Env Biol Fish* 8:81-96

_____ 1983b Mate selection by cichlid females in Lake Malawi breeding arenas. *Abst Ethol Beh Ecol Fish Conf* Illinois State Univ. Normal

McKaye K R, W N Gray 1984 Extrinsic barriers to gene flow in rock-dwelling cichlids of Lake Malawi. 169-184 A A Echelle, I Kornfield eds. *Evolution of Fish Species Flocks.* Univ Maine Press at Orono

McKaye K R, T Kocher, P Reinthal, I Kornfield 1982 A sympatric sibling species complex of *Petrotilapia* Trewavas from Lake Malawi analyzed by enzyme electrophoresis (Pisces: Cichlidae). *Zool J Linn Soc* 76:91-96

McKaye K R, M K Oliver 1980 Geometry of a selfish school: defence of cichlid young by a bagrid catfish in Lake Malawi, Africa. *Anim Behav* 28:4

Myers G S 1960 The endemic fish fauna of Lake Lanao, and the evolution of higher taxonomic categories. *Evolution* 14:323-333

Paterson H E H 1978 More evidence against speciation by reinforcement. *S Afr J Sci* 74:369-371

Powell J R 1978 The founder-flush speciation theory: an experimental approach. *Evolution* 32:465-474

Pinsker W, E Doschek 1980 Courtship and rape: the mating behavior of *Drosophila subobscura* in light and darkness. *Z Tierpsychol* 54:57-70

Ringo J M 1977 Why 300 species of Hawaiian *Drosophila*? The sexual selection hypothesis. *Evolution* 31:694-696

Rosenzweig M L 1978 Competitive speciation. *Biol J Linn Soc* 10:275-289

Sage R D, R K Selander 1975 Trophic radiation through polymorphism in cichlid fishes. *Proc Natl Acad Sci USA* 72:4669-4673

Schilcher F von 1977 A mutation which changes courtship song in *Drosophila melanogaster*. *Behav Genet* 7:251-259

Schroder J H 1980 Morphological and behavioural differences between BB/OB and B/W colour morphs of *Pseudotropheus zebra* Boulenger (Pisces; Cichlidae). *Z Zool Evolut.-forsch* 18:69-76

Smith G R, T N Todd 1984 Evolution of species flocks of north-temperate lacustrine fishes. 47-68 A A Echelle, I Kornfield eds. *Evolution of Fish Species Flocks.* Univ Maine Press at Orono

Spiess E B 1982 Do female flies choose their mates? *Am Nat* 119:675-693

Spieth H T 1974a Mating behavior and evolution of the Hawaiian *Drosophila*. M J D White ed. *Genetic Mechanisms of Speciation in Insects.* Reidel. Boston

_____ 1974b Courtship behavior in *Drosophila*. *Ann Rev Entomol* 19:385-406

Svardson G 1953 The coregonid problem. V. sympatric whitefish species of the lakes of Idsjon, Storsjon, and Hormavan. *Rep Inst Reshwater Res Drottingholm* 34:141-166

Taborsky M, D Limberger 1981 Helpers in fish. *Behav Ecol Sociobiol* 8:143-145

Tauber C A, M J Tauber, J R Nechols 1977 Two genes control seasonal isolation in sibling species. *Science* 197:592-593

Templeton A R 1980 The theory of speciation *via* the founder principle. *Genetics* 94:1011-1038

_____ 1981 Mechanisms of speciation -- a population genetic approach. *Ann Rev Ecol Syst* 12:23-48

Thornhill R, J Alcock 1983 *The Evolution of Insect Mating Systems.* Harvard Univ Press

Trewavas E, J Green, S A Corbet 1972 Ecololgical studies on crater lakes in west Cameroon fishes of Barombi Mbo. *J Zool Lond* 167:41-95

West-Eberhard M J 1979 Sexual selection, social competition and speciation. *Proc Amer Phil Soc* 123:222-234

_____ 1983a Sexual selection, social competition, and evolution. *Q Rev Biol* 58:155-183

_____ 1983b Polymorphism, speciation, and phylogeny. Unpubl ms.

Wood D, J M Ringo 1982 Artificial selection for altered male wing display in *Drosophila simulans*. *Behav Gen* 12:449-458

Zimmerman E C 1948 *Insects of Hawaii*. Univ. Hawaii Press. Honolulu

WHO'S TENDING THE FLOCK?

· Irv Kornfield and Anthony A Echelle

INTRODUCTION

Because of the geologically ephemeral nature of lakes, the frequent extinction of fish species flocks seems inevitable. The Mesozoic semionotids (McCune et al. 1984) and pliocene sculpins (Smith, Todd 1984) of North America provide convincing examples of flock extinction, but scenarios for the cichlids of the African lakes, the Sculpins of Lake Baikal and the various endemic faunas of the arid Americas can be envisioned. We consider two kinds of extinction: natural and anthropogenic.

NATURAL EXTINCTION

The large "cichlid lakes" of Africa are either deep and located in tectonically active areas or are relatively shallow and sensitive to changes in local water budgets (Banister, Clarke 1980; Grove 1983). Rapid regional faulting and warping can produce dramatic changes in littoral and sub-littoral regions by direct or indirect (e.g. sediment induced) habitat alterations. In deep permanently stratified lakes, mixing of the epilimnion with anoxic waters from below 250 m would result in massive extinctions (D Eccles pers. comm.) and might be triggered by subsurface tectonic activity. In the case of the shallow systems, the same physical processes which provide geographic isolation and promote speciation may just as easily cause the eradication of endemic species. Two examples of cichlid flock extinction have been described recently from the Miocene of Africa (Van Couvering 1982). We should generally expect local extinctions to be almost as common as radiation in such systems.

Lake Baikal, because of its apparent geological stability and huge water volume, perhaps should be relatively immune to tectonic activity or changes in water budgets. However, there is a mechanism which might

make Baikal vulnerable. The last continental glaciations which extended beyond 40°N in North America did not approach Lake Baikal at 50°N. However, isolated ice cover occurred over a large area less than 400 km west of the lake. A short period of permanent ice cover associated with glaciation or regional cooling would be sufficient to destroy the food web of Baikal and its component species. The next episode of continental glaciations, if more extreme, could extend into this region.

For most of the arid climate flocks discussed in this book, projections for the effects of natural climatic cycles can be imagined. Because of their extreme shallowness and generally small geographic areas, the duration of suitable lake environments is expected to be extremely short for most lakes in these areas. In addition, some of these regions are subject to local tectonic instability (Echelle, Echelle 1984).

What possibility is there that some fraction of endemic lake taxa would be able to survive extinction? Species could persist by dispersing to contiguous and presumably stable riverine environments, assuming that they were available. The probability of such transitions may be small. First, extant riverine fish fauna are apparently not influenced by proximity to species-rich lakes. In the rivers associated with Baikal, Lanao and the African Great Lakes, specialized lacustrine endemics are uniformly absent (Taliev 1955; Roberts 1975; Greenwood 1976; Tweddle et al. 1979). The generally reduced trophic complexity of riverine environments compared to lakes suggests that in rivers, displaced endemics would experience high levels of resource competition with native forms. Of course this depends on the existing diversity of rivers. In the canals of southern Florida, a number of exotic species, including African cichlids have established breeding populations (J Taylor pers. comm.). It is possible then that some fraction of endemic faunas may be preserved if the transition to rivers were made rapidly. The reduced volume and spatial complexity of river environments would inhibit colonization by spatially specialized lacustrine species; equivalent niches for deep-water and mid-water forms would simply not be available. To the extent that some forms are actually trophic generalists (Liem 1980; cf. Barel 1983), transition to riverine environments might be possible. However, it is difficult to envision successful displacements unless they occurred rapidly.

Based on very recent studies, it has become increasingly clear that in African cichlids, closely related sympatric species invariably differ in some niche dimension (Ribbink et al. 1983; Hoogerhoud et al. 1983). In Lake Malawi, a sizable number of allopatric species which were intentionally introduced by commercial fish exporters to already diverse communities have successfully established breeding populations. But since most of these forms already exhibited niche differences with congeners when naturally sympatric, competitive extinction would not be expected. However, there is at least one example of the successful introduction of a conspecific which apparently differs from the resident only in breeding coloration; it is too early to predict whether this introduced form will persist (Ribbink et al. 1983). Regardless, because of previously existing niche differences, coexistence in forced riverine allopatry might be expected for a fraction of these species. It is not necessary to invoke mechanisms which could permit the coexistence of several species on a limited resource (Emlen 1984). In sum, if a stable riverine habitat were available as a refuge for species from a shrinking or deteriorating lake, it is possible that some endemics could persist. However, such transitions might require the rapid introduction of a fair number of individuals.

ANTHROPOGENIC EXTINCTION

Students of species flocks should be acutely concerned with the vulnerability of endemic faunas. The geologically instantaneous effects of human activities are more significant than any associated with major natural cycles. The political and economic priorities of contemporary human affairs, particularly industrial expansion and demands for water and natural resources, strongly conflict with the persistence of endemic elements. With the exception of a few well-isolated lakes, virtually all modern endemic lacustrine faunas are at risk. In addition, many of the component taxa possess those characteristics which make them particularly vulnerable (Soule 1983).

The demise of the Lanao cyprinid flock is particularly instructive. The flock went extinct in less than twenty-five years. Virtually all possible factors which could operate to cause extinction did operate: the system suffered from unregulated exploitation (including extensive electrofishing and dynamiting), the intentional and accidental introduction of exotic predators and competitors, and human population growth with its associated pollution. Even if these factors had not been present, the fauna was doomed because of variation in water demand associated with recent hydroelectric development. Biologists who were aware of these changes early on alerted the scientific community through comments in professional publications (Kornfield 1982). Given the political instability at the time, it is not clear whether any organized effort from the scientific community could have affected the outcome, but no such efforts were made. The Lanao episode is now being repeated for other species flocks, but sufficient time potentially is available to affect the outcome.

In Cuatro Cienegas, Coahuila, Mexico, major channelizations for local irrigation have caused significant habitat reductions for *C. minckleyi* (Minckley 1969). The disappearance of isolated ponds due to increases in very local water usage has also occurred (Kornfield unpubl.) and can be predicted to increase in the future. The status of this species is currently being assessed by the International Union for the Conservation of Nature. In a similar way, the endemic cichlids of Laguna de La Media Luna, San Luis Potosi, Mexico (Taylor, Miller 1983) are now considered threatened because of habitat alterations associated with water use and exotic introductions. In the large lakes on the Mesa Central of Mexico, introductions of exotics, such as *Micropterus salmoides, Ictalurus punctatus, Sarotherodon* spp., and unregulated exploitation represent major threats to the endemic faunas. Although not substantiated with data, local fishermen note a decline in abundance of *pescados blancos*.

Biologists of Israel have voiced concern about the continuous stocking of *Oreochromis aureus* into Lake Kinneret (Gophen et al. 1983a). Stocking has resulted in declines of another native cichlid, *Sarotherodon galilaeus*, and has affected the trophic dynamics of the lake (Gophen et al. 1983b). The lake supports a number of endemic species of evolutionary interest (Kornfield et al. 1979) which potentially would be affected by trophic changes. The scientific position here is interesting because the ultimate concern about stockings is not with their impact on species abundance or diversity per se; since *S. galilaeus* is the primary consumer of the dominant alga of the lake, the decline of this planktivore may result in unacceptable water quality for human activities.

To date, the great lakes of East Africa have been spared major anthropogenic disturbances, but there is reason to believe that this condition will not persist. First, what appears to be a major oil reserve has been discovered under Lake Tanganyika and a significant hydrocarbon potential apparently exists under Lake Malawi (Cromie 1982). The long renewal time of these lakes (Fryer 1972, 1973) suggests that the effects of pollution associated with petroleum exploitation (or any environmental alteration) would be pervasive and persistent. Second, there is anecdotal evidence that the introduction of exotic species may be beginning to affect the diversity of endemic taxa. *Lates* (Nile perch) which was introduced into Lake Victoria a generation ago has increased in numbers and is now widely distributed. Collectors working the Kenyan side of the lake have recorded anomalously low diversities of cichlids in nearshore communities (M Anderson pers. comm.; R Dorit pers. comm.). Unfortunately, baseline data necessary for an accurate perspective on temporal flux in species diversity are not available. In Lake Malawi, the introduction of exotic clupeids has been advocated to boost exploitable fish biomass by shortening planktivore food chains (Turner 1981). As emphasized by McKaye et al. (1984), the effect of such introductions would probably be a

significant decrease in diversity of endemics. Because of the size and ecological complexity of the African lakes, the application of standard criteria to evaluate the possible impact of exotic introductions (Hubbs 1977; Kohler, Stanley 1984) is extremely difficult. Third, changing fishing patterns and methods in the lakes can potentially induce significant shifts in diversity. In Lake Victoria particularly, the development of a major bottom trawl fishery presents the means necessary to effect such change (Rinne 1980). Expansion from traditional artisanal inshore gillnet to mechanized trawling has occurred rapidly and can be expected to increase in the future.

Lake Baikal has an interesting history with respect to conservation. Public and professional concern over the effects of ecological perturbations to the lake in the late 1960's (Goldman 1972) resulted in major governmental commitment to control pollution associated with future industrial development in the region. While point source pollution is present (Vallentyne 1972), the fauna does not seem affected in a qualitative way; diversity of the endemic sculpins is high in recent collections (R Brauer pers. comm.). If the governmental commitment to protection is long term, the fauna of Baikal would appear in little immediate danger.

It is the professional responsibility of all who study fish species flocks to be involved in insuring the preservation of endemic faunas. The problems are politically complex, but not intractable. We strongly advocate that individuals knowledgeable about lake systems at risk be given explicit support in any preservation efforts from the membership of national and international societies of ichthyologists and evolutionary biologists. Such societies should actively solicit and disseminate information on the status of species flocks. Without the effort of organized groups of scientists, the opportunity for future studies of these evolutionarily unique faunas will not exist.

REFERENCES

Banister K E, M A Clarke 1980 A revision of the large *Barbus* (Pisces, Cyprinidae) of Lake Malawi with a reconstruction of the history of the southern African Rift Valley lakes. *J Nat Hist* 14:483-542

Barel C D N 1983 Towards a constructional morphology of cichlid fishes (Teleostei, Perciformes).*Neth J Zool* 33:357-424

Cromie W J 1982 Beneath the African Lakes. *Mosaic* 13:2-6

Echelle A A, A F Echelle 1984 Evolutionary genetics of a "species flock:" antherinid fishes of the Mesa Central of Mexico. 93-110 A A Echelle, I Kornfield eds. *Evolution of Fish Species Flocks.* Univ Maine Press at Orono

Emlen J M 1984 *Population Biology.* Macmillan. New York

Fryer G 1972 Conservation of the great lakes of East Africa: a lesson and a warning. *Biol Conserv* 4:256-262

———— 1973 The Lake Victoria fisheries: some facts and fallacies. *Biol Conserv* 5:304-308

Goldman M I 1972 *The Spoils of Progress: Environmental Pollution of the Soviet Union.* MIT Press. Cambridge

Gophen M, R W Drenner, G L Vinyard 1983a Fish introductions into Lake Kinneret--call for concern. *Fish Mgmt* 14:43-45

———— 1983b Cichlid stocking and the decline of the Galilee Saint Peter's fish (*Sarotherodon galilaeus*) in Lake Kinneret, Israel. *Can J Fish Aqu Sci* 40:963-986

Greenwood P H 1976 Fish fauna of the Nile. *Monogr Biologicae* 29:127-139

Grove A T 1983 Evolution of the physical geography of the East African Rift Valley region. 115-155 R W Sims, J H Price, P E S Whalley eds. *Evolution, Time and Space: The Emergence of the Biosphere.* Academic Press. New York

Hoogerhoud R J C, F Witte, C D N Barel 1983 The ecological differentiation of two closely resembling *Haplochromis* species from Lake Victoria (*H. iris* and *H. hiatus*; Pisces, Cichlidae). *Neth J Zool* 33:283-305

Hubbs C 1977 Possible rationale and protocol for faunal supplementations. *Fisheries* 2:12-14

Kohler C C, J G Stanley 1984 Suggested protocol for evaluating proposed exotic fish introductions into the United States. *Trans Amer Fish Soc* In press

Kornfield I L 1982 Report from Mindanao. *Copeia* 483-495

Liem K F 1980 Adaptive significance of intra- and interspecific differences in the feeding repertoires of cichlid fishes. *Amer Zool* 20:295-314

McCune A R, K S Thomson, P E Olsen 1984 Semionotid fishes from the mesozoic great lakes of North America. 27-44 A A Echelle, I Kornfield eds. *Evolution of Fish Species Flocks.* Univ Maine Press at Orono

McKaye K R, R D Makwinje, W W Manyani, O K Mhone 1984 On the possible consequences of introducing non-indigenous plankton feeding fishes into Lake Malawi, Africa. ms

Minckley W L 1969 *Environments of the Bolson of Cuatro Cienegas, Coahuila, Mexico, with Special Reference to the Aquatic Biota.* Texas Western Univ Press. El Paso

Ribbink, A J, B A Marsh, A C Marsh, A C Ribbink, B J Sharp 1983 A preliminary survey of the cichlid fishes of the rocky habitats in Lake Malawi. *S Afr J Sci* 18:149-310

Rinne J N 1980 Notes on the occurrence and biology of *Xenoclarias* Greenwood (Siluriformes: Clariidae) in Lake Victoria, East Africa. *J Nat Hist* 14:579-587

Roberts T R 1975 Geographical distribution of African freshwater fishes. *Zool J Linn Soc* 57:249-319

Smith G R, T N Todd 1984 Evolution of species flocks of fishes in north temperate lakes. 45-68 A A Echelle, I Kornfield eds. *Evolution of Fish Species Flocks.* Univ Maine Press at Orono

Soule M E 1983 What do we really know about extinction? 111-124 C M Schonewald-Cox, S M Chambers, B MacBryde, W L Thomas eds. *Genetics and Conservation.* Cummings. Menlo Park

Taliev D N 1955 *Sculpins of Baikal (Cottoidei).* Acad Sci USSR. Moscow

Taylor J N, R R Miller 1983 Cichlid fishes, genus *Cichlasoma* of the Rio Panuco basin, eastern Mexico, with description of a new species. Univ Kansas Mus Nat Hist Occ Paper 104:1-24

Turner J L 1981 Malawi lake flies, water fleas and sardines. Fl DP/MLW/75/019. FAO. Rome

Tweddle D, D S C Lewis, N G Willoughby 1979 The nature of the barrier separating the Lake Malawi and Zambezi fish faunas. *Ichthy Bull J L B Smith Inst* 39:1-9

Van Couvering J A 1982 Fossil cichlid fish of Africa. *Spec Pap Paleont* 19:1-103

Vallentyne J R 1972 A scientific look at Lake Baikal, USSR. *Biol Conserv* 4:305-306

INDEX